INESCAPABLE ECOLOGIES

Inescapable Ecologies

A History of Environment, Disease, and Knowledge

LINDA NASH

UNIVERSITY OF CALIFORNIA PRESS   Berkeley   Los Angeles   London

University of California Press, one of the most
distinguished university presses in the United States,
enriches lives around the world by advancing
scholarship in the humanities, social sciences, and
natural sciences. Its activities are supported by the UC
Press Foundation and by philanthropic contributions
from individuals and institutions. For more
information, visit www.ucpress.edu.

Portions of chapters 1 and 2 were previously published
in Linda Nash, "Finishing Nature: Harmonizing Bodies
and Environments in Late-Nineteenth-Century
California," *Environmental History* 8 (January 2003):
26–52. *Environmental History* is jointly published by
the American Society for Environmental History and
the Forest History Society, Durham, NC.

Portions of chapter 4 were previously published in
Linda Nash, "The Fruits of Ill-Health: Pesticides and
Workers' Bodies in Post–World War II California,"
Osiris 19 (2004): 203–19 (©2004 by the History of
Science Society).

University of California Press
Berkeley and Los Angeles, California

University of California Press, Ltd.
London, England

Library of Congress Cataloging-in-Publication Data
    Nash, Linda Lorraine.
    Inescapable ecologies : a history of environment,
    disease, and knowledge / Linda Nash.
        p.   cm.
    Includes bibliographical references and index.
    ISBN-13: 978-0-520-24891-5 (cloth : alk. paper)
    ISBN-10: 0-520-24891-0 (cloth : alk. paper)
    ISBN-13: 978-0-520-24887-8 (pbk. : alk. paper)
    ISBN-10: 0-520-24887-2 (pbk. : alk. paper)
        1. Medical geography—California—History.
    2. Environmental health—California—History.
    3. Public health—California—History.   I. Title.
    RA807.C2N37   2006
    614.4'2794—dc22   2006002009

Manufactured in the United States of America

15   14   13   12   11   10   09   08   07   06
10   9   8   7   6   5   4   3   2   1

This book is printed on New Leaf EcoBook 50, a 100%
recycled fiber of which 50% is de-inked post-consumer
waste, processed chlorine-free. EcoBook 50 is acid-free
and meets the minimum requirements of ANSI/ASTM
D5634-01 (*Permanence of Paper*).

For Jim and Helen

Contents

Illustrations

Acknowledgments

It is a long road from the beginning of graduate school to completed book. If I had not had help from many people, I certainly would not have reached this point. At different times during this project I relied on the excellent research assistance of Anna Bailey and Wendi Willeford. William Harrison helped select the photographs. Jim Hanford stepped in at the last moment to pull together the bibliography and check the notes.

I am indebted to many librarians and archivists who made the work possible and often enjoyable. I am especially grateful to the staff at the University of Washington Libraries, including those in the Interlibrary Loan Department. In California, I wish to thank the staff at the Bancroft Library and the Water Resources Archives at the University of California, Berkeley, the California State Archives, the California State Library, the California Historical Society, and the Huntington Library. I owe a special debt to the staff at the California Department of Health Services Environmental Health Investigations Branch, in particular Raymond Neutra and Rick Kreutzer, who took time away from their own work to talk with me and make files available. At the Kern County Department of Public Health, I am similarly indebted to Kirt Emery and Ron Talbot. Noor Tietze of the California Vector Control Association generously made available several oral histories of those involved in mosquito and vector control. Thomas Milby kindly agreed to discuss his experience at the California Department of Public Health.

My greatest intellectual debt will always be to Richard White, who, in addition to everything else, read and commented on several drafts of a very different dissertation as well as an earlier version of this book. Though he needs no one to extol either his scholarly vision or his unsurpassed commitment to graduate teaching, I am one of many who do. I was fortunate to have many outstanding teachers in graduate school whom I am privileged to call my colleagues. I especially thank James Gregory, Susan Glenn, John Toews, John Findlay, and Richard Johnson for freely sharing their knowledge over many years while also challenging me to think harder and write more clearly. In addition, several colleagues read all or parts of the manuscript and generously offered their comments and suggestions, including Nayna Jhaveri, Celia Lowe, Gregg Mitman, Janelle Taylor, Simon Werrett, Elliott West, and members of the History Research Group at the University of Washington. I also benefited from the feedback I received from audiences at the University of British Columbia, the University of Wisconsin–Madison, the University of Washington History of Science Group, and the annual conference of the American Society for Environmental History.

Material from chapters 1 and 2 was previously published in *Environmental History,* where it received a careful reading from Adam Rome and Conevery Valenčius. Portions of chapter 4 were published in an issue of *Osiris* edited by Gregg Mitman, Christopher Sellers, and Michelle Murphy; I thank the editors for inviting my contribution and for their comments. For their advice and support at different points, I also thank Warwick Anderson, William Cronon, Mark Fiege, Jonathan Mayer, Kathryn Morse, Uta Poiger, William Rorabaugh, Laurie Sears, Paul Sutter, Julidta Tarver, Lynn Thomas, and Kathryn Utter.

At the University of California Press, Monica McCormick first recruited the manuscript, and I am grateful for her interest and encouragement. In the end, all the work fell to Niels Hooper, who has shepherded the book to publication with enthusiasm, efficiency, and, not least, good humor. It has been a pleasure to work with him. At the University of Washington, this project has been supported by grants from the Walter Chapin Simpson Center for the Humanities, the Royalty Research Fund, and the Department of History. In addition, a generous grant from the National Library of Medicine provided an uninterrupted year of writing. For many years now, I have been fortunate to have the History Department at the University of Washington as my academic home, first as a student and now as a faculty member. I have benefited not only from the department's commitment to graduate education but also from its sustained support of junior faculty.

My family and friends have been extremely patient while waiting for this project to come to a conclusion, and they have all supported it in myriad ways. Perhaps above all, I appreciate their willingness *not* to ask when it would be done. I owe special thanks to Bill Harrison, Anne Semans, and Caroline Streeter for ensuring that my many trips to California involved more than just work. I also thank Nancy Hanford for her willingness to take care of the granddaughter she loves, often at a moment's notice. In the end, few outside academia understand how interminable a process writing a book such as this can be and the sacrifices it inevitably requires. In my case those who know this best are Jim and Helen. Despite that fact, they remain my most unwavering and enthusiastic supporters. Thus it is at once completely necessary and wholly inadequate that I dedicate this book to them.

Introduction

One of the major cultural developments of the late twentieth century was the reenvisioning of human beings' place in the world. In many ways the most radical notion to emerge from the modern environmental movement was the idea that people were inescapably part of a larger ecosystem. The genesis of this popular understanding is typically traced to the publication of Rachel Carson's *Silent Spring* in 1962. The influence of Carson's book is undeniable. *Silent Spring* made the science of ecology accessible to a broad public and helped galvanize a generation of environmental activists who would push through several groundbreaking political and institutional reforms. But perhaps her ideas took hold so strongly and so quickly not because they were wholly new but because she had articulated existing understandings in a new way. Although Carson's book was pivotal, "ecological" understandings of human bodies have a much longer history. This book tells that story, recounting both the marginalization and the persistence of ecological ideas of health and disease as that history unfolded in California, and more particularly within the Central Valley. By doing so, it also complicates the history of Americans' relationship to nature.

California's Central Valley has attracted the attention of many scholars, and most have invoked the region as the antithesis of both romantic agrarianism and enlightened environmental management. The valley's notoriety stems from its position as the most productive and the most industrialized of America's rural landscapes. Whether one's perspective is social or environmental, the valley's history can be told as the relentless domination of space by capital and technology. Social historians have chronicled the region's exploitative wage-labor system, its racialized workforce, and its history of violence. Environmental historians have found the valley's history equally appalling and irresistible: the large dams and fully engineered rivers constructed with no concern for their effects on fish and wildlife, the almost total obliteration of a native prairie and wetland landscape in favor of profitable crops, and the nearly unrestrained introduction of highly toxic pesticides and nonnative species. By any standard, the environmental changes have been immense. What most of these histories share, environmental histories in particular, is the idea of alienation and the theme of regret.[1]

The twin stories of capitalist exploitation and environmental conquest are not wrong. In fact, they are essential to understanding American— especially western American—history. In the words of Donald Worster, ours is "a culture and society built on, and absolutely dependent on, a sharply alienating, intensely managerial relationship with nature."[2] Worster is right. When we stand alongside one of the Central Valley's major irrigation canals, amid a landscape of engineered rivers, laser-leveled fields, and two-thousand-acre cotton ranches planted with genetically modified seed, the power of capitalism to shape the landscape, and the resulting sense of alienation, can seem at once undeniable and overwhelming.

Yet critiques such as Worster's suggest that there may be cracks in the metaphorical irrigation canal, that there are instances in which understandings of the landscape fall outside the rubrics of conquest and alienation that dominate so much of American environmental history. Perceptions of health and disease constitute just such a crack. Although they do not fall wholly outside the capitalist and managerial mentality, neither can they be fully contained within it. Thus, placing the stories of colonization and capitalist development alongside stories of health and disease creates a more complicated environmental history, one in which we can perhaps begin to see ourselves.

Such a story might be set anywhere, but I have chosen to return to the Central Valley, putting aside the classic narrative of regret to tell a less linear history of body and environment, health and place. Although California is most strongly associated with its coastline and its two major cities, physiographically, the state is more accurately described as a very big valley encircled by mountains. The Central Valley is so big that it hardly seems like a valley when you are in it; it is roughly seventy-five miles wide and more than four hundred miles long. There is no other flat area of such size west of the Rocky Mountains. Although from a topographic perspective the Central Valley is a single unit, defined by the Sierra Nevada on the east and the Coast Ranges on the west, it actually comprises three different watersheds. The Sacramento and San Joaquin Valleys are defined by their rivers. The Sacramento River emerges from the mountains just north of Redding and flows south to the city of Sacramento, where it empties into a large delta and flows out into San Francisco Bay. The San Joaquin River flows out of the Sierra Nevada near Fresno and then runs north to the same delta. Immediately south of the San Joaquin Valley lies the Tulare basin—an area defined by the now-vanished Tulare Lake and separated from the Los Angeles basin by the Tehachapi Mountains. Although these watersheds differ from one another in important ways, their similarities bind them together in most people's minds. Since the late nineteenth century, the vast majority of the region has been devoted to agriculture, and by most measures, it remains the most productive landscape in the world. That productivity has been the outcome of many things: a warm climate, good soils, large groundwater basins, massive water engineering projects, and substantial state investments in mechanization, plant hybridization, and modern chemicals.

When I set out to write a cultural history of this environment, I found ample evidence of settlers' market orientation and belief in environmental conquest. Clearly they looked upon the western landscape as a resource ripe for exploitation, and they eagerly set about to mine the region's gold, harvest its timber, and plow up its prairies. But I also stumbled across settlers' overwhelming concern with the region's effect on their health, and here I found a different way of thinking about the land. Many immigrants to the southern California coast extolled its therapeutic effects, calling attention to the purity of the atmosphere, the temperate climate, and the healing properties of local springs. Settlers in the Central Valley, on the other hand, were far more anxious. They talked about miasma and poisonous north winds that left them feeling sluggish and debilitated, about dysentery and sunstroke, and also about a host of fevers—"bilious fever," "Sacramento fever," "autumnal fever," and "typho-malarial fever," to name but a few. But whether they experienced improved health or new forms of illness, settlers wrote about the

Map 1. The Central Valley of California, with the Sierra Nevada to the east and the Coast Ranges to the west.

environment and its effects in quite personal terms. These sources date from the mid-nineteenth century, the moment in which whites first migrated to California in substantial numbers. When these early immigrants surveyed their new home, they were as likely to discuss the prevalence of disease as the fertility of the soil. Physicians were among the keenest environmental observers, and their medical papers are filled with assessments of the landscape and its features: mountains, coastlines, and valleys; swamps, soils, winds, and springs. The extent of this preoccupation surprised me, and I thought it might make for an interesting paragraph or two. Instead it reoriented my entire project, as I realized how important perceptions of health were to understanding the natural landscape in earlier eras.

These nineteenth-century discussions reveal that the development of the West's resources—the planting of the most profitable crops, the cutting of forests, the importation of cheap labor, the rapid adoption of technology—often ran up against settlers' concerns for their health. Put another way, the body, specifically the body's physical well-being, offered a powerful way of understanding local environments, a form of understanding that lay outside simple calculations of profit. These ideas had a more complicated history, and they pointed to a more intimate connection that settlers felt with the landscape. As Conevery Valenčius has put it, antebellum Americans "moved and breathed in profound connection with their environments."[3]

It is easy to dismiss nineteenth-century medical beliefs as a historical curiosity, as something that has no relevance to the more recent past. The voluminous writings on "sickly" and "salubrious" places suggest a deeply held fear of certain landscapes—swamps, forest edges, tropical climates. These fears were an outgrowth of Europeans' colonial experience and are no longer widely shared. Miasmatic disease, the quintessential environmental affliction of the nineteenth century, is now regarded primarily as an inadequate understanding of malaria, before medical science had revealed a mosquito-borne parasite as its cause. For many, the rise of "germ theory" at the end of the nineteenth century clearly separates the unscientific beliefs of earlier centuries from "modern" medicine and public health.[4]

What struck me about nineteenth-century health concerns, however, was that they were strange *and* familiar. While concerns over miasma and locally produced fevers are a product of a very different era, they seemed to resonate with and foreshadow late-twentieth-century anxieties about the environmental sources of cancer, asthma, and even the disputed disease of multiple chemical sensitivity.[5] Yet the rise of environmental health concerns in the decades after World War II has typically been portrayed as a wholly new development. Recent medical interest in environmental factors can be explained by pointing to shifting disease patterns (i.e., the decline in infectious disease and rising rate of chronic and noninfectious conditions) and to the material changes associated with America's rapid postwar growth (e.g., urbanization, rising rates of air pollution). In this account "environmental health" has no connection to an earlier environmental medicine. But to emphasize the discontinuities in medical thinking may well be an artifact of modern medicine's desire, in David Morris's words, to "burn away the memory of its pre-scientific ancestors." In this book, I turn the story around. My question is not why and how the link between environment and health was finally recognized in the late twentieth century but why it had ever become invisible. From this perspective, the narrow situating of disease in the organic dysfunction of bodies and particular pathogens begins to look like a brief period of modernist amnesia.[6]

If we set aside the scientific narrative of medical progress, as well as the environmentalist narrative of regret, it becomes possible to trace continuities between the nineteenth and twentieth centuries. That concerns over environment and health persisted throughout this period, even in a paradigmatic modern space like California's Central Valley, suggests that connections and continuities deserve closer scrutiny. My story begins with the accounts of early-nineteenth-century immigrants and travelers to California, just before the initiation of large-scale American settlement in western North America, when humoral theory still dominated medical and environmental thinking. I show how nineteenth-century understandings of the body as porous and vulnerable shaped early responses to the California landscape and to the colonial project more generally. My story chronicles how physicians, boosters, and settlers drew on understandings of bodies and health to evaluate the diverse landscapes of California. In focusing on the diseases of the valley, I emphasize the growing tension between settlers' desire to rapidly transform the landscape for profit on the one hand and ecological understandings of health on the other.

Just as most histories of medicine and public health see the rise of germ theory in the late nineteenth century as a critical turning point, the turning point in my narrative also comes with germ theory—not, however, because it signaled the end of an unscientific environmental medicine, but because it marked the institutionalization of a certain concept of both the body and the environment. Advocates of the "new public health" defined health much more narrowly than their predecessors, as the simple absence of disease. And they insisted that disease-causing pathogens were situated in human bodies, not environments. Their models seemingly, and conveniently, resolved the tension between modernization and health by exonerating the landscape from any independent role in disease. Yet the discourse of germ theory obscured as much as it revealed, and, moreover, rhetoric did not always match practice. Sanitary engineers and entomologists, among others, remained attentive to the environment's role in disease; their daily practices undermined many of the modern premises that doctors and public health officials purported to embrace.

Immediately after World War II, public health specialists had to confront head-on the environmental sources of disease. The emergence of concerns about air pollution, radioactive fallout, and pesticides forced health professionals and laypeople to consider anew the interaction of bodies and environments and their definitions of health and illness. To tell this part of the story, I focus on the introduction of organic chemicals into the valley's environment and the concurrent emergence of concerns about pesticide poisoning among farmworkers in the 1950s and 1960s. In these decades modern conceptions of the body were publicly challenged by an ecological view that—much like earlier humoral medicine—saw bodies as intimately connected to their environments. With the rise of environmental concerns and the popularization of ecological science, bodies seemed as if they might once again be permeable and porous. Fearing the health effects of recent environmental changes, many middle-class Americans began to question the trajectory of their own modernity. The seminal text of modern environmentalism, Rachel Carson's *Silent Spring,* while popularly remembered as a book about the effects of pesticides on wildlife, centered on the dangers that newly introduced chemicals posed to human bodies.[7] In the Central Valley, farmworkers and environmentalists marshaled an ecological view of the body to challenge the models of professionals and to critique existing environmental and agricultural practices as inherently unsafe. In the decades after World War II, it became increasingly clear that human beings were not simply agents of environmental change; they were also objects of that change.

Unlike classic tales of either progress or decline, this history has no definitive ending or denouement because the dilemmas and conflicts that I delineate remain unresolved, even today. However, I take the story up through the 1980s by tracing conflicting ideas about the body, health, and disease as they played out both in the field of public health and in conflicts between health experts and laypersons. The final chapter describes the discovery of more widespread pollution in the valley and the ensuing debate over several cancer clusters and their possible environmental causes. These events reveal the resilience of models derived from germ theory in certain institutional contexts—even as both professionals and laypersons pointed repeatedly to the inadequacy of those same models.

By placing the human body at the center of an environmental history, this work challenges the modern dichotomy that separates human beings from the rest of nature, a dichotomy that underwrites the very discipline of history. By definition, history is about the ways in which human beings have shaped their world. Environmental historians, for their part, have sought to complicate that story by drawing attention to the role of plants, animals, soils, and climate. Nonetheless, environmental history has typically reinforced the modern dichotomy by placing humans and their creations on one side and everything else—nonhuman nature—on the other. In all histories, the actors are assumed to be human; the rest of the world is a set of constraints that human actors must work within.

When we focus on the human body, however, the boundary between the human and the nonhuman world, the actors and their objects, becomes much more fuzzy and the distinction much more tenuous. Where does the body end and "nonhuman nature" begin? When we recognize that human bodies are directly affected by their environments, we are forced to acknowledge that humans are not simply agents of environmental change but also objects of that change. Conversely, the environment is more than an object upon which change is enacted; it is also an agent of sorts that acts upon the bodies inhabiting it. As landscapes can be investigated to uncover the effects of certain human actions, human bodies—their symptoms and diseases—become sites for investigating the quality and effects of certain landscapes. Subjects blur into objects, and historical agency becomes distributed among a multitude of entities: humans, insects, microbes, trees, groundwater, and chemicals. It is no longer so easy to separate the human from the nonhuman world, to insist that modernity represents the triumph (for better or for worse) of humans over nature. Modernization was indeed enacted on the landscapes of North America but also—often unwittingly—on the people who inhabited those landscapes. It is a question not simply of how a manipulated environment has broadly influenced society but also of how environments have shaped human flesh in minute and profound ways. The history of disease, because it lies at the nexus of the human and the nonhuman, offers a useful means for rethinking these divisions and for reconsidering how we divide and write our histories—environmental, social, or otherwise.[8]

If environmental histories typically have ignored issues of health, it is also true that histories of health and the body typically have been told without reference to environments. Scholars generally write cultural and intellectual histories, including those of medicine and public health, in nonlocal ways. To the extent that these histories are located anywhere, it is usually in those institutions where ideas are codified and recorded: universities, government agencies, disciplinary societies. But to tell the history of health without reference to specific landscapes is to assume at the outset that landscapes do not matter. In fact, that has been the dominant assumption of twentieth-century medicine, and most of the field's historians have followed suit.[9]

But the history of medicine and health is also local. Though we tend to think of knowledge as residing in minds, knowledge and ideas do not emerge from nowhere but from the interaction of human minds with specific places, materials, and things. The contexts that shape knowledge are not only social and economic but also material and environmental. To take one example, knowledge about pesticide-related illnesses emerged first in the fields, as workers fell ill for unknown reasons. It was the subsequent investigation of local orchards and individual workers that began to reveal the complex relationship between chemicals and health. Once formulated, scientific ideas travel quickly, but they also continue to interact with actual environments and messy realities. Even the most abstract knowledge is ultimately produced from human experience in particular places; three decades of work in science studies have demonstrated that the modern scientific laboratory is its own kind of local space. Thus, in addition to highlighting the tension between the imperatives of capitalist modernization and the concern with health, this book engages the tension between local and translocal knowledge.[10]

In marked contrast to the modern laboratory, nineteenth-century medicine was determinedly and self-consciously local. Immigrant physicians brought with them to California ideas about miasma and warm climates, but they also expected to reformulate those ideas based on the diseases they witnessed and the soils, winds, and swamps they encountered. Twentieth-century medicine turned away from local knowledge, at least in theory, and highlighted the role of the laboratory. Yet even today we recognize that the history of health and disease is not fully divorced from place. The material qualities of a specific landscape remain critical to the production of certain diseases: local habitats that support anopheles mosquitoes, warm temperatures that allow the survival of parasites and bacteria, the material and biological conditions of particular human communities, the ability of local rivers to carry away and dilute infectious sewage, the local history of chemical usage, the meteorologic and geologic characteristics that determine the fate of environmental toxins.

This tension between local and translocal knowledge often emerges in the materials produced by state and local public health officials, whose records provide an important set of sources for this book. From the early twentieth century on, public health has been an important arena for discussions of health and disease; moreover, public health professionals have held the power to regulate the relationship between human bodies and local environments. While committed to the professional discourses of biomedical science, public health officials have had to contend directly with local realities. Sometimes they found that local experience reinforced translocal knowledge about disease; at other times, however, they found that local experience challenged or resisted knowledge that had been generated elsewhere. California's public health experts, engineers, and environmental scientists have struggled over the decades to uphold the visions of a universal science, but again and again they have had to account for the particularity of environments and the idiosyncrasies of individual bodies. As Alan Irwin and Bryan Wynne have observed, "The local is the site of renegotiation of the 'universal.'"[11]

Any environmental history must confront the idea that the environment about which we write is, inevitably, something that we always understand through language and certain cultural practices. The chapters that follow pay close attention to understandings and perceptions: how people talked and thought about health, disease, and the environment. But like most environmental historians, I remain committed to a materialist view of the world. I am interested not only in how people talked about environment and disease but also in what happened on the ground—the changing pattern of disease, the changing uses of the land, the changing qualities of air, water, and soil. Consequently, I do not hew to either a materialist or a cultural approach, nor have I tried to separate out the two. That is precisely the point. Our understandings of environment and disease are shaped simultaneously by culture and by the material realities of the world. These stories need to be told together.

To talk about environmental changes in the valley, I rely on the understandings of contemporary science, which offer a powerful means of understanding past environmental and biological events. But while I evaluate earlier understandings of the environment or disease using modern scientific understandings at certain moments, that is not my only—or even my principal—goal. I am not especially interested in ascertaining how much malaria might have been caused by nineteenth-century irrigation, or in how effective anti-insect campaigns were in reducing the prevalence of certain diseases. In many cases the historical record is far too incomplete to allow such an assessment. More to the point, I am only too aware that contemporary scientific understandings, however powerful they seem now, will inevitably give way to different or more refined interpretations. Contemporary scientific understandings are not the principal metric against which I measure the value of earlier ideas; rather, I try to understand those ideas on their own terms. Nonetheless, it would be foolish if not impossible to ignore what contemporary science has to say about the environments and diseases of the valley. Thus, I use scientific insights to make past events more understandable but do not assume that there is only one correct way to describe what happened. My point is that, like all narratives, scientific description is a product of its own time and place, shaped as much by past practices and prevailing attitudes as by empirical "facts." Contemporary science, like any other lens, simultaneously brings the past into focus and distorts it.

I treat the human body and its diseases in a similar fashion—as at once material realities and products of language and culture. Recent scholarship has insisted that the body, like the natural environment, cannot be taken simply as a biological given. People experience their bodies differently in different historical moments, according to the languages and practices available to them. And yet, as the historian Barbara Duden suggested some time ago, the body seems to exist in two kinds of time—historical and transhistorical. Though our experience of our own bodies is constructed by both culture and practice, there are elements of our bodily experience—including birth, death, fatigue, and symptoms of illness—that undeniably connect us to those who came before. Though there is no precise equivalence between the experience of "bilious fever" in the nineteenth century and malaria in the twenty-first, certainly many of the symptoms and effects that people experienced were quite similar. And though the experience of disease or even death is necessarily different in different times and places, in some ways it is clearly not.[12]

In this work, I emphasize two competing conceptions of the body. The first is what I will call for convenience's sake the "modern" body—the body of Western allopathic medicine and American consumer capitalism, the body that is defined in medical textbooks, the body that is composed of discrete parts and bounded by its skin; in other words, the idea of the body that most of us take as so self-evident that it requires no comment. This idea of the body emerged only gradually. Certainly it existed no earlier than the eighteenth century, and in many places it developed much later or only partially. Its consolidation came only at the end of the nineteenth century with the rise of laboratory medicine. In Euro-American history, this modern body was preceded by a different, less than modern body that owed much to the writings of the fifth-century Greek physician Hippocrates. This older body was characterized by a constant exchange between inside and outside, by fluxes and flows, and by its close dependence on the surrounding environment. Using contemporary terms, I label this earlier conception of the body "ecological," though it clearly antedated the modern science of ecology. Until the late nineteenth century, the concept of health typically referred to a state of balance, or harmony, between a person's body and the larger world. In contrast, for the modern body, "health" came to connote primarily the absence of disease; it implies both purity and the ability to fend off harmful organisms and substances. Above all, health became a quality possessed (or not) by an individual body rather than a dynamic relationship between a body and its environment.[13]

To label different conceptions of the body "ecological" and "modern" is obviously an oversimplification—but a useful one. It calls attention to certain continuities and differences. The critical difference between the two in my account is the quality of permeability: that is, to what extent and by what means a body is closed off from the larger environment. While I track the occurrence of certain diseases, I also trace the construction of bodies as more or less permeable to their environment in different times and places. The move toward bacteriological theories of disease in the late nineteenth century marks an important if incomplete shift in this respect. Initially, those who embraced germ theory melded the new ideas with their long-standing environmental beliefs; ecological and modern concepts of the body coexisted openly and easily for some time. Nineteenth-century physicians were likely to see germs everywhere and capable of penetrating the body in many ways. However, the increasing dominance of bacteriology encouraged doctors to narrow their definitions of health and to limit their focus to the specific pathogenic agents that were revealed under the microscope. Gradually, they reinterpreted health as the absence of disease and disease as something confined to discrete entities. By the early twentieth century, advocates of the "new public health" were confidently proclaiming that the skin protected a body from its surroundings and urging their colleagues to shift their attention away from the environment; what mattered now was shoring up the self-contained body.[14]

What the existing scholarship has overlooked is how this shift from an ecological to a modern conception of the body affected understandings of the environment. In the rhetoric of public health, local environments were no longer understood as active components in the production of health and disease; instead, they were recast as homogeneous spaces that were traversed by pathogenic agents. In this formulation, the environment itself (aside from pathogenic bacteria) had no agency of its own in the production of disease. But that narrowed focus was always belied by actual public health practices and by the persistence of other, more environmentally oriented medical subspecialties. In the years after World War II, the modern concept of the body was strongly challenged by new types of illness. When investigators began to research pesticide poisoning in the 1950s, they found that bodies were far more porous than they had previously acknowledged and the local environment far more implicated. Once again, the skin was less a boundary than a zone of transfer and connection. The singular and self-contained human body of the early twentieth century came, by the end of that century, to seem distressingly porous and vulnerable to the modern landscape.

Any discussion of the body and its diseases cannot be understood without reference to the complex history of race. The relationship between bodies and environments in Euro-American culture has always been racialized, like the bodies themselves. Although race is an arbitrary category that has no firm basis in modern biological science, it has been an overwhelmingly powerful and persistent idea with vast material effects; the very indeterminacy of the concept has spurred tremendous efforts to make it seem more real. In the eighteenth and nineteenth centuries, ideas of race, scientific and otherwise, stood at the center of Euro-American understandings of body and place. Colonial discourse insisted that races were associated with particular places and that "race" itself was both an outcome of location and the principal determinant of one's fitness for any particular land. Until the late nineteenth century, most Euro-Americans believed that it was the very permeability of the body that created its race and that a person's race was liable to change in a new location. Careful scrutiny of the land and its bodily effects provided an important arena for the ongoing construction of race and racial ideology. As Europeans moved into new areas of the world, they often found themselves highly susceptible to disease and death, considerably more susceptible than the natives of those areas. That physical vulnerability would find expression in the European fear and denigration of hot, tropical climates as places inherently hostile to civilization. In the nineteenth century, as Americans embarked on their own project of expansion, many still feared that the western portions of North America might threaten their racial identity and undermine their colonizing project. The heat and diseases of the Central Valley in particular suggested its similarity to "tropical" locations and thus its inherent hostility to white bodies and Euro-American culture, and the process of white settlement generated a far-reaching discourse about environment, race, and health that linked the environments of California with those of other colonial frontiers in Asia, Africa, and Latin America.[15]

Although climatic theories of race would give way in most quarters by the turn of the century, the category of race still remained central to discussions of disease and health. In the first half of the twentieth century, the link between race and disease was reformulated through appeals to bacteriology. People of certain races were no longer perceived as more or less suited to specific environments; nonetheless, certain races were deemed more likely to "carry" disease into otherwise healthy locations and to be more susceptible to certain kinds of illness.

After World War II, overtly racist thinking lost much (though certainly not all) of its power in biomedical discourse, but the social and economic inequality fostered by racism meant that race and disease would continue to map onto one another. Those who were not "white" were more likely to suffer from specific illnesses and to lack many of the characteristics that signified good health. For some, this remained evidence of the susceptibility of certain kinds of bodies and the bad habits of certain kinds of people. For others, however, it pointed to the fact that socioeconomic inequality had a geographic, and environmental, dimension. A long history of residential and employment segregation meant that nonwhites were far more likely to live in badly polluted communities and to have more hazardous jobs.[16] In the late twentieth century, the health status of minority communities became a crucial political issue that would underwrite calls for "environmental justice." Whereas colonial science had understood the relationship between disease and race as one that potentially set limits on the colonial project, late-twentieth-century activists argued that that relationship required Americans to acknowledge and contain the unequal effects of their modernization. Although the relationship between race and disease has been continually reformulated, the linkage remains remarkably and distressingly persistent.[17]

Of course, my own intellectual practices and concerns influenced the narrative long before I entered the archives. And while I believe that all understandings are contingent, partial, and situated, underlying my own practice is, admittedly, the belief that our bodies powerfully affect and depend on the environments they inhabit, that the changes we make in the land will ultimately be registered in ourselves and in those who follow us in very material ways. This is the belief that underwrites my entire project; otherwise, this book would make no sense. And yet those linkages may be understood in multiple ways, or even not at all. A major part of the story here is about when and how the links between environment and disease are made visible or invisible, a tracing of the cultural and intellectual practices, my own and others', that connect or disconnect the health of bodies and the condition of the landscape, and the ensuing social and political implications.

1

Body and Environment
in an Era of Colonization

A knowledge of the *etiology of diseases* can best be attained
by studying the affections of different localities in connection with every condition and circumstance calculated to
operate prejudicially or otherwise upon the health of the
inhabitants. Such philosophical investigation is particularly
useful in tracing the modifications diseases may undergo
from the agency of causes of a local or special character;
and being also calculated to elucidate the relationship of
diseases to climate, to the prevailing geological formations—
the fauna, the vegetables, the minerals, the waters, which
vary with the earth's crust, wherever man can make his
abode, commends itself to the pioneer physician of our
extended territory.

> Dr. Thomas Logan in *Transactions of the*
> *American Medical Association,* 1859

It is typical to think of the colonization of western North America as a
process in which Europeans and Americans remade the land by
reworking natural environments into forms that were both aesthetically
pleasing and materially useful. This is surely true, but it is also true that
in earlier eras Americans understood colonization as involving bodily
transformation as well. The process could work both ways. Places could
alter bodies as much as bodies could alter places. Despite the political
and cultural rhetoric of conquest, those engaged in colonizing western
North America recognized that the effort often brought substantial physical risks. Western immigration was a gamble in physical, as well as economic, terms.

Historians of American expansion have not neglected concerns about
health; however, they have overwhelmingly emphasized the disease experience of Native Americans. The story of "virgin soil epidemics"—the

transmission of European diseases to Indian populations with no previous exposure and thus no acquired or inherited resistance—is now quite
well known. That the consequences of European disease were horrific for
most Indian peoples is certain. But Indians were not the only people who
suffered extensively from illness in the eighteenth and nineteenth centuries. The processes and exchanges brought about by the colonial
endeavor of that period created what one scholar has labeled a "global
epidemiological crisis." Everyone was more vulnerable to illness, even
those who stayed put. Diseases that were already familiar to white
colonists were not necessarily less debilitating or frightening on that
account. Accordingly, concerns about disease and disability permeated
much of nineteenth-century European and American culture.[1]

The focus on the disease experience of Native Americans is justified by
the unprecedented scope of Indian depopulation and the role that illnesses played in that catastrophe. But to ignore the disease experience of
white immigrants is problematic. Such a selective focus can in some cases
serve to retrospectively naturalize Euro-American colonization. What
was historically contingent—European dominance in North America—
can come to seem biologically predestined, and the centuries-long struggle between native peoples and Euro-Americans, which was marked by
incredible violence, can too easily be rewritten as a passive and unavoidable conquest. The historical "forgetting" of disease, other than the diseases of Indians, may itself be part of a centuries-long process of normalizing white colonization in the western United States.[2] Moreover, by
failing to acknowledge the perceived vulnerability of white as well as
nonwhite bodies in earlier periods, we run the risk of reading those periods through the lens of later demographic transitions. By contrast, those
engaged in colonization were often far less certain of its ultimate outcome, particularly as they waged their own struggles with Native
Americans, unfamiliar landscapes, and a host of lethal diseases: cholera,
malaria, dysentery, typhus, yellow fever, tuberculosis.

Understanding the health concerns of nineteenth-century settlers in
western North America requires that we put aside more recent understandings of both the human body and the environment. The one-sided
focus on the disease history of Indian peoples can have the effect of rewriting white bodies in contrasting and somewhat ahistorical terms—as clearly
bounded, always resilient, and unproblematically cosmopolitan. But this
modern understanding of the body cannot be found in early- or even mid-
nineteenth-century sources. In fact, the very idea of a distinct and bounded
body, clearly separate from its environment, and able to move unproblem-

atically from one location to another, is a relatively recent historical development. Nineteenth-century bodies, white and nonwhite, were malleable and porous entities that were in constant interaction with the surrounding environment, an environment that retained a complex agency of its own. Disease in the nineteenth century, even when acknowledged to be contagious, was not reducible to specific pathogenic agents or person-to-person contact. Contemporaries understood the causes of disease as spread widely across both bodies and landscapes. Consequently, prospective settlers approached new environments with caution, recognizing that the land itself could be either a font of health or a source of illness.

For those who moved west, human bodies were the most sensitive and reliable indicators of place.[3] The presence or absence of certain illnesses, rates of birth and death, and the course of epidemics—all these were important clues to the qualities of an unfamiliar landscape. Settlers and travelers alike were typically attuned to the reactions of their bodies and to the appearances of those they met. Their physical reactions—the onset of fever, a new sense of vigor, a persistent cough, the timing of menstrual cycles—became important means to understand new places. As settlers set about to alter the landscape, they recognized that the landscape, in turn, might also alter them. Settlers' bodies were thus instruments of colonialism in a double sense—in that they both facilitated the colonial project and registered that project's physical effects. Nineteenth-century American medicine eagerly addressed itself to this project, assessing both bodies and landscapes with an eye toward preserving health and whiteness in new locations.

COLONIZATION AND HEALTH

Today California is commonly, even prosaically, associated with health. In our own health-obsessed time, California stands out as an especially health-obsessed place. But the rhetorical association of California and health was largely a creation of mid-nineteenth-century western boosters. Firsthand accounts of the period offer a much more equivocal and sometimes negative picture. Until the late nineteenth century, California, in European and American minds, was a distant frontier about which little was known, a "terra incognita" as more than one source referred to it. Although California may not have raised the same level of fears among Euro-Americans that southern Africa or the Caribbean did, we should not then assume that early migrants to the Far West understood their relocation in trivial terms.

In the Spanish and Mexican colonial periods, Alta California's colonizers and explorers did not consider it a particularly healthful place. There is no obvious reason why they should have. The existence of disease among Indians in California is indicated in part by their extensive knowledge of therapies, which early European observers simultaneously derided and recorded. Among those native remedies that Americans adopted were *Eriodictyon californicum* (yerba santa), a treatment for bronchitis; *Rhamnus purshiana* (cascara sagrada), a well-known cathartic; and *Grindelia robusta,* used for both lung and skin diseases. By the eighteenth century, Indian peoples were also dealing with an onslaught of new diseases. Scholars have typically assumed that European diseases emerged in California only after the establishment of the first Spanish mission in 1769, but some diseases may have preceded colonization. There is no question, however, that disease arrived anew with the Spanish. Contemporary scholars concur that venereal diseases (both syphilis and gonorrhea) were rampant among the Spanish and the mission Indians and had spread to the tribes of central California by 1814.[4]

Venereal diseases were the most prevalent but hardly the only old-world illnesses in colonial California. In the early nineteenth century Franciscan missionaries reported the presence of consumption, dysentery, and various fevers. A devastating measles epidemic swept the missions in 1806 and may also have spread beyond. Smallpox probably arrived in 1828. In 1837 a smallpox epidemic broke out at Fort Ross on the northern California coast and moved south, killing more than 2,000 individuals mainly among the Pomo, Wappo, and Wintun. Another epidemic began in 1844 among settlers in the Central Valley town of Stockton; it subsequently spread through the valley and foothill regions, affecting mostly the Miwok. In addition to smallpox and measles, pneumonia, diphtheria, scarlet fever, and tuberculosis were recorded in California prior to the 1840s. Disease undoubtedly played a critical role in the decline of the California Indians. The demographer Sherburne Cook estimated that Indian numbers dropped by 21 percent between 1770 and 1830, from more than 300,000 individuals to approximately 65,000. Declines were far higher in the missions than elsewhere, a reflection of both a more concentrated population and the oppressive and often violent nature of mission life.[5]

The few medical men who attempted to assess the health of California in the Spanish and Mexican periods were circumspect. In 1786 the physician Henry Rollin accompanied a French expedition to California and published an account of the voyage in Paris eleven years later. Rollin cataloged the various diseases suffered by California Indians, which he attributed largely to the "great changes in temperature" during the year. He listed several diseases as prevalent in the region, including "ephemeral and intermittent fevers," "digestive disturbances," "putrid fever," "petechial fever," "bilious fevers," and dysentery, neuritis, rheumatic "affections," scabies, opthalmias, pox, and epilepsy. Rollin laid special emphasis on the "high fevers" and "bilious fevers," which he noted were widely feared and frequently fatal.[6]

Among the Spanish, the only professional physician in Alta California was the surgeon general stationed at the provincial capital of Monterey, a position that was evidently difficult to fill. Of the eight men who occupied this position between 1769 and 1824, few left significant records. However, in 1804, at the behest of his superiors, who were concerned by the exceedingly high mortality among mission Indians, Dr. José Benites wrote a lengthy report summarizing the medical condition of the province. He reported that syphilis, scrofula, and tuberculosis were common illnesses. He also made reference to the region's unfavorable climate: the humidity, heavy fogs, and great cold, all of which he believed were contributing to the prevalence of disease. Authorities in Mexico City had little interest in supporting Benites's principal request—that they establish a hospital at Monterey. Instead the Royal Medical Board noted somewhat fatalistically that disease in Alta California was unavoidable because of "the extreme cold, the lack of shelter, the bad water, lack of vegetables, and badly prepared meats," as well as the "voluntary indiscretions" of the inhabitants. Impressions recorded at about the same time by George Heinrich von Langsdorff, a surgeon accompanying a Russian expedition to California, were more favorable. While Langsdorff found the west coast of Mexico unhealthy in the extreme, he reported that the climate of Alta California was "better and more salubrious." But he was hardly enthusiastic on that point, noting that the local Indians were often afflicted with fevers, measles, venereal diseases, and a mysterious palpitation of the heart.[7]

As these sources indicate, disease was a constant presence in the region by the early 1800s, if not before. Several epidemics swept through California in these decades, including at least three severe outbreaks of smallpox.[8] Yet, by all accounts, a different and especially devastating illness appeared in the California interior in the 1830s. Indian tribes throughout central California were catastrophically affected, as were the few white settlers and travelers in the region. John Work, an Irish immigrant who had settled in Canada and the leader of a Hudson's Bay trapping expedition to central California, was one of those who fell ill in the summer of 1833; Work's journal offers a firsthand account of the disease among both local Indians and members of his party.

WEDNESDAY 31 [JULY 1833]

Several of our people have been for some days unwell and some symptoms of the fever breaking out among them.—Indeed for a length of time back, the weather has been very unfavorable for health. The heat, except for a few days back excessive during the day and a heavy chilly dew in the night, so that our blankets would be completely wet in the morning as we slept in the open air. Besides we often had very bad water.

TUESDAY 6 [AUGUST 1833]

Some sickness prevails among the Indians on feather river. The villages which were so populous and swarming with inhabitants when we passed that way in Jany or Febry last seem now almost deserted & have a desolate appearance. The few wretched Indians who remain seem wretched they are lying apparently scarcely able to move. . . . We are unable to learn the malady or its cause.

TUESDAY 20 [AUGUST 1833]

Our sick people get no better, nine more have fallen ill within these two days, making in all 61 that are ill, a good many of them attacked with trembling fits. . . . Our condition is really deplorable, so many of the people taken ill and no medicines, fortunately not many of the men are yet ill, but is is to be apprehended they soon will fall and that we will soon become so weak that we will not be able to raise camp, and I am afraid to stop lest we die like the Indians the most of the people completely disheartened, and indeed well they may.—I endeavour to keep up their spirits as well as I can but it is become now of little effect.[9]

Four days later Work reported that seventy-two persons were ill out of a party of one hundred, and over the next two and a half months, several died. "Our whole party is now become exceedingly helpless," Work wrote on September 7. At the same time, the death toll among resident Indians was almost incomprehensibly high. Of the Indians in the northern Sacramento Valley, the Wintun, Work noted that "the villags [sic] seem almost wholly depopulated." Later accounts confirmed the magnitude of the epidemic. An American trapper, J. J. Warner, recalled of the once densely populated region that every native village along the rivers had been abandoned and his party saw "but one living Indian." A member of the Yokuts tribe told the ethnographer Stephen Powers in 1872

Figure 1. A rancheria near Yuba City, in the area where the 1833 malaria epidemic devastated Native American populations. From *Gleason's Pictorial* 13 (27 March 1852): 96. Courtesy California Historical Society, FN-4341.

that a plague had raged throughout the San Joaquin Valley several years earlier, destroying "thousands of lives." On October 31, 1833, Work finally made it back to Fort Vancouver. As he described his condition some years later, "I was so much exhausted by this debilitating disease that I was reduced to a perfect skeleton and could scarcely walk."[10]

Modern scholars interpret this event as an epidemic outbreak of malaria and typically trace the origins of the disease in central California to Work's own party, though malaria may have appeared in conjunction with influenza, which could explain the dramatically high death rates. While malaria is not generally believed to have been endemic to California before the nineteenth century, at least four species of anopheles mosquitoes were. Once what we now understand as the plasmodium parasite was introduced into California, it could spread through those regions that supported large numbers of anopheles. Moreover, the temperate climate and long, hot summers of California were conducive to an epidemic outbreak, as they fostered multiple cycles of mosquito reproduction. The disease, or diseases, that reached California in 1832 were the southern extension of an epidemic, most likely of vivax malaria, that had begun on the lower Columbia River in 1830, at Fort Vancouver. The epidemic had a devastating impact on the Chinook and Kalapuyan peoples in the Pacific Northwest, prompting one contemporary scholar to label it "the single most important epidemiological event" in the recorded history of the region. By the time Work's party left Vancouver, malaria had apparently infected most of the white population in the Northwest, and, in fact, Work reported that several of his party became sick with "intermittent fever" en route to California. Disease then traveled south from the Columbia in the bodies of the trappers and their families. Its overall effect on the Indians of California was as terrible as it had been on the Indians of the Northwest. In 1955 Cook estimated Native American mortality in California at 20,000; however, he later revised that number upward to 50,000, or what he estimated to be one-half the entire Native American population in central California. Though the numbers cannot be determined with any accuracy, it is clear from contemporary accounts that the epidemic radically disorganized California Indian societies, leaving Indian peoples ill prepared to resist or adapt to the dramatic invasion of their territory that came a decade and a half later with the discovery of gold.[11]

But the impact of the epidemic on Indians should not obscure the fact that whites themselves were highly vulnerable to malaria and often incapacitated by it—as the fate of Work's expedition attests—though they were typically less likely to die.[12] Malaria, which remains a significant global problem, was the preeminent disease of the nineteenth-century frontier. Fear of the intermittent and remittent fevers were shared by all western colonists, as well as by those who were long settled in the southern states. Even where death rates from malaria alone were not high, it often debilitated much of the population and complicated other, more fatal illnesses. The disease spread rapidly in mosquito-ridden areas such as central California and could easily infect an entire community.[13]

Such contemporary diagnoses were, of course, unavailable to either the Indians or John Work. Mosquitoes were an ongoing nuisance to Indians and whites in California, but no one had reason to think them a source of illness. Indians generally interpreted disease as a foreign or hostile object that had entered the body, the result of an offended or malignant spirit. To cure the ill, these disease objects or "pains" had to be extracted from the body. For most Indian groups in California, the preferred cure was a bloodletting ceremony performed by a shaman, often in combination with specific medicines.[14] Work, on the other hand, drew on Euro-American frameworks of disease. He believed his party suffered from two diseases—a mysterious fever that caused violent headaches and intense pain in the bones and the more familiar "ague," or intermittent fever. European cures for fever and ague were similar to those of many Indian tribes and included both bloodletting and quinine.

Yet having neither medicine nor a doctor in his party, Work sought a change of climate and location. He urged his exhausted and discouraged companions to continue, believing that once they reached the mountains they "would experience a difference of climate which would most likely effect a change for the better."[15]

Work was clearly desperate to save himself and his party, and in seeking a "difference of climate," his response was consistent with the medical advice of his day. In the nineteenth-century world, bodies were understood differently than they would be in the next century. Professional and popular beliefs about health derived from humoralism—a system of medicine that held health was the outcome of balance among the essential bodily fluids, or "humors." Though the idea of humoralism may seem bizarre to most contemporary Americans, in slightly different forms humoralism was the basis for Western medicine from the time of ancient Greece until nearly the end of the nineteenth century. In contrast, the history of "modern" Western medicine and the corresponding "modern" body is far more brief. These ideas would come only with the bacteriological discoveries of the late nineteenth century and their institutionalization in the twentieth.[16]

Before the late nineteenth century, a healthy body was a body in equilibrium, and disease signified that balance needed to be restored. Composed of flows and fluxes—of blood, mucus, saliva, feces, perspiration—the body could easily be either over- or understimulated. The result was imbalance, and the likely outcome was illness. An improper diet, poor habits, a shock to the system, mental anguish—any of these might push a body out of kilter. But especially important were changes in the external environment. Changes in temperature, winds, humidity, or simply an unfavorable landscape could alter the body's normal functioning and leave it prone to a variety of ailments. Work connected the onset of disease among his party with weather that was "very unfavorable for health." Similarly, an army surgeon reporting on the health of the U.S. troops stationed in the California desert in the 1850s wrote that the "unusual mortality" witnessed at his post was "attributable to but one cause, viz.: their transfer from a comparatively cold climate to one so much warmer and more debilitating."[17]

For nineteenth-century Americans, the body itself was not a clearly bounded entity, separate and distinct from its surroundings; rather, it was porous and permeable. The skin did not close off an individual, separating him or her from the larger world. The body flowed into the environment, and the environment seeped into an individual body—through the air one breathed, the food one ate, the water one drank. These inter-actions were not only unavoidable; they were critical to health as well as illness. External surroundings could shape the body in both subtle and profound ways. A given environment inevitably left its mark in a body's shape, color, and strength, while radical changes in a person's environment could effect wondrous cures or induce sudden illness. The prevailing winds, the onset of floods, a local earthquake, a distant volcanic eruption—these might all be factors affecting an individual's condition. Local surroundings might be managed for better health, but they could never be kept at bay—nor would one want to do so. Health was not the product of successfully closing a body off from external influences but of intelligently managing the relationship between an individual and his or her surroundings.[18]

This ongoing concern with the environment does not mean that nine-teenth-century individuals were oblivious to the process of contagion. Quite the contrary. By the end of the eighteenth century, Spanish measures to prevent the spread of smallpox included isolation, quaran-tine, and inoculation—which may well have stemmed the spread of sev-eral epidemics known to have ravaged Mexico and the Southwest.[19] Moreover, throughout most of the nineteenth century, a heated and sometimes vitriolic debate raged between those physicians who advo-cated "contagionism" and those who held to the doctrine of "anticonta-gionism." And though it is easy to interpret this debate through the lens of later scientific developments—and thus to read the "contagionist" position as a forerunner of germ theory—such a reading erases an ear-lier context. The very ideas of contagion and infection held different meanings than they would in the following century, and they were not necessarily incompatible. Part of what makes eighteenth- and nineteenth-century sources so opaque to a modern reader is that the categories that are so meaningful to us—contagious versus noncontagious, infectious versus chronic—were neither crucial nor discrete distinctions in earlier eras. Even when these words were employed, their meanings differed from contemporary usage. Although the most extreme anticontagionists, such as the well-known Philadelphia physician Benjamin Rush, suggested that there were but two opposed ways of understanding any given dis-ease, most medical men embraced a more complex position. For instance, a Spanish directive on smallpox issued to the governor of California in 1786 emphasized the need for quarantine. But while acknowledging that most professional men believed the disease was transmitted by "contact with the victims or the houses in which they are treated," the writer nonetheless insisted that victims should be quarantined "in a healthy location [that] shall be situated so that the prevailing winds in the region cannot communicate the contagion to the villages and farms of the vicinity."[20]

A report on disease in California published in the 1860s reveals the slipperiness of medical categories. The writer divided epidemic disease into three classes: "contagious" (which included smallpox, scarlet fever, and measles), "meteoratious" (diseases that were contagious to a limited degree), and "infectious" (diseases "assumed to possess the property of propagating . . . by means of a vitiated or poisoned atmosphere emanating from and surrounding the diseased person, without contact of the body or clothing"). Moreover, he elaborated the difference between contagious and meteoratious epidemics in the following way: "The great majority of cases [of contagious disease] spring immediately from specific poisons, generated in primary or atmospheric cases, and communicable from one individual to another. Whereas in meteoratious epidemics, excepting in one or two of them, every case is of atmospheric *origin;* and, in the exceptionable instances, as in the cholera (and to which may be added diphtheria), which is believed to possess the contagious attribute, the great *majority* of cases manifestly arise, not from the diffusion of its contagious virus, but from the existing meteoratious influence." Implicit in this description was an acknowledgment that the distinctions between these categories were anything but firm. Certain diseases, notably smallpox and syphilis, were widely held to be transmitted from person to person. Even so, contagious diseases might have an atmospheric origin, while meteoratious epidemics might have contagious "attributes." Contagious disease shaded into environmental disease and vice versa.[21]

Despite the recognition of contagion, the local environment was always regarded as critical to health or illness. In the words of the historian Charles Rosenberg, "Disease entities played a relatively small role in a scheme that emphasized the body's unending transactions with its environment."[22] On the other end of the spectrum from smallpox were various fevers, nearly all of which were understood to originate from local places and were thus labeled "endemic." Yet epidemics were also understood to have local causes, for that offered the most logical explanation for why so many people in one place became ill at the same time. Still other diseases such as diphtheria and yellow fever were believed to emanate from environmental causes but were liable to become contagious, depending on the circumstances. And most physicians concurred that even "contagious" diseases, such as smallpox and plague, had important "climatological relations." When smallpox broke out in California in 1868, leading California physicians readily acknowledged its contagious character and argued for vaccination, but nonetheless they believed that both local and global climatic conditions were relevant to the course of the epidemic. As Thomas Logan put it, "There is some peculiar, but as yet inscrutable condition of the climate which favors its development." Another California physician, Frederick Hatch, offered the hypothesis that climatic conditions, "having brought about such modifications in the constitutions of our people as to renew a susceptibility to the agent," might explain the failure of vaccination to protect against smallpox that year.[23] These and other writings reveal understandings of causality that are multiple; environmental explanations easily overlap with theories of contagion. Disease always had many potential sources, both human and nonhuman.

MAPPING THE DISEASE ENVIRONMENT: MEDICAL GEOGRAPHY

Nineteenth-century writings about disease offer a window into earlier conceptions of the body. Perhaps less obviously, these same writings speak to earlier conceptions of the environment. Different conceptions of illness point to differences in how people have understood the nonhuman world. When viewed from the perspective of health, the nineteenth-century environment was neither passive nor necessarily benign in its natural state. To the contrary, the "natural" environment, especially those environments least touched by the processes of civilization, acted on settlers' bodies in sometimes aggressive and unpredictable ways. Consequently, untested landscapes were always physically threatening. This fear of distant and unfamiliar places generated reams of popular advice for would-be settlers and travelers. At the same time, existing medical and scientific practices brought the environmental sources of disease into focus.

Interest in the medical effects of certain environments has a very long history, dating at least to the Greek physician Hippocrates and his treatise *Airs, Waters, and Places,* written in the fifth century B.C. Theories of environmental causation gained particular prominence in seventeenth- and eighteenth-century Europe. In that period several European intellectuals drew on Hippocratic ideas to articulate a discourse that denigrated warm places and their inhabitants. Among the most influential was Montesquieu, who, in *The Spirit of Laws* (1747), famously argued that hot climates produced sloth, excessive sexuality, and despotic forms of government. This view would be widely held in Europe for at least the next century. Only in the 1900s, however, would the professions of medicine and geography scientize these beliefs. For both Europeans and Americans, the project of colonial expansion fostered the new disciplines of medical geography—which studied the large-scale distribution of diseases across continents—and medical topography—which cataloged the physical factors that affected health in certain localities.[24]

What motivated these inquiries was the desire to explore and colonize new environments. Medical geography implicitly and often explicitly served the needs of European colonialism. Many of the earliest medical topographies emerged from various militaries out of the concern for troop mortality in distant regions, and, not surprisingly, it was British physicians who did the most to systematize the geographic approach to disease. The British colonial project had generated an obsessive interest in the "tropics" as a zone of overabundant nature that was inherently inhospitable to European "civilization," and these emerging ideas about "tropical" environments owed much to European fears of disease. The canonical English texts on environmental medicine in the nineteenth century were James Lind's *Diseases Incidental to Europeans in Hot Climates*, published in 1768, and James Johnson's *The Influence of Tropical Climates on European Constitutions*. Johnson's volume chronicled the diseases experienced by Europeans in so-called tropical lands: India, Asia, Batavia, southern Europe, western Africa, and the West Indies. First published in 1812, the book had reached its sixth edition by 1841, a testament to its influence. It would remain the principal reference on the subject for two more decades as it was expanded and edited by another British colonial physician, James Ranald Martin.[25]

Though their work on climate and disease was separated by more than seventy years, Lind and Martin occupied a similar intellectual milieu. For both authors, health was the result of humoral balance in the body, and warm climates were likely to overstimulate the temperate European constitution. Excessive heat, especially temperatures that exceeded that of the body, predisposed an individual to all kinds of diseases. Prevailing medical opinion held that the greatest effect was felt on the liver, which produced irregular secretions until, exhausted, it ceased to function adequately. The texts themselves were both diagnostic and prescriptive, offering not only a chronicle of disease and its symptoms but also suggestions on how Europeans might lessen the impacts of hot climates on their selves (through rigid temperance and prophylactic measures). The message was that European bodies were highly sensitive to relocation and required careful observation and intensive self-management in unfamiliar and inherently hostile places. In the case of serious illness, however, the best and often the only hope lay in returning to a more temperate climate. As Johnson wrote, a change of locality was frequently "tantamount to a transition from almost hopeless disease to rapid recovery."[26]

German and French physicians also contributed to the development of a global geography of disease. Especially important to the continental versions of medical geography was the work of Alexander von Humboldt, who is considered the founder of scientific geography. Humboldt sought to understand the natural world by collecting quantitative information about various landscapes and then seeking mathematical correlations among the variables he had measured. His most significant contribution to physical geography was the isothermal map—a cartographic representation that linked regions by their average temperatures. Humboldt noted that these lines of average temperature, along with altitude, set limits on the occurrence of certain plants, and he produced numerous maps of the world that charted distributions of flora. He also suggested that, like plants, certain diseases were produced under specific conditions of temperature, humidity, and altitude. It was this observation that medical geographers, Germans in particular, took as their starting point. They sought to map the spatial distribution of disease in the same way that Humboldt had mapped the distribution of plants. Like their mentor, medical geographers looked for correlations between the occurrence of disease and measured characteristics of the landscape. Practitioners held out the hope that with the collection of enough data—temperature, barometric pressure, rainfall, and so on— they would be able to predict the response of human bodies to diverse environmental conditions.[27]

At root, the primary concern of nineteenth-century medical geography lay in preserving the health of the white race in unfamiliar lands. Behind the desire to uncover the relationship between bodies and landscape lay the belief that the success of Europeans had always hinged, to some undetermined extent, on climate and, moreover, that climate might ultimately set limits on their continuing colonial ambitions. Nineteenth-century Europeans and Americans understood race in multiple and contradictory ways—as variously a sign of biology, nationality, and culture. A concept anchored in incoherence, race necessarily eluded precise definition. Yet it was quite clear to contemporaries that race was associated in some way with place. Whites came from Europe, blacks from Africa. The yellow race originated in Asia, the red in America. Race always had a geographic component, and thus it is hardly surprising that ideas of race played a central role in nineteenth-century medical geography. After all, the question that most interested European medical geographers was whether those of northern European descent (i.e., whites) could survive and prosper in climates that they associated with "other" races. While there was general agreement that strange environments had negative effects on European bodies, contemporaries debated the extent to which those same bodies might adapt and acclimatize to their new surroundings.[28]

Proponents of acclimatization believed that human bodies could, over time, adjust to new surroundings. As Europeans had succeeded in introducing plants and animals to unfamiliar regions, they argued, the same would be true for transplanted peoples. But theories of human acclimatization had many opponents. For most who argued against acclimatization, the central issue was racial malleability. If European bodies could in fact physically change to survive in a new climate, would they still be European? More to the point, would they still be white? As James Johnson put the question in his medical treatise of 1820, "Will it be said, that the fair complexion of Europeans, may, in two or three generations, acquire the sable tinct of the inter-tropical natives, by exchanging situations?"[29] By answering with an unequivocal "no," Johnson adhered to a belief in racial stability; whites would remain white no matter where they resided. But Johnson's was hardly the last word on the issue. The question would continue to preoccupy European and American intellectuals for the rest of the century.

The concern with whiteness and its potential malleability was paramount in European settler societies. The "frontier," whether in Africa or North America, was never a zone that separated empty from populated lands; it was, however, a zone that separated lands dominated by those identified as "white" from those whom they deemed nonwhite. By definition, frontiers posed challenges to racial identities; their miscegenated populations only underscored the problem. In a period in which place helped to produce ideas of race and bodies were perceived as porous and 5 permeable, migration always threatened racial identity as well as health.[30]

Even the suitability of North America for European immigrants had been an issue of long-standing debate. In the seventeenth century, British settlement in America was accompanied by considerable fears over health, especially in the southern colonies but also, to a lesser extent, in

New England.[31] In the late eighteenth century European and American elites intensely debated theories of climate and civilization. Among the most influential authors on this point was Georges-Louis Leclerc, Comte de Buffon, who argued that the cold, humid climate of North America could not support plants and animals of the same size and quality as those in Europe. Pointing to the absence of large native mammals—such as the giraffe, the hippopotamus, and the lion—and to the degeneration of European livestock in the new world, Buffon argued that North America produced neither the same quality of person nor the high level of civilization that existed in Europe. The evidence for this lay in Buffon's description of the physical inferiority of Native Americans—their lesser strength, their low fertility, their "lack of ardour." Though transplanted Europeans might survive in North America, they would not flourish. So influential was this thinking among elites that Thomas Jefferson felt compelled to mount a detailed defense of the North American climate and its plant and animal species in his only book, *Notes on the State of Virginia*. For Jefferson, establishing the existence of the mammoth—a beast larger than any found in Europe—and defending both American livestock and the sexual prowess of Native Americans were crucial to predicting a healthy and fertile white population in America. His vision of a republican civilization ultimately depended on the natural environment's ability to support properly European bodies.[32] Like the mammoth, the native body was a "production" of nature and a testament to the New World's inherent virility.

Discussions of the North American climate's effect on both health and fertility would continue. However, by the nineteenth century whites had proven themselves capable of prospering in both New England and the South. Colonists' bodily fears had gradually been replaced by a sense of their physical fitness for the eastern regions of North America.[33] But at midcentury, as Americans embarked on the effort to colonize the western half of the continent, western climates remained something of a wild card for white settlers. The regions west of the Mississippi were unfamiliar and relatively untested. Much of the landscape of the West was treeless and arid, in contrast to the humid and well-forested lands of the East. The initial American settlement experience in the Mississippi Valley had not been encouraging for whites, who had sickened and died in large numbers. "It is to be suspected," wrote the English physician John McCulloch in 1829, "that no changes and no cultivation will ever bring it into a state of salubrity." Moreover, these regions were still filled with nonwhite populations. The different climate and environment in the

West were subjects of constant commentary, although contemporaries acknowledged their lack of information. "The arid climates of the interior and the cool Pacific coast have been occupied so recently, and so little observed, that is difficult to trace the climatological geography of disease there," wrote J. W. Blodgett in a massive work on the American climate.[34] While most American elites espoused confidence that white Americans would eventually populate western North America, the region's suitability for white bodies and Euro-American civilization was a subject of ongoing debate. Like their European counterparts, Americans feared that the environmental characteristics of new lands might frustrate their desire for expansion.

Consequently, knowledge of the relationship between climate and disease became as important as geologic or agricultural assessments to furthering the colonization of western North America, and the period saw several important American contributions to medical geography. In 1842 the army surgeon Samuel Forry used data collected by the military to compose the first complete medical geography of the United States. Forry emphasized the need to move from anecdotal accounts of climate and disease to the quantification of climatic features, and his work was widely lauded as an original and important contribution to medical science.[35] Like most American physicians, Forry drew heavily on Humboldt, whose mapping of isothermal lines had disrupted an older reliance on latitude alone as the principal factor determining climate. When latitude was considered in isolation, the more southerly location of North America relative to northern Europe was cause for concern. Humboldt's isothermal maps, on the other hand, helped to draw attention away from differences between the old and new worlds over which Jefferson and Buffon had argued. He reassuringly linked North America with Europe by redefining the United States as an unambiguously "temperate" region.

The publication of Daniel Drake's massive work, *A Systematic Treatise on the . . . Principal Diseases of the Interior of the Valley of North America as They Appear in the Caucasian, African, Indian, and Esquimaux Varieties of Its Population* (1850–54), drew international attention to the medical geography of North America. In many ways, Drake was an unlikely person to make such a contribution. He was raised on the Kentucky frontier at the turn of the century, far from the centers of medical knowledge. His medical training consisted of an apprenticeship with the most prominent of Cincinnati's four physicians. After establishing his own practice, Drake published a pamphlet describing the climate, topography, and diseases of his town. The work was well received, and that publishing success inspired him to undertake an exceedingly ambitious effort to chronicle the diseases of the entire West. Drake embraced as his region of study the area between the Allegheny Mountains in the east and the Rockies in the west and running from the Gulf of Mexico to the Polar Sea. He was explicitly Humboldtian in his approach, emphasizing the primary importance of latitude and altitude to disease and the local character of both symptoms and cures and insisting that the watershed was the proper unit of medical analysis. All told, Drake spent more than ten years collecting firsthand information and traveled more than thirty thousand miles. When published, his book ran to more than 1,800 pages. It immediately became a seminal publication in American medicine and inspired much more work along the same lines. That growing interest in medical geography also helped drive an interest in meteorology among American intellectuals, doctors in particular. In 1848 the newly formed Smithsonian Institution enlisted physicians across the country to assist in the systematic collection of national weather data. Lorin Blodget, a former employee of the Smithsonian, would publish much of that work in his 500-page *Climatology of the United States* (1857), which American physicians embraced as a critical reference.[36]

Nineteenth-century medical geography was an elite, scientific, and transnational discourse, but it emerged at a time when even scientific knowledge was acknowledged to be profoundly local. The goal of medical geographers was not to erase local particularity but to quantify and systematize it. Thus the most valuable work in the field could, and often did, emerge from the periphery—as in the case of Drake's *Systematic Treatise*. It was on the periphery where new and unusual relationships might be uncovered and where existing theories could be tested against new circumstances. Medical geography was a science, but it was a science of local experience.

EVALUATING THE ENVIRONMENT OF CALIFORNIA

As Americans set their sights on colonizing the Pacific Coast, discussions of the far western environment and its effects on health appeared in a variety of places: newspaper articles, medical periodicals, emigrant guides, almanacs, personal letters, and government reports. The region of California generated no shortage of medicoenvironmental commentary, and among the most prolific writers were those who most enthusiastically and uncritically advocated settlement: western boosters.

Booster literature on California proliferated along with the push for American expansion in the 1840s. Despite the profusion of writers, the tracts themselves have a formulaic quality. Boosters routinely discussed those details that had some bearing on settlers' economic prospects: land availability, soil fertility, the length of the growing season, the size and character of towns, and the availability of transportation. These same writers also almost invariably addressed the region's effect on health.[37] And despite the devastating experience of illness in the 1830s, California boosters made much of the region's "salubrity" in the following decade. Perhaps this is not surprising. Like modern tabloids, booster literature is notoriously unreliable. History and environments alike can too easily be rewritten to further social and political goals, and booster accounts of California were motivated at least as much by desires for American colonization as by empirical observation. Those inspired to write about the region in the 1840s typically sought white settlement in the Far West, America's "manifest destiny," regardless of any potential costs in human suffering. And though what they wrote was not necessarily or even likely to be true, they tell us something about prevailing cultural vocabularies: what boosters wrote about presumably held some meaning for their intended readers. Thus while the repeated, almost obligatory, insistence on the region's healthfulness might say little about the actual prevalence of disease, it suggests that perceptions of health were important, even critical, to understanding a foreign place. Nineteenth-century boosters did not invent the connection between climate and health; they did, however, wield that connection freely, often with considerable flair.

Richard Henry Dana, one of the most widely read popularizers of California in the 1840s, claimed that the region was "blessed with a climate, than which there can be no better in the world; free from all manner of disease, whether epidemic or endemic." John Marsh, who settled in California in the 1830s and offered his services as a physician (although he, like many other physicians in this period, had no formal medical training), wrote, "It is much the most healthy country I have ever seen, or have any knowledge of. There is no disease whatever than can be attributed to the influence of the climate." Yet Marsh himself maintained a thriving medical practice, treating both Indians and whites for fever and ague and other diseases. The author of a popular emigrant guide published in the mid-1840s made a point of denying that the virulent "fevers" known to have killed thousands of Indians in the previous decade were attributable to any "local" causes but instead blamed the mortality on the habits of the Indians themselves. He went on to claim that "there is no country in the known world, possessing so fertile a climate, of such mildness and uniform salubrity."[38] Victor Jean Fourgeaud, a physician with connections to the expansionist politicians Thomas Hart Benton and William Gilpin, went considerably further in his account of the region's effect on health. In a piece intended for eastern audiences and potential immigrants, Fourgeaud asserted that "the general salubrity of California has justly become a proverb. The surgeons of California have remarked that wounds heal here with astonishing rapidity, owing, it is supposed, in a great measure, to the extreme purity of the atmosphere." For those who sought American colonization of California, it was critical to establish the region's healthfulness, and among this set Fourgeaud's claim for the wound-healing properties of the atmosphere would be frequently repeated.[39]

What boosterism alone could not accomplish the discovery of gold did. In 1849 a massive migration to California began that decisively shifted the region's racial and ethnic demographics. Perhaps as many as 90,000 immigrants arrived in California in a single year, and between 1848 and 1860 the population rose by almost 300,000.[40] With so many people suddenly in the region under such extraordinary circumstances, accounts of the local environment proliferated. Those who traveled to California during the gold rush were not of one mind regarding the region's healthfulness, however. Individuals evaluated the environment's effects through both their personal experience and their hopes. Some, such as the physician John Baker who came to California from New Hampshire in 1853, managed to stay relatively healthy and attributed their vigor and success in part to the positive effects of the local climate. But many more wrote of illness and disease. Sickness seemingly surrounded and enveloped miners and travelers in the early 1850s. The journalist Bayard Taylor visited the interior of California in 1849 and claimed that "three-fourths of the people who settle in Sacramento City are visited by agues, diarrhoeas, and other reducing complaints." The experience of most miners confirmed this claim. As the miner George Kent confided to his journal, "Almost all of us had severe attacks of the diarrhoea or dysentery either before or after our arrival at this place." "Never had I been so ill before," wrote Thomas Kerr after he came down with "the ague" at Sacramento. Still another miner, John Gunnell, wrote, "[I] had not been in good health since I bin in Calaforn" and on that account advised others not to make the journey. "Gold was not a sufficient recompense," another physician and failed miner wrote, "for disease and broken constitution."[41]

From the early 1850s on, the incongruence of booster accounts and bodily experience was a common theme in California writings. "I am satisfied," wrote George Kent, "that the ideas we had formed of California before leaving home were very incorrect, and people who come out here must form their opinions of this new state independent of any home notions derived from Fremont, newspaper accounts &c." The writer Alonzo Delano repeatedly maligned the popular account of Edwin Bryant, in particular his claims about the salubrity of the climate and the purity of the atmosphere. Delano wrote, in contrast, "I never saw so much suffering and misery from disease in all my life as I have seen during a five months' residence in California." Immigrant doctors were particularly apt to attack the booster literature of the period for its inaccuracies. Jacob Stillman, a physician from New York, joined the rush to California in 1849 and subsequently wrote, "I was deceived in some respects; the healthfulness and beauty of the country was exaggerated by the early explorers." Dr. Thomas Muldrup Logan, a native of Charleston who arrived in San Francisco in 1850, wrote after four months in the state, "As to the health and climate of California, I now speak from experience when I affirm that we have all been grossly deceived. . . . [Since my arrival] I have not passed one perfectly well or pleasant day." In fact, Logan concluded a particularly gruesome depiction of the ravages of cholera in California with the ironic comment, "[This was] a land where I had been led to expect an Italian clime—an Archipelagian salubrity, and El Dorado harvest!"[42] The early reports of the California State Board of Health, first published in 1871, similarly undercut any consensus on the region's healthfulness; instead the authors went to considerable length to catalog the diseases associated with every region of California and condemned as injudicious the "extravagant" portrayals of the state's healthfulness promulgated by "non-professional travelers." As one leading California physician wrote, "The most erroneous statements have been circulated, either by travelers or by interested residents. It is our imperative duty as medical men to correct such error and to disseminate the truth."[43]

Such cautious and often negative assessments seemed to be borne out by the arrival of cholera in northern California in 1850, smallpox in 1852, and the rapid spread of dysentery and various "malarial fevers" in the ensuing decade—not to mention the high rates of insanity reported among recent immigrants to the state. Disease spread rapidly in the havoc of colonial invasion, though it is difficult if not impossible to assess the material prevalence of disease in contemporary scientific terms. The

only available statistics on death and disease from the period are frag-
mentary and unreliable by contemporary standards. Moreover, nineteenth-century categories of disease do not correspond neatly to contemporary ones. The most systematic accounts of disease from the period
appear in army reports on the health of troops stationed in California;
yet even these reveal more about the cultural gulf that separates the nineteenth from the twenty-first century than about disease as we might now
understand it. What are we to make of the category "fevers," one of the
most common causes of death in early California? The reports do contain death rates for the army, and by themselves these do not suggest that
troop mortality was especially high in California—at least as compared
with the southern United States or the tropical regions of the world. Yet
disease in California was acknowledged to be highly localized. In some
regions, rates of illness and death rivaled the most disease-ridden sections
of the South. Army surgeons noted that the prevalence of illness, if not
death, at certain posts was disturbingly high by any contemporary standard. Camp Far West, located in the Sacramento Valley, was abandoned
in 1849 on account of its unhealthfulness, and a second fort, established
some distance farther north, was similarly abandoned in 1856.[44] The city
of Sacramento, a center of gold rush activity, was known to be especially
sickly. Death rates calculated for the city in the 1850s are considerably
worse than those for the state as a whole: 39 per 1,000 persons in 1851;
74 per 1,000 in 1852, when cholera was at its height; and 27 per 1,000
in 1855.[45]

Yet physicians routinely commented that death rates alone failed to
tell the story of illness adequately. Many diseases were prevalent in early
California, and the incidence of disease seemed to be increasing in some
regions. Cholera appeared again in 1860, and smallpox struck the state
three times in the first two decades of American occupation. Of equal or
even greater concern were the various fevers. "Malarious" diseases
reached epidemic proportions in central California in 1858, and some
physicians regrettably acknowledged that this class of diseases was
endemic to their new home. Fifteen years later, the State Board of Health
noted with resignation that "throughout the whole of the State there
must continue to be more or less of malaria for centuries to come, if not
for all time." Aside from malaria, California physicians recorded the
presence of scarlet fever, measles, diphtheria, influenza, typhoid, phthisis
(consumption), and various forms of dysentery and intestinal disease.
Insanity also elicited deep concern. California had a much higher pro-
portion of supposedly insane individuals than other regions of the coun-

try, as high as one in every 490 persons. Insanity had been a local concern since the gold rush, but in the 1870s the State Board of Health noted somewhat anxiously that it might become an epidemic. Explanations focused on the "pace" of life in California, the heterogeneous social climate, and even "nostalgia," but most acknowledged that the local climate was at least partly responsible.[46]

Observers of California often pointed to bad habits and poor social conditions that were exacerbating disease, especially among the miners. Whereas the environment was a critical factor, the characteristics of an individual body were certainly relevant. Even in an unhealthy climate, not everyone succumbed to illness. Disease and death were the result of the "combined influence of the meteorological and physiological conditions modified by temperament."[47] Dr. J. P. Leonard of Rhode Island, who arrived in California in 1849, immediately wrote to the *Boston Medical and Surgical Journal,* noting the region's general healthfulness and downplaying the existing diseases. However, just four months later he was unable to maintain the same sanguine assessment. Writing to the same publication, he now acknowledged a "vast amount of sickness in San Francisco during the past summer," much of it fatal. He listed dysentery, diarrhea, pulmonary disorders, and fevers as the most prevalent diseases. Yet Leonard was reluctant to change his overall assessment of the California environment; instead he, like many others, called attention to the "intemperance, dissipation, disappointment, privations, exposure &c." that complicated recovery.[48]

How one understood disease causation had potentially enormous implications for the future of the region. Americans such as Leonard worried that the environmental characteristics of new lands might frustrate their desire for expansion, and mass outbreaks of disease were particularly disillusioning. While the British in India could resort to rotating new recruits into unhealthy districts, the American project of settlement depended on the ability of settlers' bodies to remain healthy, and reproduce themselves, in their new locations.[49] Only relatively healthy lands could be colonized through settlement in the long run. Thus to the extent that disease was the result of human action (intemperance, poor diet), the health of the community could be restored; accordingly, the prospects for settlement remained good. But to the extent that disease was the outcome of local environmental factors, it was largely outside human control; settlement, in turn, was threatened. By insisting on the role of "intemperance, dissipation, disappointment, privations, exposure &c.," Dr. Leonard and many others asserted somewhat hopefully the ability of immigrants to manage their own well-being in California.

The health concerns of nineteenth-century colonizers were inextricably connected to their obsession with race, and American immigrants to California were no exception. Although white immigrants spoke of health in general terms, the question that actually interested them was whether the region would foster the health of Euro-Americans, specifically those of northern European descent, the "Anglo-Saxon race."[50] The answer varied, but the intensity of the discussion indicates that both doctors and laypeople remained concerned about white racial health in the Far West. White settlers saw themselves as *more* vulnerable to certain diseases because of their race. Conversely, they believed that nonwhites, with the important exception of Native Americans, were less susceptible to the "tropical" diseases encountered in California, such as malarial fever. Writing of malaria, Thomas Logan paid particular attention to its differential effects among the races, noting "the insusceptibility of negroes and of those of mixed blood, born and bred in hot climates." He offered the tentative conclusion that "the susceptibility of the different races of mankind to malarial fevers appears to be in direct ratio to the whiteness of the skin." Others made frequent note of the seeming insusceptibility of Chinese immigrants: "The Chinese seem to be constituted something like the negro; they are not affected by the malaria as the Anglo-Saxons are."[51] Medicine and public health provided a scientific arena in which concepts of race and place were simultaneously constructed. And in an ethnically heterogeneous society such as post–gold rush California, vulnerability to specific diseases such as malaria could itself be a sign of whiteness. Illness could reaffirm one's race in the Far West, even while it raised questions about the suitability of the region for white settlement.

At the same moment, the growing crisis over slavery intensified both popular and medical interest in the debate over races and their proper places. Among Anglo-American intellectuals, racial categories hardened over the course of the 1840s and 1850s, and several leading British and American scientists argued the evidence for "polygenesis"—the belief that different human races had separate origins in different parts of the world and were thus biologically distinct. These ideas cast human migration—including American expansion—in threatening terms. In 1850 the Edinburgh anatomist Robert Knox published his treatise, *The Races of Men: A Philosophical Enquiry into the Influence of Race over the Destinies of Nations.* Knox, a former British army surgeon stationed at the Cape of Africa, argued that migration from east to west was as dangerous as that from north to south; he held that Europeans in America and Australia had, in fact, degenerated. His outlook on the American future was equally dim: "A *real native* permanent American . . . race of pure Saxon blood, is a dream which can never be realized." The American physician and committed raciologist Josiah Nott concurred, writing that the races of the temperate zones had already "paid dearly for their migratory propensities." At the same moment, America's foremost scientist, Louis Agassiz, drew on Humboldt's geography to articulate his theory of zoological provinces and "natural racial zones." Agassiz argued that the various races of men could maintain and reproduce themselves only in distinct regions of the world.[52] In his view, human migration across these zones was doomed to failure.

For some, these debates over races and their proper places cast American westward migration in ambiguous if not completely negative terms. For Euro-Americans engaged in Western colonization, the concern with the health of white bodies took on particular urgency. In California, Euro-American perceptions of Mexican society only served to intensify the question of degeneration. In a passage quite typical of the period, Lansford Hastings, author of a popular emigrant guide, described the Mexican inhabitants of California as "scarcely a visible grade, in the scale of intelligence, above the barbarous tribes by whom they are surrounded." It was an open question whether the small population and what Hastings saw as the backwardness of Mexican California could be ascribed to the physical and moral inferiority of the inhabitants or whether the climate and landscape were in some way responsible. Common, as well as professional, knowledge held that the numbers and characteristics of the local people were important indicators of the quality of the land. As the Englishman James Martin had written in his medical treatise on the tropics, it was "an axiom of medical topography . . . that a slothful, squalid-looking population invariably characterizes an unhealthy country."[53] Euro-Americans were already convinced that the mild tropical climates of Latin America fostered degeneration and debility among their own kind, and accounts of American forty-niners frequently made anxious reference to the degenerate Europeans and Americans encountered in Panama and Mexico. As one prospective miner wrote of Chagres, the port of disembarkation in Panama, "Idleness and sloth meet you at every turn; you feel that you are in the midst of an inferior race of men, enervated by the climate, whom bountiful nature has made stolid and indolent." In contrast, harsh northern

European climates supposedly bred vigor. But what the relatively mild climate of California would yield was, in 1850, still unknown. Dr. John Baker expressed this mixture of hope and anxiety after he arrived in San Diego in 1853. "The people here were the first specimen of Yankeedom that we had seen since leaving New York," Baker declared, "or at least those who manifested in their appearance the healthy and active life which the Yankee is accustomed to do in our section of country. If we found a man from the States on the Isthmus [of Panama] (where there were many) they had the appearance of sickness and debility about them. But at San Diego they seemed to be healthy."[54]

Early white immigrants like Baker hoped that the environment of California would be more like Europe and eastern North America and less like South America, but that was merely a hope. In 1850, firsthand information about the region was still relatively scarce, and its vast distance from centers of civilization underscored its unknown character. Although the presence of gold made California irresistibly attractive, immigrants had almost no idea what to expect when they arrived. Moreover, in intellectual circles the rise of Humboldtian geography had cast California in an ambiguous light. Though North America existed securely within Humboldt's "temperate" zone, maps of average temperatures revealed local and regional anomalies. Much of California was anomalous in just this way; some localities stood out as exceedingly hot. In fact, at the time the nation's highest recorded temperatures came from Fort Miller in California's Central Valley. As Lorin Blodget noted in his *Climatology,* these summer temperatures "exceed those measures in [the] humid tropical climates," a fact that was not reassuring to prospective immigrants.[55]

California stood apart not only as a result of its summer temperatures but also because of its diverse society. Contemporary observers almost always commented on the state's racial and ethnic heterogeneity, and the need to attract more immigrants of "northern European stock" became a paramount concern of the state's American boosters. At the same time, however, Euro-Americans could not help but wonder whether there might be some underlying relationship between the social diversity that they feared and the regional climate. Many immigrant physicians viewed California as an experiment in racial health, and neither the social conditions nor the overall state of health in gold rush California initially inspired white confidence on this point. The editors of the *California Medical Gazette* noted in 1857 that recent immigrants to California were "peculiarly susceptible of disease," and the *Second Biennial Report* of the State Board of Health acknowledged the popular sentiment that whites were degenerating in their new home and called for further scientific study.[56]

Among other things, concerns about degeneration prompted the close scrutiny of California's native inhabitants. Implicit in Jefferson's argument with Buffon had been the belief that Native Americans symbolized the quality of the land that whites now sought to colonize. Several decades later Euro-Americans still held Indian bodies as proxies for the natural environment. Whites understood Indian bodies as even more permeable than their own and thus as especially sensitive indicators of the region's healthfulness. As the doctor Frederick Hatch remarked in an early account of California, Indians were "by birth and hereditary impress . . . the peculiar subjects of the climatic influence and serve . . . to illustrate its features." Hatch, much like Thomas Jefferson before him, tried to read Indian bodies as healthy and resistant to disease. Similarly, John Griffin, an army surgeon in California during the war with Mexico, wrote that the Indians in southern California were "fine large, healthy looking fellows—and speak well for the salubrity of the climate." In the 1890s, when California health boosterism was at it height, Dr. Peter Remondino would extol the appearance, longevity, and endurance of the southern California Indians, comparing them to that paragon of physical and moral perfection, the ancient Greeks.[57]

Yet in the early decades of settlement, whites were also anxious to confirm their own biological superiority and their physical, as well as moral, fitness for the land they were appropriating. Indians presented something of a conundrum in this regard. It was necessary to understand Indians as both physically superior and physically inferior. The answer that many would settle on was to assert, as Dr. Logan put it, the remarkable "viability of the native Indian race . . . so long as he is not subject to the habits of civilized life." Native Californians were healthy and long-lived, but they were also ill adapted to progress and civilization. From this perspective, Native Americans were robust indicators of California's natural environment even while they were doomed to extinction.[58]

White women's bodies, like those of Indians, were understood as relatively more permeable than those of white men. Thus women were the most sensitive indicators of the environment's effect on white immigrants. Moreover, avoiding racial degeneration depended on the ability of female immigrants to produce able-bodied and unambiguously white children. Manifest destiny hinged not only on conquest and migration but also on reproduction. The success of colonial settlement hinged on the ability of whites to outproduce indigenous peoples. Reproduction, in turn, depended on female health. Consequently, the diseases of women drew particular attention because of their potential effects on fertility and childbearing. Though opinions on women's health in California were mixed in the mid-nineteenth century, writers were more likely to declare their anxiety than to express optimism. In a description of California's diseases written in 1852, Dr. James Blake asserted that "there can be no doubt that the climate is conducive to fertility in the female." Yet in that same year the army surgeon at Monterey remarked that the "diseases peculiar to women" were more common than any other malady in that region. James Hittel, in an otherwise promotional account, admitted that women's diseases were common in the state, that fertility was low, and that women began to "wither" at the age of twenty-five. In the late 1850s the state's newly formed medical society splintered over a paper on women's health in California prepared by one of its members. Dr. Beverly Cole had written of the moral and physical degeneration of white women in the state, claiming that "in no place of civilization do the causes [of ill health among women] exist or prevail to the same extent as in California." Several members of the society walked out in protest, claiming that Cole, by his references to immorality, had disgraced California's white women and, not incidentally, had impugned the suitability of the California environment for white immigrants. In response to Cole's paper, a medical colleague argued that children born in the state to immigrant women were remarkably healthy and would constitute "a highly improved variety of the human species."[59]

Despite the harsh reaction to Cole's paper, the debate over women's health in California continued for the next three decades. In its first report the State Board of Health corroborated the popular perception that "females [were] more susceptible to all kinds of disease, especially in California," and a book titled *Female Health and Hygiene on the Pacific Coast,* published in 1876, began by referring to the "unusual prevalence of disorders affecting the reproductive organs among ladies on the Pacific Coast." Others, doctors and boosters alike, continued to argue to the contrary that the local climate fostered both female fertility and healthy children. Dr. Thomas Logan appealed to mortality statistics to demonstrate that the proportionate mortality of women in California was lower than that of men. Charles Nordhoff, in his popular booster tract written at the behest of the Southern Pacific Railroad in 1874, included the obligatory reference to the attractive forms of women and children: "The climate is most kindly to little children, which is perhaps one its best tests.

One cannot travel anywhere in California without noticing that the forms of the women who have lived some years here are more full and robust than [in the East]; while the children are universally chubby and red-cheeked." But the attention devoted to the issue only underscored the anxieties of early white immigrants.[60] The ways in which the California environment might alter their bodies—and those of their children—was, at best, an open question.

Over time, as white dominance became an established reality in the region, California promoters such as Nordhoff, as well as some of the more boosterish doctors, would turn concerns about degeneration around and claim that the mild climate of the Pacific Coast produced an even healthier breed of white Americans. Already in 1869 the author of a popular tract on California, Charles Loring Brace, had acknowledged the effect of climate on the human "type," but he also assured his readers that the result was a sturdier and more attractive breed of Anglo-Saxons: "One sees great numbers of fine manly profiles, with full, ruddy cheeks, and tall, vigorous forms." However, Brace still felt the need to reassure his readers that while climate could improve health and vitality, it could not alter race. As he put it, "Blood is stronger than isothermal lines." But even these remaining reservations about racial malleability would soon disappear. Dr. Peter Remondino, who became one of the foremost boosters of southern California (as well as the owner of a popular health resort in San Diego), would pen several articles on climate and health in the 1880s in which he refuted the widely held idea that humidity was bad for health and only harsh climates bred vigor. Instead, as Remondino put it, California's "moist marine air and equable temperatures produce the most perfect specimens of physical development."[61]

COLONIAL MEDICINE AS ENVIRONMENTAL SCIENCE

Nineteenth-century immigrant doctors were among those who wrote most prolifically about the environment of California. In contrast, contemporary medicine is not much concerned with the landscape; physicians generally confine themselves to the terrain of the human body, while the natural environment is left to a host of other disciplines. This narrowing of professional perspective and the intellectual parsing of environmental and medical sciences is largely a product of the early twentieth century. Nineteenth-century understandings of health required physicians to pay close attention not only to the sufferer's body but also to the surrounding landscape. It was only logical that among the early European exploring expeditions, the same person typically served as both doctor and naturalist.[62] And that the issue of health and environment drew the sustained attention of professionals as well as laypersons should caution against reading the large popular literature on the subject as merely the invention of boosters or the writings of medical eccentrics. Rather, nineteenth-century science underwrote and sustained widely held beliefs that melded human health and the natural environment into an inextricable whole. Even while popular and professional writings diverged in their particulars, they reinforced a view of the body as an entity that was both porous and environmentally sensitive.

But for professional medical men who saw themselves as serious scientists, the relationship between health and environment remained frustratingly vague and qualitative. In the national and even transnational intellectual debates over health and environment, several early California physicians saw themselves at the forefront of an empirical effort to answer questions about climate, environment, and racial fitness in a more definitive way. As the reception of Daniel Drake's treatise on the Mississippi Valley attests, professional interest in scientific medical topography was high in the 1850s. Consequently, at the moment of its colonization California formed a rich field for the extension of medicoenvironmental studies, and colonial physicians worked hard to institutionalize and scientize the study of the local environment. Several gold rush–era physicians brought environmental interests with them to California, quite conscious of the fact that they were encountering a new environment to which bodies might react in unforeseen ways. Immediately on their arrival, several physicians committed themselves to the close study of the local environment in their adopted home and to the "patient and laborious accumulation of exact statistics."[63] From the outset, professional medicine in California had a strong environmental cast.

In many ways, the critical figure in early California medicine was Thomas Logan, a devotee of Humboldt who arrived in San Francisco with an established interest in climatology and the environmental basis of disease. Logan was born into a family of physicians in Charleston, South Carolina, in 1808 and was educated at the Medical College of South Carolina in a period in which probably less than half of all physicians actually took a medical degree. He supplemented his formal training with a tour to Europe and wide reading in his field, and he subsequently practiced in Charleston and New Orleans. But the mid-nineteenth century was a difficult time to be a doctor, and there is no

Figure 2. Portrait of Dr. Thomas Muldrup Logan, one of the most prominent and prolific American physicians in nineteenth-century California. Courtesy of the California History Room, California State Library.

evidence that Logan ever established a profitable private practice in the South. In 1850 he left Louisiana and the South for good. He joined the gold-inspired migration to San Francisco and quickly settled in Sacramento where he would practice medicine for the next twenty-five years. Having brought along meteorological instruments on loan from the Smithsonian Institution, Logan immediately commenced recording weather statistics. Eventually he would become one of California's most prominent physicians, a professor at the University of California, and president of the American Medical Association. As a leading figure in the California Medical Society and later as the first secretary of the State Board of Health, Logan lobbied strenuously for more meteorological study, arguing that "every city, village, and settlement should have its meteorologic record."[64]

Logan was joined by several other physicians who shared his professional interests and environmental orientation. Henry Gibbons immigrated to California in 1850, arriving from Philadelphia where he had been a faculty member at the University of Pennsylvania. In San

Francisco, Gibbons supplemented his medical practice with the study of native plants and of meteorology. As the editor of the leading California medical journal, he urged every physician to "train himself as an observer of meteorological phenomena. The thermometer, the hygrometer, the currents of wind and cloud, should be as familiar to him as the stethoscope, the microscope and the speculum." Frederick Hatch, a graduate of New York University's medical school and a successful physician in Wisconsin, immigrated to Sacramento in 1853. Hatch would become a close observer of the California environment and would write several papers on the subject; and, like Logan, he would serve as a meteorologic correspondent for the Smithsonian. His observations on climate and health in his adopted state were included in Blodget's 1857 *Climatology.* Hans Herman Behr, an immigrant from Germany, had trained in medicine at the University of Berlin, where he had been a student of Humboldt and another well-known German geographer, Karl Ritter. In California, Behr combined his interests in climate and health with the close study of native plants and insects.[65]

These and other individuals would succeed in institutionalizing the study of environmental medicine at an early date. In a speech before the newly formed California Medical Society in 1856, Logan insisted that a key aspect of the organization's mission was "to work out the problem of climatic influence on the physical condition of man—to investigate the nature and causes of endemics and epidemics—to show how far man's agency has to do in the matter." At the second meeting of the society, Logan was instrumental in establishing the Committee on Medical Topography, Meteorology, and Endemics and Epidemics. The following year, he chaired the committee and wrote a lengthy report on the subject. When the California State Board of Health was founded in 1871, Henry Gibbons served as its first president and Logan as the first permanent secretary. Under their leadership that organization would make the study of the physical environment and its relationship to health a priority, and medical topographic studies would proliferate in California for two more decades.[66]

This emphasis on the environmental causes of disease may well reflect the fact that until the last decades of the nineteenth century California remained very much a settler society, and settler anxieties about relocation made western physicians especially attentive to their surroundings. Given that nineteenth-century bodies were permeable, new and diverse environments required both wide and meticulous scrutiny. While indigenous bodies were powerful indicators of the land, immigrant bodies required careful monitoring and care in a new place. Doctors such as Logan and Gibbons saw their studies of environment and health as indispensable to securing the successful colonization of California by white Americans. However, since it was not yet clear what aspects of the environment were critical to health, Logan advised his colleagues to collect as much data as possible—not only temperature and altitude but also dew point, quantity of clouds, timing of frosts, depth of ground frozen, temperature of wells and springs, timing of animal migrations and fish runs, presence of ozone in air, and causal phenomena such as thunderstorms, tornadoes, hailstorms, the aurora borealis, meteors, shooting stars, and earthquakes.[67] Discourses on medicine and health were thus not only discussions of the human body but also important realms of environmental understanding.

The approach of nineteenth-century immigrants to questions of health and disease reveals a world in which the very concept of agency was understood in nuanced ways. Disease was not simply contained within certain pathogens. Discussions of causality, whether carried on by physicians or laypeople, embraced theories of environment and contagion, individual constitution and moral rectitude, personal habits and social progress. In this world, the local environment was sometimes healthful and sometimes threatening—but it was always active, contingent, and relevant to the bodies that resided there. Agency, moreover, was not necessarily confined to human beings, nor were the causes of disease discretely located in certain microorganisms, at least not yet. Rather, disease was only the most obvious sign that humans were part and parcel of a larger whole, a world that, though not completely opaque to scientific methods, often responded in unpredictable ways. Certainly American immigrants who came to California in the mid-nineteenth century did not doubt the virtues of white settlement. But when Logan wrote of the need to study "the modifications diseases may undergo from the agency of causes of a local or special character," he, like many others, acknowledged at the outset that the history of that project would be the outcome of nonhuman as well as human forces.

2

Placing Health and Disease

There can be no question . . . that the extent of territory,
and variety of climate and soil, within the limits of the State,
render it a peculiarly favorable one for gaining valuable
and comprehensive knowledge of the influence of various
conditions upon the rate and causes of mortality. There is
an opportunity to compare, in degrees of latitude; sea levels
with elevations of eight thousand to ten thousand feet; and,
what affords an unusual contrast, seacoast valleys chilled
by an Arctic current, with vast interiour prairies of almost
tropical temperature. Doubtless, when sufficient time shall
have been given to the study of these conditions . . . the result
will be a demonstration of important relations between them.

*Second Biennial Report of the California
State Board of Health*, 1873

Understandings of environment and health in the nineteenth century
were shaped by broad cultural and political currents—European
medical geography, debates between contagionists and anticontagionists, transatlantic racial theories, American expansionism, the political crisis over slavery. But they were also shaped by the physical experience of individuals and the material realities encountered in specific places. Laypersons and physicians alike believed that both disease and cure could be understood only in their specific contexts. Diseases were often unique to their localities, and a treatment that worked in one place could all too often fail in another.

Although considerable emphasis has been placed on the supposed triumph of germ theory in the last third of the century, the theory took hold quite slowly among professionals as well as nonprofessionals. In brief, germ theory held that disease could be traced to singular and discrete etiologic agents that penetrated the body rather than to the much vaguer and more nuanced concept of imbalance. However, nineteenth-century medicine was intellectually capacious, and most physicians had no diffi-culty mixing germ theories with long-standing environmentalist beliefs. Many of those who professed their allegiance to the new bacteriological views still maintained that environmental conditions either "multiplied" germs or brought them into activity.[1] At least until the mid-1880s the overwhelming majority believed that the source of health as well as dis-ease lay primarily in the local environment and that the only way to gauge an environment's effect on health was to live in it. Thus to recover nineteenth-century understandings of bodies and environments and their interrelationship requires more than an explication of national and transnational intellectual developments. It requires a return to the social and environmental conditions of specific places.

Local conditions in California were rapidly changing in the late nine-teenth century. The gold rush faded quickly, but larger mining operations continued, often with devastating environmental effects. At the same time, immigrants realized that farming offered potentially greater opportunities than mining ever had, and they eagerly exploited the agricultural poten-tial of the state. Settlers in this period wielded their power to remake the Central Valley's natural landscape—through clearing, plowing, draining, and planting—and the results were obvious. The lightly settled landscape that immigrants entered in 1849 looked vastly different by 1900, when hundreds of thousands of acres had been converted to grain and fruit crops, and both irrigation and reclamation were proceeding apace.

Yet even as immigrants eagerly commodified and transformed the environment that they occupied, they still felt themselves intimately con-nected to the land. Americans came to understand the western landscape in many different ways—through the lens of the market, through their own labor, and through their experience of health and disease. The prac-tices of medicine, both lay and professional, assumed and made explicit connections between the human body and its external surroundings. Discussions of health and disease suggest that the rhetoric of economic development and environmental conquest, so familiar in historical accounts of the period, always existed alongside the knowledge that white immigrants depended on a landscape that could be neither com-pletely dominated nor fully understood. Health was an arena in which the assumed separation between the human and the nonhuman world broke down. In their discussions of miasma and other endemic diseases, settlers acknowledged that they could not fully control or even predict the results of their environmental interventions; those interventions, moreover, would be registered in their own bodies.

The felt connection between bodies and environments generated a persistent tension over the processes of modernization. In a period of profound environmental change, the environment's perceived effect on bodies underwrote a debate about what kinds of transformations were desirable and how best to judge their effects, a debate that historians have largely overlooked. Nowhere was that tension more evident than in California's Central Valley, which, within the state's emerging medical geography, stood out as a disturbingly "insalubrious" region. While the warm temperatures, long growing season, and rich soils recommended the region for farming, those same characteristics, along with others, also indicated its physical dangerousness for immigrant bodies. The landscape itself was a contradiction. Seen in economic terms, it offered enormous potential and invited transformation; seen in physical terms, it threatened disease and even death, and consequently its transformation was a risky undertaking. As physicians and lay observers came to understand the valley as both an undeveloped resource and a pathogenic space, they would debate the implications and their own responses. Not only did the environment affect health; ideas about health and disease sometimes shaped landscapes in quite material ways.

MEDICINE IN NINETEENTH-CENTURY CALIFORNIA

In the writings of California doctors, it is impossible to discern any sharp turning point in medical thinking during the late nineteenth century. The disjuncture supposedly created by the rise of germ theory is largely the retrospective creation of twentieth-century medical historians. Although they were well aware of new discoveries in bacteriology, early California physicians greeted them as developments to be debated rather than as an intellectual watershed. Moreover, most viewed germ theory as something of a fad. Already in 1868 members of the Sacramento Society for Medical Improvement were debating the prize-winning theory of James Henry Salisbury, which held that spores of palmella, blown from field to field, were responsible for ague; yet the members agreed that Salisbury's position was too simple and failed to account for the varying environmental conditions that gave rise to malarial disease in most of California. After Salisbury's theory fell into disrepute some years later (the palmella spores were found in areas with no history of malaria), several California physicians would point to the enthusiasm for palmella spores as a cautionary tale.[2]

California physicians took up discussions of germ theory more stridently in subsequent decades. In October 1881 Dr. Ira Oatman presented a paper before the Sacramento Society for Medical Improvement, and a lively discussion of the new theories ensued. Though several physicians announced their allegiance to germ ideas, they had not relinquished their environmentalist beliefs. They were willing to acknowledge that microorganisms might play a role in disease but insisted that so too did the individual constitution and the local environment. As one doctor described his position, "I believe in the germ theory, [but] the germ may differ somewhat in different localities, thus producing various grades of the disease." Another argued that there was no reason to assume a disease had only one cause; the malarial poison that they were discussing might well be "a gas as well as a germ." In a lengthy article on miasmatic disease published in the *California Medical Gazette,* Dr. Campbell Shorb of San Francisco explained that "the evolution [of the germ] depends on contingencies. It is developed when subjected to some exciting cause, or when the energies of resistance are reduced or destroyed by morbid processes, which have no association or connection with this pre-existing germ."[3]

Although some physicians positioned themselves as vociferous germ skeptics, most of California's nineteenth-century physicians cannot be easily divided into those who accepted germ theory and those who remained miasmatists. Most embraced a complex approach to questions of etiology and would do so until the end of the century. For instance, in 1889 Dr. Marshall Chipman wrote an article on advances in bacteriology for the California Medical Society that demonstrated his wide knowledge of the field and its recent developments. Though he was a proponent of the new laboratory approach, Chipman saw bacteriology as merely "coadjutant" with older branches of medical science, including medical topography in which he maintained a keen interest. Even those who embraced the basic tenets of germ theory saw the uneven distribution of disease as pointing to local environments as a critical variable. As one doctor put it, "Now that the various germ theories have obtained such a fixed footing, the question presents itself to the thoughtful man, where can I reside that the conditions so far as any germ infection is possible may be minimized?" Even with the advent of germs, place remained critical to both disease and health. Thus medical topography remained an emphasis of California doctors (and a standing committee of the California Medical Society) from the 1850s through the end of the century.[4]

That disease varied from one region to another had been a focus of the state's leading physicians since the 1850s, an interest that emerges clearly in the early reports of the California State Board of Health. In the *Second Biennial Report* (published in 1873), Dr. Thomas Logan drew on census data to map mortality rates from consumption, intestinal disease, and malaria by region. These maps immediately invited speculation about the local causes of disease. For instance, consumption was shown to be far more prevalent in the northern coastal regions, while southern California and the southern desert regions were nearly exempt from it. Logan admitted that the distribution was difficult to explain, but he hazarded that it might be related to the fact that in the cooler regions homes were more shut up so that residents had less exposure to pure air. He suggested that the higher mortality from intestinal disease observed in coastal regions was the result of "chilling Summer winds of the ocean [that] prevail in all their intensity—driving the blood upon the vital organs."[5]

Physicians recognized that disease varied over time as well as space, and here again the local environment was believed to have a critical role. Both doctors and laypeople correlated the onset of certain disease-prone periods with annual changes in temperature, rainfall, and atmospheric conditions. Almost every observer noted that malarial disease was more prevalent in late summer and fall, and one doctor went so far as to call autumn in California "the season of death."[6] Lay advice manuals recommended that anyone capable of doing so should migrate away from miasmatic regions in August and September. Barring that, extra precautions (proper clothing and diet, adequate sleep, abstinence from stimulants) were advised.[7] During certain seasons, strong winds could bring disease into an otherwise healthy area, while unusually high floods might "wash" a disease-prone landscape and improve the health of its inhabitants.[8] Some observers connected seasonal outbreaks of pneumonia with the onset of the valley's north winds.

Physicians did not confine their interest to the immediate environment; they also paid attention to the ramifications of distant environmental and astronomical events.[9] The year 1868 was medically noteworthy in California for the amount and diversity of disease. The "unusual epidemic prevalence" included outbreaks of pneumonia, whooping cough, measles, and smallpox. Logan hypothesized a connection between the health of Californians and volcanic eruptions at Mount Vesuvius and Mauna Loa and a series of earthquakes that had occurred throughout the Pacific. While acknowledging that no one understood the precise relations between these global phenomena and pestilence, Logan suggested that volcanoes impregnated the upper atmosphere with an "invisible gaseous poison" and that earthquakes were typically coincident with abrupt and dangerous changes in the weather.[10]

To explain disease required not only education and intelligence but also immense amounts of local knowledge. Physicians openly acknowledged the critical role of local experience to their etiological debates. Residents' bodies served as instruments of a kind, which measured a region's unhealthfulness and offered important clues to the sources of disease.[11] Local people, moreover, were best positioned to observe the subtleties of a given landscape and its effects, and thus professionals valued lay experience and hypotheses. As Logan reminded his colleagues, "an army of observers" would be needed to generate accurate understandings of disease in a new place.[12] In that vein, Logan issued a circular on behalf of the California Medical Society in 1856, asking local physicians to report "authentic information respecting the topographic and climatic characteristics of the several localities or regions in California."[13] Several years later, as secretary of the State Board of Health, Logan took it upon himself to travel throughout much of California, observing the topography and the residents in different locales and canvassing both physicians and laypeople about health. In his official capacity he also canvassed local doctors by mail, asking about the frequency, characteristics, and likely sources of specific diseases. In 1877 the board queried local physicians about the incidence of malaria and consumption and reported their answers, along with their assessments of climate and topography, in detail. Their responses make clear the presumed connections between local environments and disease. On malaria, Dr. W. H. Patterson of Cedarville reported that "a chain of three shallow lakes extends through this (Surprise) valley, and when these dry up the affluvia from their muddy beds causes remittent fevers." Dr. Kunkler, writing from Placerville, reported that malarial diseases "prevail to a moderate extent in some parts of this vicinity," which he attributed to the "the partial obstruction of waters from the creeks by the ditch owners, and also to excessive or injudicious irrigation upon some farms." Dr. J. S. Jackson of Modesto reported that his locale was bad for consumptives, owing "to the dust and dryness of the air."[14] These responses, and the fact that Logan published them, reveal the importance of local knowledge. Medical expertise was the outcome of experience in a given landscape.

Environmental theories of disease encouraged physicians to seek cures as well as prevention in California's landscapes. Through their writings, many of the state's leading physicians emerge as strong advocates of what John Harley Warner has called the "therapeutic revolution" of the mid-nineteenth century, eschewing aggressive therapies and strong medicines in favor of allowing diseases to run their course and the body to heal itself with minimal intervention.[15] As one physician reminded his colleagues in the Sacramento Society for Medical Improvement, the physician was merely "the assistant of nature"; another stridently criticized the profession for giving "too much medicine and rely[ing] too little upon the great natural forces which are constantly at work in organic life."[16] If anything, the controversial medical doctrine of "trusting to nature" was stronger in California than elsewhere, removed as it was from the centers of medical orthodoxy and staffed by immigrant physicians with diverse training and backgrounds. In place of intrusive practices such as bleeding and purging, California physicians were more likely to urge the moderate use of drugs and passive environmental cures. One observer had already noted in the 1850s that "hydropathy is the popular treatment [in California], and a good bath is thought to be far more conducive to health than bleeding or calomel." As the century wore on, California would become well known for the "wilderness cure" and recognized as a center for altitude therapy in the treatment of consumption.[17]

Their embrace of environmental etiologies and natural cures demanded that California's early physicians study and measure the local environment. In addition to collecting weather statistics, they digested the reports of geologists, railroad engineers, and surveyors. Logan himself became an authority on the hydrology of the Sacramento River; his statistics were incorporated into the reports of the California surveyor-general. Both Henry Gibbons and Marshall Chipman became experts on California flora.[18] This dual focus on land and bodies, made most explicit in the field of medical topography, encouraged physicians to connect environmental change quite concretely to human health—even when their knowledge was admittedly incomplete. Writing on behalf of the State Board of Health, Logan insisted that it was "only by the careful and comprehensive study of the laws of nature and the correlation of forces operating throughout the physical world, that the highest department of the physician's art can be brought into operation."[19]

Writing of contemporary biomedicine, the philosopher Annemarie Mol has argued that the realities of disease do not exist apart from the practices through which disease is enacted. Disease is never solely the outcome of practices, and yet it is only knowable through practices—and practices may differ. In other words, there is no single reality of disease that controls how we go about describing and defining it. Even in modern Western medicine, different practices often yield quite different "diseases" when they are closely scrutinized.[20] Because of the gulf that separates our own understandings of disease and health from those of the nineteenth century, it is perhaps easier to see that the practices of Thomas Logan and his California colleagues created both environmental disease and environmental cures. Nineteenth-century medical practice already located disease *and* health in the landscape—and so physicians gathered environmental data, solicited local comments on the landscape, and mapped disease prevalence. These practices then helped confirm the critical role of the environment; the only question was how to determine and explain that role with greater precision. As Logan put it, "We do not yet understand the various ways in which climatic conditions and changes cause sickness and death, although we do know that such is the fact."[21] To be able to predict and contain disease required ever more careful scrutiny of healthy and unhealthy landscapes. Nineteenth-century medicine took every opportunity to make visible the links between bodies and environments.

HEALTHY AND UNHEALTHY LANDSCAPES

What immigrant physicians had immediately found striking about California was the sheer variety of its landscape and climate. In contrast to the more homogeneous landscapes of the Midwest, California struck doctors as a place comprising many different environments, each with its own effect on health. "In every other respect than the long rainless season, the climate of California has numberless modifications, according to locality," wrote Frederick Hatch for the State Board of Health.[22] While the potential of certain localities, especially the coastal resorts of San Diego and Santa Barbara, to foster convalescence for certain types of patients was recognized, such associations could not be generalized to the whole state. Instead each region needed to be studied for its own healthy and unhealthy qualities. "If we would make our work and our statistics of any true or permanent value," wrote Dr. Joseph Widney of Los Angeles, "climatic belt must be differentiated from, and contrasted with, climatic belt. It is only thus that our work will lead to a clear understanding of the varied pathological peculiarities of the State." Such a complicated landscape could shape disease in unforeseen ways. The presence of a given ailment might be the product of a region's seasonal climate, the local soil, a particular swamp, or a periodic wind. Nature itself was inherently neither good nor bad so far as health was concerned. Rather, it was local and particular. A complicated geography posed a host of potential threats while offering new possibilities of cure.[23] Thus it was incumbent on physicians to construct a geography of disease.

To assess scientifically the relationship between environment and health, California physicians divided the state into three principal regions: the coast, the Sierra Nevada, and the Central Valley. From the early days of settlement, physicians and patients alike noted the therapeutic benefits of certain landscapes. The state had numerous local hot springs and mineral waters that were quickly recognized for their healing properties, particularly in the treatment of rheumatism, chronic constipation, and skin diseases. Writing for the State Board of Health in 1873, Logan listed some two dozen springs believed to have therapeutic value, along with what was known about their chemical composition, average temperature, and surrounding climate.[24]

But it was as a treatment for consumption, or "phthisis," that the California landscape gained most attention. Tuberculosis was the most widespread disease in the nineteenth century, and the available drug therapies were woefully ineffective. By contrast, the "travel cure" offered some hope to consumptive patients, and by 1834 the leading American medical text was promoting emigration to the western prairies as a treatment. According to contemporary accounts, the critical factor promoting healing in western environments was the purity and dryness of the atmosphere, but invalids were also believed to benefit from exercise and a changed diet. Already in 1860 immigrant physicians were touting the therapeutic properties of the California coast for consumptives, and by the 1870s both coastal and mountain locations were being actively promoted by boosters and physicians as health resorts. Health seekers soon followed. The low incidence of epidemics in these areas was typically attributed to an unspecified "antiseptic property in the climate" that prevented contagion. In addition to the benefits of pure air, doctors pointed to the equable and warm temperatures of the southern coast and the low humidity of the interior region of Los Angeles. Some advocated the benefits of the Coast Ranges, because these mountains offered a climate intermediate between that of the moist and sometimes foggy coast and the dry, overheated valley.[25]

While portions of California were curative, it was also clear that the vast interior regions were an unrelenting source of disease. The problem of sickness in the Central Valley had been widely noted by travelers during the gold rush, many of whom insisted that illness was a certainty for

Figure 3. A health resort (Montecito Hot Springs) in the mountains above Santa Barbara, circa 1880. The landscape of the southern California Coast Range was identified as curative for those suffering from phthsis (tuberculosis) and other respiratory conditions in the late nineteenth century. Courtesy of the California History Room, California State Library.

those who spent much time in the lowlands. "I can conceive of no part of the Mississippi Valley more prolific of disease, than the valley of the Sacramento must be," wrote one physician in 1850. Newspapers reported the problem of disease in both the foothills and the valley, and the extant diaries and letters of miners confirm the presence and virulence of "fever and ague."[26] Dr. Jacob Stillman remarked in 1849 that the settlement of Reading's Springs (present-day Redding) in the Sacramento Valley was "little better than a fever hospital," and soon after visiting the region he fell ill.[27] In his reconnaissance of the Central Valley for the army in 1850, Lt. George Derby reported that at Sutter's farm and all along the nearby Yuba River most of the occupants suffered from periodical fever, to which several deaths had been attributed. A U.S. Army post established in the eastern Sacramento Valley in the same year was moved two years later on account of its unhealthiness. As Assistant Surgeon Robert Murray wrote, "In common with the whole Sacramento valley, this post is very sickly from June till October." John Audubon, visiting Sutter's Fort (present-day Sacramento) in 1850, wrote that fever and ague were "very prevalent" and that dysentery was "feared by all." Many farmers informed Audubon that "nothing would induce them to settle" in such a place.[28]

Figure 4. The mining regions of the state were known to be particularly sickly in the mid-nineteenth century. Deforestation coupled with the diversion of rivers and streams increased the habitat of anopheles mosquitoes and helped spread malaria through the immigrant population. Photograph by Eadweard Muybridge, "The Heathen Chinee with Pick and Rocker." Courtesy California Historical Society, FN-13890.

The fact of disease made the Central Valley a focus of medical concern and etiologic debate. Although California's latitude placed the entire state well within the temperate zone, Humboldtian isothermal lines immediately called attention to the valley's extraordinary heat. In fact, both physicians and lay travelers frequently equated the Central Valley with the tropics. Victor Fourgeaud, a physician-turned-miner who came to California from St. Louis, wrote that the temperature of the valley was "unsurpassed by anything [he] had conceived of the heat of the tropics." An 1855 army medical report on a post in the San Joaquin Valley referred to the "torrid heat" and the "burning rays which through several months unintermittingly pour down from a relentless sun." "The vertical rays of the summer sun fall upon it with full intensity," creating an "oppressive heat," wrote a physician on behalf of the State Board of Health. Dr. Frederick Hatch, in an address to the local medical society, referred to the state's "vast interiour prairies of almost tropical temperature." In his summary of the meteorological data for the region in 1859, Logan could not fail to note that summertime temperatures far exceeded anything encountered in his hometown of Charleston. He conceded that the valley had "some features of a tropical rather than of a temperate climate."[29]

The labeling of the Central Valley as "tropical" in the mid-nineteenth century was more than simply a rhetorical flourish; it expressed serious doubt about the region's potential for white settlement. Heat threatened health, and Europeans had long associated warm places with early death. As the British physician James Martin wrote of the tropics in his seminal medical text, "From *heat* spring all those effects which originally *predispose* to the reception or operation of other morbific causes."[30] For Europeans who found themselves in tropical climates, it was essential to adopt prophylactic measures and, above all, to avoid both vigorous exercise and hard labor. In such a climate, the body was already "overstimulated," and fatigue could easily push it past its limit. In a world where bodies were understood as porous, whites feared that they would not be able to maintain their health or reproduce themselves in any region labeled "tropical." Such an environment acted upon the body in negative ways. An immigrant physician noted that in the bodies of white immigrants "the cutaneous surface is almost constantly in a state of hyperaemia—of congestion under the influence of an exalted temperature, a condition only partially relieved by the free perspiration which ensues."[31]

All whites, not just physicians, felt vulnerable in this landscape, and many were hyperconscious of their physical sensitivity. It was not only medical discourse but also immigrants' own experience that pointed to the environment's harmful effects on health. As the miner Thomas Kerr complained in his journal, "The heat of this country is really intolerable for any European to bear with for any length of time." In 1850, while surveying the valley for the army, Lieutenant Derby reported that he "found the heat most intense during the mid-day." "I was obliged to work merely during the morning and evening," he wrote, "but in spite of this precaution my assistant and two employees were taken sick, apparently from the effect of exposure to the sun."[32]

Long-standing fears of hot climates and the experience of illness, however severe, still competed with American desires for colonization. Although immigrants might fear the health effects of the valley, they needed to make a living, and most wanted to make a profit. From the outset, those who immigrated to California were committed to finding success in their adopted state. Consequently, immigrant physicians struggled to read the valley's climate as habitable by whites if not ideal. The heat encountered in the valley was undeniable, but many physicians argued that the region was not as inhospitable as the temperature alone might suggest. Debates over the effects of heat and the possibility of acclimatization reveal an ongoing effort to reconcile the desire for colonization with environmental understandings of both health and illness.

Thomas Logan was but one defender of his adopted home, arguing that the "tropical tendency" of the regional climate should not be misconstrued, that "in the aggregate of its constituents" the climate was on the whole more temperate than tropical.[33] Similarly, Frederick Hatch dismissed the assumption that the region's high temperatures made white settlement impossible:

> It might very naturally be inferred by those familiar only with the hot
> weather of the Atlantic States, that the exaltation of temperature we have
> described would necessarily be attended by the most disastrous consequences
> to those exposed to its influence; and that, during some of the days of the
> past summer, the climate must have been almost uninhabitable by the Anglo-
> Saxon race. Yet such impressions are far from being justly entertained. . . .
> Even during the hottest days, with the sun pouring out the concentrated heat
> of the tropics, there is an elasticity and freshness maintained by the system.[34]

Hatch was vague about the reasons that such heat was not oppressive or even prohibitive to "Anglo-Saxon" settlement, but others argued that both the dry atmosphere and the presence of strong winds mitigated against "the wretchedness of a purely tropical state."[35] Still others pointed to the rapid decrease in temperature at the end of the day and the coolness of the evening; moderate nighttime temperatures gave the body a chance to rest and recover its balance, the thinking went. In the valley temperatures could vary by as much as forty degrees in a day, but this also was a cause for concern. A rapid change could throw a body still further out of balance; the prevailing assumption held that "equability of temperature is the great desideratum." Thus, while some argued that the valley's large variations in temperature were an advantage in that they offset the daytime heat, the same phenomenon was more commonly blamed for the prevalence of certain diseases, such as dysentery and malaria.[36]

Still others argued that the problem posed by heat was only temporary and that settlers could adapt to it. Drawing on widely held beliefs about acclimatization, many California immigrants found reason to remain optimistic. "It is not uncommon for fresh comers to suffer from the heat the first season," one settler near Bakersfield wrote, "but, becoming acclimatized, they do not experience the same inconvenience afterwards."[37] Having immigrated from the South, Logan aligned himself with the theories of Josiah Nott and initially opposed any suggestion that whites might acclimate to their new surroundings. But after a decade and a half in California, even Logan began to modify his opinion on the issue, claiming that experience suggested that residence in California gradually bred resistance to the region's climate and diseases.[38]

It was not merely heat that made the valley dangerous. Physicians and laypeople alike associated the valley with numerous illnesses including dysentery, diarrhea, neuralgia, sciatica, pneumonia, "malarial neuroses," constipation, pleurisy, bronchitis, temporary blindness, "dysmenorrhea," and even paralysis. All of these were linked to some extent to local causes. But it was various forms of "fever" that were most disturbing. In some localities, at least three-fourths of the white population were believed to be ill with some form of fever, with entire families often afflicted. Patent medicine ads for ague tonics and liver pills filled the back pages of local agricultural publications, and many observers claimed that the incidence of fever was so common in certain localities that it went unremarked. Indians interviewed in the late nineteenth century noted that summer and fall were sickly seasons for their people and that they migrated away from the main rivers at this time of year. The ethnographer Stephen Powers, who interviewed several Yokuts Indians in the late 1870s, reported that they were prone to both fevers and lung complaints. And all observers agreed that the number of deaths directly attributable to fever in the valley conveyed "but a feeble idea of the amount of sickness, suffering, loss of time, of the impairment of health of body, of the enfeeblement of intellect produced by them."[39]

Fevers had been epidemic in the region during the gold rush, a fact that many attributed to the "unsettled" nature of the country and the privations of the mines. Though the prevalence of fevers waxed and waned over the next several decades, they often reached epidemic proportions in the valley. The year 1874 was particularly devastating. Disease raged throughout the region, but the hardest hit area was Oroville in the Sacramento Valley, where reportedly more than four thousand Chinese miners abandoned the region out of fears for their health. Logan attempted to map the prevalence of the disease that year for the State Board of Health, drawing on both local reports and his own observations. He correlated the distribution of disease with the summer isotherm of sixty degrees. To no one's surprise, he concluded that the entire Central Valley was a source of malarial fevers while the California coast was relatively exempt. On Logan's map—as in the accounts of miners, farmers, and travelers—the valley emerges as a pathological space.[40]

In the usage of the time, "fever" was something of an umbrella term that has no precise correspondence with contemporary diagnoses. Nineteenth-century physicians recognized several variants of "fever"— "intermittent," "remittent," "typho-malarial," "bilious"—and classified them in a variety of ways, not all of them consistent. If we put aside the inadequacies of retrospective diagnoses, however, the most prevalent forms of fever in California likely corresponded to what we would understand as malaria.[41] Though malarial fevers were often not fatal, their effects on the body were severe. A typical attack was characterized by an initial "cold" stage, or "chills," accompanied by excessive languor, aversion to food, a pale or yellowed complexion, and intense joint pains and nausea; a "hot stage," which brought intense thirst, a flushed complexion, violent headaches, still more "frightful pains in the bones," and, often, delirium; and finally, profuse sweating. Sufferers were likely to experience a "congestion" of the liver, bladder irritation, constipation, nausea, shaking, and a general and often long-lasting "debility." Believing that their systems were not only out of balance but also "blocked," sufferers often resorted to strong purgatives. The organ most severely affected was the liver, which, along with the stomach, experienced "extensive derangements."[42]

The prevalence and virulence of disease in the Central Valley prompted physicians to take a keen interest in the region's environmental details. Like farmers, physicians incessantly discussed and attempted to classify the local ecology. In order to better prevent and cure disease, they documented temperature, rainfall, and humidity and carefully scrutinized the air, soils, water, and vegetation. Immigrants immediately took note of the valley's atmosphere; it obviously lacked the "purity" for which physicians celebrated the state's coastal regions. Dr. W. W. Taylor, who traveled through Sacramento in 1850, found the air "dense and sultry and saturated with dust and impurity."[43] Logan referred to the "vapory condition of the atmosphere," yet another quality that he associated with the tropics.[44] Settlers also feared the valley's dense winter fogs, which settled on the land for days and often reduced visibility to almost nothing; they felt certain that these fogs harbored the seeds of disease. Even more disturbing were the strong winds that descended on the valley from the north in summer and fall. Known locally as "northers," these winds were said to quickly desiccate crops and kill small birds. Cows reportedly produced less milk, and sheep sheared during a norther quickly lost weight. Humans suffered similarly, as the winds were known to debilitate even healthy persons and to exacerbate the problems of those who were already sick. In 1869 Dr. H. W. Harkness published an article in the *Pacific Medical and Surgical Journal* in which he linked the north winds with several diseases, including rheumatism, neuralgia, sluggishness, depression, respiratory problems, eye diseases, and an increased incidence of sunstroke. As one local doctor summarized it, "All the organs of the body suffer . . . from the effects of this, the most noxious of all air currents to be encountered upon this portion of the American continent." [45]

Physicians also cataloged the region's diverse soils. The "internal nature" of local soils was held to influence climate and the quality of local waters. Some soils retained heat; others did not. Some exhaled poisonous gases; others might absorb them. And the composition of the soil influenced the form (remittent, intermittent, etc.) that local diseases were likely to take. Throughout the continent, poorly drained soils were strongly correlated with the production of malaria, and, along these lines, the California State Board of Health noted that "retentive clay subsoils" found in the San Joaquin Valley kept "the air over large districts . . . always more or less damp and unhealthy." The valley's alluvial soils, moreover, were believed to be "charged with the elements of malaria, which only await a summer temperature above 60 degrees and an upturning of the surface to induce that poisonous fermentation which destroys health and endangers life." In his investigations of epidemic dysentery, Dr. Walton Todd of Stockton observed that the alluvial "black lands" in the delta region appeared to be associated with the most malignant cases of disease. The western side of the valley held mostly alkaline soils, which, though not associated with malaria, nonetheless left the local waters mineral laden and unhealthy.[46]

The local topography also invited medical study. In some areas, mountains and hills trapped unwholesome vapors and fogs, but they could also block disease-carrying winds. Of course, most of the valley was exceedingly flat, but nonetheless there were important differences within that landscape. The outer reaches of the valley, the areas farthest from the main rivers, were grasslands, with few trees, which settlers found off-putting. The land's barren appearance not only signified its potential infertility but also offered no barrier to the aerial transport of disease. Under such conditions, miasmatic diseases could spread long distances from their source and infect otherwise healthy localities. Not surprisingly, immigrants found the eastern edge of the valley most appealing. Here the land rose gradually up to the Sierra Nevada foothills and was generally well watered, with cottonwoods and willows lining the riverbanks. The western side of the valley was drier, especially in the southern San

Joaquin Valley—parts of which were nearly a desert and attracted few settlers. Yet it was the interior region that generated the most comment. These were the most fertile and well-watered, and in many ways the most desirable, lands—and also the most sickly.

Much of the inner valley was a seasonal swamp. Before the reengineering of the valley that settlers and engineers undertook in earnest in the 1880s, rivers and streams were not confined to their banks but dispersed into a variety of sloughs and channels on the valley floor, the number and size of which varied seasonally. The major rivers—the Sacramento and the San Joaquin—as well as their tributaries were shallow, aggrading streams that easily overflowed their banks in many places and dispersed into multiple ill-defined and slow-moving channels as they approached their outlets. The volume of water in the rivers varied dramatically over the year, rising precipitously in spring and early summer when the snows of the adjacent Sierra Nevada melted and then declining over the long dry summer and fall. During the high-water season, flooding was a matter of course; river waters spread into any low-lying land. In years of above-average rainfall, most of the region, excepting only the highest points, would lie underwater for a period of days or even weeks, creating "one immense sea, leaving only a few scattered eminences which art or nature have produced, as so many islets or spots of refuge."[47] In localized patches, the land could remain flooded for a year or more. Historical estimates of seasonal swampland place it at one to one and a half million acres, and accounts of the valley from the 1840s and 1850s repeatedly emphasize the difficulty of moving through the swampy and overflowed environment and the confusing nature of the landscape, especially during the wet season. As one surveyor wrote, "The number and intricacy of the winding sloughs and channels that traverse this wide area of low marshy land is worthy of notice." Boats that sailed through the delta region were known to head the wrong way in the slow-moving currents and to become lost in the maze of crisscrossing channels.[48] In every sense, it was a disorderly and chaotic environment, even in the eyes of those who desperately wanted to colonize it.

Aside from the condition of the rivers, the feature that most caught the attention of European and American observers was the vegetation. Extremely large rushes lined most of the streambanks and sloughs. Called "tulares" by the Spanish and "tules" by the Americans, these plants grew thickly in many areas. Though some equated tules with the common bulrushes found in the East, the California plants grew "to an enormous size, attaining a height of from 8 to 15 feet, and sometimes a diameter of three-quarters of an inch." In the 1850s these plants, "a remarkable feature of the vegetable physiognomy of California," covered most of the aptly named Tulare Valley, the banks of the San Joaquin River, and the banks of the lower Sacramento River; they also extended across the valley for a distance of one to fifty miles. During winter and spring, the tules trapped and held floodwaters; for miners and early travelers, they were yet another obstacle to movement. Lieutenant Derby found the tules a "thick, tenacious, quagmire, which it is difficult, not to say dangerous, to attempt crossing." The surveyor of San Joaquin County had a similar complaint about the landscape: "It is almost one solid mass of tule, with frequent deep ponds and sloughs."[49]

Swamps and tule lands were not merely an unsightly landscape, an inconvenience to travel, and a hindrance to agriculture—though they were all these; they were also a frightening source of disease. That swampy environments were unhealthy was axiomatic to nineteenth-century settlers and physicians alike. The chaotic, overgrown vegetation along with the meandering and changeable river channels surely called to mind the tropics and their associated diseases. "Where the tule grows the rankest, we shall always find the most positive type of intermittent fever," wrote Dr. Washington Ayer for the California Medical Society. During the gold rush, Dr. James Tyson had referred to the tule marshes as "nurseries of disease," and even popular writers who boasted of California's healthy climate often advised newcomers to avoid the riverbanks. John Audubon wrote of Sutter's Fort, located on the Sacramento River, "The swampy neighborhood, bad atmosphere, and malarial conditions must render this section of the country unhealthy of a great degree for half the year."[50] Dr. Thomas Logan, the reigning expert on the health of the Sacramento region, declared that the Sacramento River was the principal factor, for better or for worse, in the region's health. Elsewhere Logan described the Sacramento basin and its associated tule marshes as a continual source of illness:

> The watercourses, in time of high water, do not run *into* but *from* the river, and consequently they carry off into the *tulares*, or marshes, and lowlands— animal, vegetable, and mineral matters capable of solution, suspension, transportation, and putrefaction. Hundreds of miles above the city of Sacramento, the tributaries of the river—every swollen brook, rivulet, and stream reach the larger affluents charged with these matters, which during its journey towards the south, under the influence of a hot sun, undergo great modifications. New chemical compositions and decompositions must occur. . . . Hence every condition exists favorable to the dissipation of deleterious gases arising from the putrescible matter in the water.[51]

What Logan was describing was the formation of miasma, or marsh poison. Contemporaries typically understood miasma not as a disease but as a quality of certain environments that was responsible for many diseases, including malarial fever, typhoid and typho-malarial fevers, diarrhea, dysentery, and diphtheria.[52] Physicians debated the nature and cause of miasma; many noted that miasmatic diseases, including malaria, were among the most poorly understood medical phenomena, and numerous theories were expounded to account for them. There was widespread agreement, however, that miasmatic diseases were most typically associated with low-lying, swampy places, as well as with fogs and bad smells. Most believed they were the product of organic matter and warm temperatures, and contemporaries coined the term *zymotic* to describe the process of decomposition and fermentation they believed led to disease. Dr. Daniel Drake, the medical authority on the Mississippi Valley, acknowledged his uncertainty on the issue but surmised that either organic matter was somehow transformed from a benign solid material into a poisonous gas or, alternatively, that it might already harbor "animalcular or vegetable germs" that could be released into the air under certain environmental conditions. Logan, following the leading theories of his day, associated malarial fevers with alluvial soil, abundant vegetation, seasonally flooded rivers, and high summer temperatures.[53]

Although nineteenth-century physicians had developed a general understanding of miasma and the conditions under which it produced disease, they were forced to acknowledge its profoundly local and unpredictable nature. In a survey of the occurrence of malaria in southern California, Dr. Joseph Widney noted "the lack of regularity in the action of certain well-established laws." After studying the characteristics of places afflicted with disease, he concluded that the "results do not, at first sight, seem altogether such as we might naturally expect." Widney was especially surprised to find malaria prevalent in irrigated coastal lands, despite a relatively low summer temperature and the fact that they were "swept daily by the full force of the sea breeze." Such perplexing realities necessitated even closer local study.[54]

As medical discussions of the Central Valley make clear, the assumptions and practices of nineteenth-century medicine insisted on the careful study and subsequent differentiation of landscapes. Environments were local and particular. California was not New York; the Central Valley was not the coast; the Sacramento River was not the San Joaquin. Health and disease were specific to place. And where disease was present, even closer environmental scrutiny was necessary in the interests of public health. The sometimes atypical patterns of disease in California only reaffirmed the importance of place and the need for local environmental knowledge. That physicians were among the keenest observers of not only disease but also the environment in nineteenth-century California was only to be expected at a time when both environmental etiologies and cures held sway.

CIRCUMSCRIBING DISEASE, REMEDYING THE LAND

Even though sensitive to disease, most Americans were far more intent on settling the environment than studying it. As valley communities competed fiercely to encourage immigration and commerce, boosters feared that any suggestion of disease would undermine their own town's financial prospects. There were obvious reasons for immigrants to deny that the environment was the source of their health problems, and many did. Some denied the presence of disease altogether. One writer from Bakersfield complained that concerns about disease in that locality were greatly exaggerated: "The climate of this State is so salubrious and life-inspiring that if a locality happens to be found within its borders where any sickness whatever occurs, it is regarded with a sort of horror." Another view held that the valley's malaria was not endemic at all but "an offspring of the tropics," which immigrants acquired while passing through Latin America, only to have the illness manifest itself after their arrival in California.[55]

But it was less common to deny the presence of disease altogether than it was to stress the importance of individual factors in its control. The interplay of environmental and human factors in the onset of disease was generally acknowledged and was a staple in the medical advice literature. Colonial British physicians, for instance, had long stressed the health of European troops in tropical environments hinged partly on personal behavior. James Martin's medical text advised specific prophylactic measures, including proper dress, appropriate food and drink, limited exercise, cold baths, and adequate sleep. But above all, he advised Europeans to exert strict control over the "passions" and any temptation to vice. Self-discipline could be the key to maintaining one's health, as well as one's moral superiority, in dangerous environments.

A similar discourse circulated in California. Boosters and their supporters often attributed ill health in the valley to bad habits rather than to the local environment. Predictably, disease among nonwhites was easily and frequently dismissed as personal rather than environmental in origin, the obvious outgrowth of their vice and unclean habits. Sanitarians repeatedly condemned the Chinese neighborhoods of valley towns as the source of "miasmatic effluvia" that wafted into unsuspecting white bodies. In discussing the fever epidemic that struck the Chinese in Oroville in 1874, Dr. P. B. M. Miller acknowledged the malarial influences but also noted that "China towns" of the region were "the most offensive, pestilential, nay abominable, hot-beds of disease in the State." At the same moment, disease was one often-repeated rationale for Chinese exclusion. But the illness of white settlers could also be individualized. For instance, H. S. Orme, president of the State Board of Health in the early 1880s, admitted the environmental sources of miasma in the valley and yet displaced the dangers of the environment onto settlers themselves, arguing that "in many instances, where diseases are attributed to malarial influences, . . . the true source of the evil lies in the habits of the people and the disregard of sanitary laws." Another prominent physician and San Joaquin Valley booster insisted that the underlying causes of illness in the region were "impure water, excessive and indiscreet drinking of water, [and] careless use of fruits"; and the *Pacific Rural Press*, a California agricultural paper, claimed that seasonal illness was often due to "dietetic error." Similarly, a lay health manual acknowledged the presence of endemic disease in California while advising its readers that "careful attention to bathing, a rigid dietary, and proper clothing will do much to protect" a person against disease.[56] Lay publications also inveighed against overwork and insufficient sleep. By attributing the cause of disease at least in part to individual behavior, such advice helped exonerate the landscape, rendering it appropriate for further settlement. The tendency to link disease to individuals rather than environments would become more pronounced as the century wore on. Yet as much as settlers might want to view the landscape as benign and welcoming, they operated in a biological and cultural world that did not fully allow them to do so. Whatever immigrants might want to believe about the Central Valley, many of them were still falling ill, and most agreed that the local environment was implicated.

Even the most committed boosters and hygiene advocates could not completely deny the importance of local environments to health, but they stressed the variability of disease across space. When physicians described the medical topography of the state, they often spoke of the valley as a homologous region; the crucial distinction they drew was between the state's more sickly interior and the healthy, even curative coast. Valley residents, however, were more likely to focus on the character of specific localities. Stockton was not the same as Fresno; the San Joaquin Valley was not to be confused with the Sacramento Valley; the plains differed from the riverbanks. Settlers drew a microgeography of disease within the generally afflicted space of the California interior. A medical columnist for the *Pacific Rural Press* noted that the valley's local environments produced their own characteristic diseases: low and marshy areas produced the severest cases of chills and bilious fever; the southern areas of the valley produced congestive and yellow fevers; the healthier localities produced merely dysentery, pleurisy, and attacks of indigestion.[57] Though no areas of the Central Valley attained the healthful reputation of the California coast, some were clearly superior to others, and boosters argued with one another over the relative "salubrity" of their own locales. They sought to draw a much more detailed medical topography of the valley, in which disease and diseased environments were highly localized. One strategy was to insist that diseases came primarily from the water rather than from the atmosphere and thus could be avoided through careful settlement such as shunning bad sections of river and properly placing wells. In this way unhealthy environments could be circumscribed, and pieces of the valley at least could be defended as hospitable to white settlement.[58]

Even while wary of unhealthy places and practices, American immigrants remained committed to transforming the valley. Although some immigrants came west seeking health, for most the primary motivations were economic. The disappointing opportunities afforded by mining and the existence of a strong export market for wheat and grains encouraged even the earliest immigrants to try farming. Even if they came to California for gold, most of those who immigrated in the late nineteenth century had been farmers, and they brought with them an agrarian background and an ideology that held farming to be the foundation of a moral and democratic society. The move toward farming was also supported by the state. Both the California legislature and the federal government encouraged the development of swamplands and provided funds for early irrigation surveys. The federal Homestead Act of 1865 provided smallholdings to individuals, and the state of California offered financial incentives for the development of new agricultural products. Not surprisingly, the number of farms rose rapidly, from 19,000 in 1859 to 73,000 by the end of the century.[59]

For the most part, conceptions of health merged easily with agrarian ideology because the principal means of curing a pathological space in nineteenth-century America was through the introduction of Euro-

American agriculture. Americans shared the belief that agriculture represented civilization and that civilization produced health.[60] Throughout the second half of the nineteenth century, local agricultural publications urged on valley farmers with the insistence that cultivation would eliminate the valley's worst afflictions. And the crusade for swampland reclamation in California invoked concerns over bodies and health as often as appeals to economic benefits, a fact easily overlooked by a modern reader. California's promoters insisted that Euro-American agriculture would inevitably improve the health of central California as it had improved health in England, where intermittent fevers had disappeared "in consequence of the high cultivation and careful drainage of the land." As one irrigation engineer argued, reclamation was "demanded by all the beneficial results which must ensue from [the valley's] restoration to a habitable, improvable, and *healthy* condition."[61]

American-style agriculture was seen as generally healthful, but the details were important. Specific plants were mediators of environmental disease. In California, settlers regarded alfalfa, which they planted widely for livestock, as curative, and residents in the San Joaquin Valley insisted that its widespread cultivation would curb the region's malaria problem. In his medical assessment of the state's landscape, Dr. Thomas Logan urged settlers to plant both alfalfa and grasses near the watercourses.[62] And like many Californians, he believed that the turn toward horticulture promised even greater health benefits. Of course, there were many reasons to grow fruit. International competition made wheat a risky crop, while fruit garnered higher prices. Moreover, Americans placed social value on fruit growing as an inherently family-oriented activity that would foster small farms and decrease the need for migrant (especially nonwhite) labor.[63]

But settlers also consistently noted horticulture's physical benefits. Its proponents argued that the labor involved was "light," and therefore the white body would be appropriately stimulated by the work but not overly taxed, an important consideration given the valley's summer temperatures. In fact, horticultural work was considered appropriate and even beneficial for invalids and white women—who otherwise appeared dangerously susceptible to disease in California. "The labor is of that character just suited to eradicate their [women's] too frequent physical debility," declared the *California Farmer*. Moreover, the consumption of fruit might act as a prophylactic against the regional diseases. Fruit was believed to lessen constipation and provide "an outlet to all febrile and bilious 'humors.'" Dr. Widney attributed the health of southern California residents to the fact that fruit was "a leading article of diet" that cured "many ills of the digestive systems [that] we fell ill to in our Eastern homes." Strawberries were reportedly a cure for gout, insanity, and certain forms of consumption; olives alleviated indigestion, kidney and liver "derangements," rheumatic disease, and diseases of the glands and throat.[64]

Trees had a well-recognized potential to modify the local atmosphere and climate, thus providing another rationale for planting orchards. Trees interrupted violent winds, lessened the daily variation in temperature, and halted the spread of disease through the air. "Like a large mechanical sieve," wrote one physician, the forest sifted the air, leaving it "free from all impurities and disease." Dr. William Gibbons, one of California's physician-meteorologists, argued that if a line of forest trees from three to five miles wide were planted along the valley floor, it would both moderate the temperature and decrease the prevalence of malaria. Another physician pointed out that planting trees in the valley would help contain the noxious north winds. In addition, "emanations" from certain species of trees could mix with poisonous vapors and neutralize them. For this reason, evergreens, especially pines, were believed to benefit phthisis sufferers.[65]

From a medical standpoint, the most important tree in nineteenth-century California was the eucalyptus, in particular the blue gum *(Eucalyptus globulus)*. Californians were following the lead of European authorities who had begun importing the trees into disease-prone colonies such as Algeria, southern Africa, and Cuba. Some insisted that the species' prophylactic effects stemmed from its strong aroma; others pointed to the fact that it protected the soil from undue heat and also absorbed excess water and humidity; but few questioned its ability to render healthy otherwise "uninhabitable districts." Both physicians and boosters advocated the massive planting of eucalyptus trees in the valley, and by 1874 Californians had planted roughly a million of them. After its workforce became incapacitated by fever, the Southern Pacific Railroad planted 1,000 blue gums near the station of Delano in the southern valley. In Tipton, locals reported a significant decrease in disease following the planting of more than 120,000 trees. In addition to the eucalyptus's effect on health, its advocates argued that it provided a supply of timber, moderated the summer temperature, increased the limited streamflow, and broke up the monotony of the landscape.[66]

Appeals to improve the valley's environment in order to foster health—through the building of towns, the draining of swamps, the cultivation of fruit, and the planting of eucalyptus—were assertions that immigrants could direct if not control the effects of nature. Consequently, immigrant physicians were keenly interested in how rates of sickness and death might have changed in the period since American settlement. Improvements in health—such as a decrease in the death rate or the continued absence of diseases like smallpox and cholera—were always read as evidence that white occupation was improving the quality of the land. In 1873 the State Board of Health remarked somewhat hopefully on the "diminution in mortality" witnessed in the state over the previous two years, crediting this change to "the advanced general intelligence, the multiplication of comforts, the hygienic and other measures for the better protection from the elements, as well as from the causes of disease, the abundant supply of wholesome food, and all the inventions for exhaustive labor-saving, as well as the improvement of morals, we are accustomed to sum up in one phrase, as 'the progress of civilization.'"[67] A local physician observed that the north winds had become less violent, and he attributed the change to the American presence. In southern California, Joseph Widney credited "Anglo-Saxon" settlement with several climatic improvements: less variation in temperature, milder winds, and increased summer rainfall. And though no one could deny the continuing prevalence of malaria in the Central Valley, many physicians argued that the disease was becoming far less virulent.[68] Popular understandings of the health effects of white settlement were similar, if based on even less evidence. The booster Titus Fey Cronise confidently informed his readers that "no epidemic has prevailed in California since its settlement by Americans." In the Sacramento Valley, a newspaper editor and local booster claimed that "now since people begin to live a little more like white folks, they find [the region] is not so sickly as they were at first led to imagine."[69] In other words, white settlers believed—or at least hoped—that they were successfully adapting the land to the needs of their bodies. From this perspective, a cultivated and healthy landscape was less a symbol of conquest and domination than a sign of the synergism, in the most literal sense, between white Americans and their new home.

It is not assuming too much to read anxiety as well as arrogance in such claims. After all, most of those who came to California were only too aware of the lives they had left behind in other, more familiar and less risky places. They undoubtedly realized that their own future in the Far West, and that of their families, depended not only on their ability to prosper economically but also on their ability to remain healthy.[70] Nineteenth-century discourse on health was always marked by ambivalence. It asserted the agency of white colonizers while acknowledging their vulnerability to the complex agencies of a nature they did not fully understand. While moral vigilance and American agriculture were certainly assets, they did not necessarily outweigh the effects of an insalubrious place, and those who moved into the Central Valley in the nineteenth century often did so despite their fears. "I doubt the wisdom of bringing my family to this confounded sickly place," the engineer William Hammond Hall confided to his diary in 1878 as he contemplated a move to Sacramento. "Typhoid fever, intermittent fever of all kinds—the regular [old?] ague appears to be not only prevalent but severe. . . . I can only try to keep my wife and little ones away from here in the most unhealthy times and trust to God and good care for their protection at other[s]."[71]

IMPROVEMENT OR DECLINE?

Although American immigrants hoped that civilization would bring health to California, the transformation of the landscape only raised more questions. Not all environmental changes that white immigrants initiated appeared to generate health. To the contrary, some changes had demonstrably increased illness, or so contemporaries suspected. Nineteenth-century immigrants were keenly aware of the limits of their environmental knowledge, and physicians were particularly attuned to the unintended consequences of settlement. From the moment of his arrival, Thomas Logan was concerned with how the changes wrought by settlers might already be affecting health. Though he was anxious to see white settlement as a positive influence, Logan saw himself as a scientist, not a booster, and he ruefully noted many instances in which recent environmental changes seemed to correlate with increased disease. For example, he suspected that the initial outbreak of cholera in 1850 might have been caused by the overturning of the soil to build levees around the city of Sacramento. Dr. William Gibbons similarly noted that the plowing of land along many of the valley's rivers and streams had seemingly "liberate[d] the subtle poison which engenders disease." Members of the Sacramento Society for Medical Improvement hypothesized that improper farming practices had contributed not only to the increase in malaria but also to the rising incidence of sunstroke.[72] In 1858, when malaria in the valley and foothills reached epidemic proportions, Logan attributed the region's sickliness to the environmental changes associated with mining and farming:

Prior to 1858 the plains as well as the mountains of California were proverbial for their salubrity. . . . But how stands the case now? An extensive system of irrigation for mining and agricultural purposes has been resorted to—canals seven thousand miles in aggregate length have been dug to lead the water in innumerable serpentine courses from the rivers into the placers, and almost every valley that can be dammed on the line of these ditches has been appropriated as reservoirs to hold water. The action of an almost tropical sun upon the decaying vegetable matter that remains in these canals and reservoirs, when they have been drained[,] . . . has been manifested in its effects. Not only in the plains and agricultural regions, but along the whole range of the foot-hills . . . nothing but fever at one time was heard of.[73]

Logan's critique of human changes to the land drew on well-established medical theories. As early as 1775 Benjamin Rush, the preeminent physician in colonial America, had emphasized the difference between clearing and cultivating the land: "While clearing a country makes it sickly, . . . *cultivating* a country, that is draining swamps, destroying weeds, burning brush, and exhaling the wholesome and superfluous moisture of the earth, by means of frequent crops of grain, grasses and vegetables of all kinds, render it healthy."[74] In this formulation, one kind of environmental transformation—clearing—yielded disease, while another—cultivation—yielded better health. Contemporaries struggled to reconcile these observations by arguing that clearing was an intermediate stage. Illness in such cases was the result of failing to complete the necessary work. It was not simply the presence of white settlers that improved the health of a region but also the specific changes that they imposed on the land. The realities of disease, however, made it far too difficult to predict with any confidence which changes would bring health and which disease, especially in the untested landscape of California. Even as Californians embraced the "development" of their region, tensions over the processes of modernization emerged again and again in discussions of health.

The overriding concern in the Central Valley was not forest clearing, however, but the effects of irrigation. As irrigation expanded in the valley in the 1860s and 1870s, so did malarial disease, and most saw an unfortunate connection. This was especially true in the southern San Joaquin Valley. Whereas in the 1850s the Sacramento Valley had been the most sickly, by the 1870s portions of the more southern San Joaquin Valley, around the towns of Fresno and Bakersfield, held that distinction. Over that period, irrigation in California had proceeded haphazardly. The first areas to be irrigated were those in the southernmost section of the valley, which averaged less than twelve inches of rain per year and whose natural topography made irrigation relatively easy. The first projects were nothing more than cuts in the riverbank, which allowed water to run into shallow ditches or old sloughs. Though the first canals were dug in the mid-1850s, it was not until the 1870s that irrigation expanded rapidly—spurred both by several dry years and the move toward fruit and vegetable crops, which required more water than grains. The engineer Carl Grunsky, surveying the San Joaquin Valley in the 1880s, cataloged forty-six irrigation ditches along the Kaweah River, thirty-four on the Kern, and forty-nine on the Tule. Ditches ranged in size from very small—only a few miles long and irrigating a few acres— to the giant Fresno Canal, which irrigated more than 160,000 acres. Total estimates of irrigated acreage in the period are sketchy, but by 1890 California had more than one million acres under irrigation, the majority in the San Joaquin Valley.[75]

As irrigation expanded, so too did concerns about disease. The *Pacific Medical and Surgical Journal* reported in 1871 that an irrigation project in northern California had been "disastrous in the development of malarial disease." Two years later the State Board of Health noted the prevalence of malarial disease in the valley and asked rhetorically whether "the extensive systems of irrigation, now under discussion, will not add to their insalubrity." In retrospect, we can understand the perceived rise in disease by pointing to the creation of new mosquito habitat as crudely constructed canals were laid across soils with poor natural drainage. In fact, the land in much of the San Joaquin Valley was underlain by an impermeable layer of clay, and irrigation had noticeably raised the water table in some localities by the 1880s.[76] In addition, irrigation canals were typically shallow, and ditch owners tried to keep water in their channels as long as possible for the benefit of livestock. These practices created numerous small pools on the surface—prime breeding habitat for anopheles mosquitoes. Contemporaries, however, believed that ill-considered irrigation projects were making the land more sickly, breeding not mosquitoes but miasma. In an attempt to ascertain the relationship between environment and disease, A. G. Warfield, an engineer working for the state, surveyed settlers along the San Joaquin River in 1879, asking everyone whether irrigation had affected health. Several people told Warfield that the region had in fact become "more subject" to fever and ague since the introduction of irrigation. A California medical almanac published in the late 1870s informed its readers that irrigation had converted "perfectly salubrious" areas into "hot-beds for malarial poison, giving rise to ague, congestive chills, remittent or bilious fever, and typho-malarial fever," and it warned that those who did the irrigating were the most likely to suffer. Engineers, doctors, and ordinary farmers all noted the link between environmental change and human health. When a committee of the U.S. Senate studied the question of irrigation in California and the West, its members were keen to query local residents about their own experiences of irrigation and its bodily effects.[77]

Not everyone was willing to attribute local diseases directly to irrigation, however, particularly those who were most committed to the modernization and rationalization of the landscape. In testimony before the Senate committee, an irrigator from Fresno claimed that when he moved to the region, people told him he would surely sicken and die, but in defense of both his adopted home and the practice of irrigation he pointed proudly to the fact that he had gained seventy pounds. In his survey of the Tule River area, the engineer Carl Grunsky, a strong supporter of irrigation works, acknowledged the prevalence of fevers in the region but emphasized that there might be other causes; in his words, it was not possible "to determine to what extent the healthfulness of the country has been affected by the practice of irrigation." Defenders of irrigation, especially engineers like Grunsky, routinely downplayed the role of environmental change; they argued that it was not irrigation per se that made the land unhealthy but only improper methods, in particular the failure to address the need for drainage. Others hoped that the ill effects of irrigation could be countered by the planting of trees and improved personal conduct. A few even claimed that in some localities irrigation had made the climate "more temperate."[78] William Hammond Hall, the state engineer and an architect of California's environmental modernization, acknowledged the environment's ability to affect health (and worried in private about his own family's well-being in the sickly landscapes of the valley), but he also insisted that irrigation could have positive bodily effects. Hall told the Senate Committee on Irrigation in 1890:

> I do not know whether this ought to be recorded, but I was struck with the very curious appearance of shape of the younger ladies . . . in that section [near Bakersfield]. It seems that I could not account for it at the time, but it turned out that they had a little affection of the spine, and there was a rotundity of the trunk which was astonishing in ladies so young. I tell you [this] to illustrate the fact that before irrigation was regularly introduced, there was a malarial influence which pervaded the population . . . , and it showed not only upon the countenance of the people, but upon their persons. But now that the slough has been cleaned out . . . [and] fresh water . . . applied to the land . . . , the atmosphere and the climatic

conditions appear to have changed to that extent that I believe Bakersfield . . . is about as healthful a locality as you will find in any portion of
this State. And I attribute it to irrigation.[79]

While engineers such as Grunsky and Hall were likely to downplay
the agency of nature in disease, and to assert their ability to manage the
landscape for both productivity and health, physicians were far more circumspect. As the pace of environmental change grew more rapid, leading physicians weighed in on the possible long-term health effects. One
physician pointed to the environmental disruption caused by construction of the Central Pacific Railroad in the southern San Joaquin Valley,
which had left "nearly every laborer . . . a victim" of malaria. An article
in the *California Medical Gazette* noted anxiously that "the spirit of
progress has unwittingly opened up new fields for the more extensive
growth of [miasmatic disease]."[80] Drawing on the work of George
Perkins Marsh, Dr. Marshall Chipman was one of several Californians to
turn his attention to deforestation of the state's riparian and mountain
landscapes. In Chipman's view, deforestation was clearly a medical issue
because it both destroyed the earth's ability to sustain the human population and altered the state's meteorology and climate, rendering vast
regions "insalubrious and less wholesome for human residence." Writing
for the state medical society in 1878, William Gibbons warned that
"while there is a steady destruction of timber going on throughout the
State, there must necessarily be a gradual increase of area over which
malarial fevers will be developed."[81]

Critics of irrigation likewise interpreted the rising incidence of illness
as a failure on the part of settlers to attend and adapt to the local environment. As one doctor residing in the valley town of Merced remarked,
"As far as I have been able to learn, this was a healthy locality till the
water of the river was turned out of its proper channel."[82] In the *Third
Biennial Report* of the California State Board of Health published in
1875, Logan—as secretary of that organization and now California's
most prominent physician—reiterated his earlier warnings about the
environmental changes under way, indicting irrigation, deforestation,
and mining for their effects on health in the Central Valley. "In no other
country or epoch in the world's history than in California at the present
time has man's action ever been known to change so rapidly or so permanently the face of nature," Logan wrote. "Unless, therefore, some
effort is made towards correcting and providing against the evils resulting . . . the most valuable portion of California will become more and

more obnoxious to the health of the inhabitants during the autumnal months." In his opinion, settlement might not necessarily bring health but its very opposite, particularly if nature's patterns were not understood or adequately respected. Both land and bodies were at risk from the effects of an "outraged nature." Writing of the uncontrolled spread of irrigation across the valley, Logan warned Californians that if they proceeded on their current course, they might well end up with an environment as sickly as that of certain sections of the South with their "notoriously miasmatic rice fields."[83]

In 1881 Marshall Chipman chaired the California Medical Society's committee on medical topography, and in that capacity he wrote a lengthy article describing the deleterious effects of mining debris that had clogged the Sacramento River and ruined farmland throughout the lower Sacramento Valley. He informed his colleagues that the long-term effects of mining were not simply economic or social but also medical, and he urged the profession to take a stand on the issue. In contrast to the beneficial effects on health that "civilization" was supposed to bring, mining had actually increased the extent and frequency of malaria. He cited several local physicians who claimed that the mining debris had engendered typhoid-like diseases that were more severe and less treatable than those experienced previously. In fact, another physician, William Robert Cluness, would testify to that effect in the famous trial concerning the effects of hydraulic mining, *Woodruff v. North Bloomfield, et al.* Elsewhere Chipman blamed human "ignorance, improvidence and wastefulness" for deteriorating the "earth's general salubrity."[84]

The concerns of Chipman, Gibbons, and others attest to their recognition that health and disease were complex outcomes of human actions and a multitude of environmental factors. It was not merely a question of transforming the landscape in accord with Euro-American norms but of attending to local conditions and the undeniable connections between land and bodies. The future of human beings in the Central Valley, as in any region, was closely interwoven with the development, or disturbance, of local animal and vegetable life. No one understood this better than physicians. Although knowledge of the precise connections between bodies and environments remained obscure, physicians were certain nevertheless that the connections were real. A disruption of external nature would lead to a disruption of internal nature.

The practices of nineteenth-century medicine—medical topography and local observation, the assiduous collection and correlation of environmental data and health statistics, theories of miasma and climatotherapy—connected health and disease to the landscape in concrete ways. These practices helped produce an understanding of the body as permeable and an awareness of the environment as an active agent shaping health. An environmentally oriented medicine combined with the human and nonhuman nature of the valley to produce certain forms of illness: malarial and typho-malarial fever, dysentery, neuralgia, and many others. Although elite discourses of health were not equally available to all, the practices of nineteenth-century physicians supported and explained what most laypeople already believed to be true.

Although environmental medicine antedated the modern science of ecology, it nonetheless prefigured aspects of that later discourse. Nineteenth-century physicians insisted on placing human bodies squarely in their local environments, and they took the relationship between the two as their object of study. They stressed the unending interactions between bodies and their environments and the environment's importance to both health and disease. In retrospect, what is most striking is that nineteenth-century medicine could for so long meld this ecological perspective with the quite different understanding of the body and disease that was then emerging in the laboratory. Through the end of the century, most California physicians embraced both germ and environmental theories of disease, contagionism and anticontagionism, personal hygiene and the concept of an epidemic atmosphere. Germs might contain the seeds of disease, but they still existed in a complex environment that had to be taken into account.

However, it would prove increasingly difficult to sustain such an ecological view, with its inherent conservatism, in a culture committed to rapidly colonizing and radically transforming the landscape. Ecological ideas would persist—but not as the dominant understanding. Much of the environmental data that Thomas Logan had dreamed of gathering and analyzing would soon be rendered obsolete by the massive changes under way. Mining, agriculture, and lumbering were transforming California's ecologies at an unprecedented pace.[85] Those who were anxious to develop and use the land were not inclined to wait for medical topographers to draw out their conclusions. This tension—between the desire for modernization and environmental understandings of health and disease—had been evident from the moment of Americans' arrival in California, and it had only increased as the century wore on. In the 1870s and 1880s the desire to disconnect health from the landscape was clearly present but not yet fully possible. In discussions of bad habits and poor hygiene, in the dismissal of the contribution of irrigation to disease, in the rising interest in germ theories lies the growing desire to individualize and to further racialize disease.

Of course, since this is history, we know the end of the story. The environmentally oriented doctors did not prevail, either politically or historically. In comparison to the power of capital and capitalist enterprise, the medical profession was weak. Despite their efforts, nineteenth-century physicians had no broad authority to impose their views. By the end of the century, politics would not favor public health as a state enterprise, and doctors themselves would focus more intently on policing the boundaries of legitimate practice. Moreover, the rise of a new generation of practitioners, schooled in the successes of bacteriology, would hasten the decline of medical topography and environmentally oriented medicine. As laboratory work succeeded in locating the "cause" of illness in specific pathogens, the focus of medicine narrowed. Those practitioners who remained attentive to environmental factors would increasingly confine themselves to cures, ceding the territory of disease etiology to the new bacteriologists. Meanwhile, California's early physicians would be largely erased from the state's progressive history, written off as hopelessly unscientific "miasmatists" and relegated to the annals of medical antiquarians. Environmental understandings of disease, however, would not be so easily or completely dismissed.

3

Producing a Sanitary Landscape

Malaria is a disease not indigenous to our soil, nor to our
sunshine, nor to our luscious fruit, nor to the clear, cool
waters that flow from the Sierras: it is an alien that has crept
in stealthily and has occupied the length and breadth of
these fair valleys.

William B. Herms, *Malaria: Cause and Control,* 1913

In 1903 the incoming secretary of the California State Board of Health,
Dr. N. K. Foster, complained that his organization had neither a permanent office nor adequate furniture. The only equipment the board
could claim as its own was a single bookcase, two file cabinets, a mimeograph machine, and a typewriter. The secretary himself had but a bor-
rowed desk in the lieutenant governor's office. Given this situation, it is
hardly surprising that few of the board's records from the period survive.
As Foster no doubt recognized, his own lack of office space was symptomatic of the board's lack of power. Although Thomas Logan and his
colleagues had begun to build a strong state health establishment in the
1870s and 1880s, the position of secretary was nothing more than a
political sinecure by the turn of the century. The State Board of Health
had lost all credibility in the wake of San Francisco's first plague epidemic (1900–1901) when, fearing the economic impact of a quarantine,
both the governor and the board denied the existence of the disease.
Foster's predecessor advised him to pocket the $1,500 budgeted for
expenses and get someone else to open the mail.[1]

All that was about to change, however. Foster would remain secretary
for six years, and he and his immediate successors, William Snow and
George Ebright, would oversee a vast expansion of the board's power. In
1905 the state legislature enhanced the board's regulatory authority,
directing it to collect and summarize vital statistics and authorizing the

establishment of a bacteriological laboratory. In 1907 the board was given responsibility for enforcing pure food and drug laws. Two years later another laboratory was added. In 1910 the election of the Progressive Hiram Johnson to the governorship led to the appointment of an activist board, and in 1911 the board received authority to actively enforce laws against stream pollution. In 1913 the legislature strengthened stream pollution laws, created a bureau of tuberculosis, and granted the board new enforcement powers. The board's authority over water pollution was expanded again in 1917. Finally, in 1927, as part of a broad reorganization of state government, the California State Board of Health became the California Department of Public Health and was again expanded. These institutional changes reflected the emergence of public health as an important state function.[2]

The rise of professionalized public health between 1900 and 1930 was not limited to California. It was one aspect of the Progressive reform movement that took hold across the United States at the turn of the century, a period in which government at all levels became more involved in both the economy and the daily lives of citizens—through the provision of services, the enactment of regulations, and the collection of information. The so-called golden era of public health was one facet of this reformist agenda, and one that was readily embraced in Progressive California, which built one of the strongest and most activist health departments in the nation. But the rise of government-funded public health represented more than simply the growth of the state administrative apparatus; it also signaled the growing authority and influence of a new group of public health professionals who would promulgate new understandings of bodies and environments in California and elsewhere.[3]

For these professionals, the most important intellectual development of their lifetimes was the widespread acceptance of germ theory and the corresponding decline of environmental medicine. This shift cannot be precisely dated; retrospectively, it is located somewhere between the first important bacteriological discoveries in the 1870s and 1880s and their consolidation in the decades around the turn of the century. By locating the source of disease in microscopic entities, bacteriology helped lay the basis for "modern" medicine and a corresponding modern conception of the body. An emphasis on a singular etiologic agent for each disease gradually replaced the idea that disease marked a disequilibrium between the body and its environment. Consequently, it became possible to think of disease as a discrete entity that entered an otherwise self-contained body through specific, traceable pathways.[4] The elevation of germ theory to medical dogma allowed its adherents to separate and compartmental-ize diseased bodies and their environments to an extent that had not been possible in previous decades.

California's Central Valley was just one place that would be both rhetorically and materially transformed by these new formulations. By localizing disease in specific pathogenic agents and individual bodies, germ theory made it possible to assert that no landscape—at least no landscape in North America—was inherently or permanently unhealthy.[5] It is no surprise that several Central Valley boosters would be among its most ardent advocates. Modern medicine and public health would now cast pristine landscapes as physically harmless while portraying disease as an invader, an "alien" that needed only to be eradicated. If disease were indeed "carried" into a region such as the Central Valley, then any lin-gering fears that the valley was naturally inhospitable to white immi-grants could be pushed aside. Although diseases remained, modern dis-courses of health now located the "causes" of those diseases in specific bodies (insect, animal, human) that could be exterminated, reformed, or tightly controlled.[6]

Although the "apostles of the germ" portrayed these developments as revolutionary, they marked less a total break with nineteenth-century environmental medicine than a radical relocation of emphasis.[7] At the level of practice, understandings of bodies and places remained deeply intertwined. The uneven distribution of disease and the recurrence of epi-demics indicated that certain places still seemed to harbor illness while others did not. In other words, germ theory could not and did not com-pletely disconnect the body from its environment. Any attempt to con-tain disease and health within a simple modernist narrative was bound to fail.[8] Consequently, the rhetoric of germ theory existed alongside the environmentally oriented practices of sanitary engineers and marginal-ized subfields of biomedicine that still took the broader environment into account. In fact, engineers and others would invoke germ theory as their rationale for a much more widespread effort to reorganize landscapes along "sanitary" lines. Somewhat paradoxically, modern public health experts could deny the relevance of the environment to disease even as they embarked upon a vigorous project of remaking local environments.

ENVIRONMENT AND DISEASE IN THE CENTRAL VALLEY

By the 1920s rural California was a vastly different place than it had been in the 1870s, when Thomas Logan died. What drew no end of con-

temporary commentary in the intervening half century were the intensification of agriculture and the shift away from large-scale wheat farming
and toward fruit and nut crops. Statistics tell part of the story. Statewide
the number of farms grew by more than 400 percent (from 24,000 to
136,000), while the average size of farms fell from 475 acres in 1869 to
220 acres in 1929, as many of the large wheat farms were broken up;
even casual observers noted a "decided tendency toward small farms."
Even more striking was the change in what farms produced. From 1889
to 1919 California's production of citrus fruits increased twenty times
over, and its production of almonds, peaches, plums, and prunes rose by
a factor of ten. Grape production rose from 550 million pounds per year
to more than 2 billion; orange production, from 1.2 million boxes per
year to more than 14 billion. The Sacramento Valley was further transformed by the planting of rice; first grown commercially in 1912, rice
was well suited to the region's adobe soils and soon became the region's
major crop. Because it could be grown quickly, alfalfa became a popular
crop among the region's small farmers, and the subsequent availability of
hay helped spur the growth of dairying. Several factors encouraged these
shifts—including reclamation and irrigation, which made more land
available for intensive production (between 1900 and 1920 irrigated
acreage in the state rose from 1.4 million to more than 4.2 million acres);
the subdivision of bonanza wheat farms; and the declining cost of capital, which encouraged farmers to invest more heavily in their land and to
plant higher-value crops.[9]

By the 1920s the vast wheat fields and cattle ranches had been
replaced by thousands of acres of orchards, rice, and alfalfa, watered by
hundreds of miles of canals. Agriculture had differentiated the space of
the valley in new ways, and as farmers became familiar with the regional
soils and microclimates, localities became identified with the crops they
supported. Yuba and Sutter Counties were known for their orange
groves. Glenn and Colusa Counties were associated with rice. Turlock
was a center for melons and sweet potatoes, Modesto and Selma for
peaches. Oakdale and Davis were known for their almonds, Clovis for its
figs, Fresno for its grapes. In the San Joaquin Valley, farmers discovered
that the foothills of the Sierra Nevada sheltered trees from killing frosts;
consequently, a citrus belt developed here that included the towns of
Orange Cove, Lemon Cove, Exeter, Lindsay, and Porterville.[10] Although
California agriculture would continue to change, the basic elements of
this pattern would persist through the end of the century.

Intensification and diversification did not represent the triumph of

Figure 5. The horticultural landscape in the early twentieth century. Shown here is the famed citrus belt in the eastern San Joaquin Valley, with the Sierra Nevada in the background. Courtesy California Historical Society, FN-36328.

agrarianism in California, however. To the contrary, they were accompanied by the industrialization of farming, especially after 1910. Although the average farm size continued to decrease, California agriculture as a whole became more horizontally and vertically integrated. Large growers and corporations owned many thousands of acres in the valley, which were worked by labor gangs or leased to tenants. And even among the region's small farmers, prices and crops were increasingly dictated by the large packing corporations.[11]

Intensification did bring significant numbers of people to central California. The valley counties held 54,000 people in 1890; by 1930 their population had risen to more than 840,000. In the Sacramento Valley, the number of residents increased by 60 percent; the San Joaquin Valley grew even faster. Dozens of new towns were founded, while established towns—Sacramento, Fresno, Bakersfield—experienced a tremendous influx of residents. Some of these new residents were farmers and merchants, but many were migrant laborers—a cornerstone of California's intensive agriculture, which depended on a large and mobile labor force that could be available at harvest.[12]

If these changes represented the coming of "civilization" to the Central Valley, then civilization alone had not produced health. In a travel account of California published in 1913, the Englishman Arthur Johnson introduced the valley by denigrating the appearance of the region's residents:

I met dozens of families, fever-stricken men, women, and children, with pale, thin faces and trembling limbs, with all their worldly possessions on board a ramshackle buggy, or some such vehicle, drawn by a broken-hearted horse or mule. They were fleeing to the sea from the plague which wrought havoc with their constitutions. . . . In the disease-stricken, interior valleys of California, every summer adds more sacrifices to the multitude of wrecked constitutions and devitalised lives for which . . . malaria and other diseases are responsible.[13]

Johnson's observations indicate that the valley retained its "unsavory reputation for unhealthfulness" well into the twentieth century, and public health officials generally concurred. The most serious problem was typhoid. Between 1909 and 1914 the State Board of Health reported 2,647 deaths statewide from the disease, but these were not distributed evenly; typhoid was three times more prevalent in the valley than in the state's large cities. Even so, public health authorities believed that most urban cases had their origins in the countryside. Death rates in the Central Valley's main towns were nearly double those in San Francisco, while the most afflicted localities had death rates more than four times greater.[14]

In addition to typhoid, malaria remained a widespread and debilitating illness in the region. The state's leading expert on the disease asserted that California in the 1910s was the most malarious state in the country, with an estimated five thousand to six thousand cases per year, almost all of which were localized in the Sacramento and San Joaquin Valleys. When the Central Valley was viewed as a separate region, the statistics were sobering. Thirteen valley communities had a death rate of 14.2 per hundred thousand as late as 1916—substantially higher than that of Mississippi, in which whites had a malaria death rate of 5.9 and blacks 8.4. In fact, the death rate from malaria rose in the period from 1906 to 1912, but where malaria was concerned, everyone acknowledged that mortality was far less of a concern than the perpetual sickness and debility that the disease caused. The prevalence of the disease remained difficult to gauge quantitatively, however. Understandings of disease were not yet standardized among rural physicians, and misdiagnoses were common. Perhaps as few as 10 percent of all cases were reported. In 1914 a physician from the town of Wheatland claimed that the malaria parasite could be found in every resident "at one time or another," and the State Board of Health reported that the prevalence of malaria in the Sacramento Valley was so well recognized that doctors administered quinine to anyone undergoing a major operation.[15]

Perhaps the most well known malarious community in the state was an English colony established in the Penryn-Loomis area in 1888, which would become the site of the state's first antimosquito campaign. The original colony had failed completely by 1895, largely as a result of disease. The entire community was reportedly affected by malaria, forcing farmers to sell out to large landholders, who then rented to tenants. In 1919 the State Board of Health undertook a comprehensive investigation of malaria in the northern California town of Anderson and reported five hundred cases of malaria, which represented an estimated infection rate of more than 70 percent. Even so, investigators noted that reports of malaria among residents were so universal that they had only bothered to record as positive those individuals who had experienced "chills" within the past three months. In other words, the actual infection rate was probably 100 percent. [16]

Neither typhoid nor malaria were, from a biological perspective, new concerns in the region, though they were newly, or at least differently, understood by public health professionals. Much like malaria, what is now diagnosed as typhoid was, in the nineteenth century, but one of many "fevers" with an unknown etiology. However, bacteriology had made possible more specific diagnoses. Typhoid fever emerged as a distinct diagnosis with the identification of the typhoid bacillus in 1880. The malaria parasite was isolated that same year. Even so, the medical profession remained divided for more than a decade over whether either the typhoid bacterium or the malaria parasite was itself causative or whether these entities merely accompanied disease. More critical in many ways than the discovery of microbial agents were new understandings of disease transmission. In the 1890s researchers tentatively established the role of polluted water supplies in the transmission of typhoid. Then, in 1897, scientists established the mosquito's role in malaria transmission, effectively laying to rest the miasma theory among the well informed. [17]

Many professionals saw these scientific developments as nothing less than a revolution in public health. In 1910 Charles V. Chapin, head of Rhode Island's Public Health Department and a leading advocate for a new, laboratory-based public health practice, published his landmark work, *Sources and Modes of Infection*. Chapin's book vigorously repudiated old ideas of filth and miasma while stressing that the environment played at most a minimal role in the spread of disease. Instead Chapin insisted on the role of human-to-human transmission. A few years later Hibbert Winslow Hill authored a similarly influential book, *The New Public Health* (1916), in which he succinctly explained the recent change in thinking for his readers: "The old public health was concerned with the environment; the new is concerned with the individual. The old sought the sources of infectious disease in the surroundings of man; the new finds them in man himself." In these accounts, the environment no longer seeped into a body and shaped it in countless ways; instead the skin was now seen as an all-important barrier between the body and the larger environment that was breached only in certain, well-defined instances—through the bite of a mosquito or the consumption of contaminated water. In 1923 the State Board of Health published an article by Hill that went even further in its insistence on the separation of human beings from their environment. "We do not struggle with the universe at large to save us from disease, as did our ancestors," Hill declared. "We do not fear or dread anything from our skins out. Nothing outside can hurt us until it gets into us; and often, not even then. Only from our skins in can anything harm us; and this is why we have turned from regarding the environment and doctoring it, to regarding ourselves and keeping ourselves diseaseless."[18]

As espoused by Chapin, Hill, and others, germ theory fundamentally redescribed both bodies and disease in the early twentieth century. The healthy body was no longer a body in equilibrium with its environment, a "body in balance," but a pure body, one that was free from germs and parasites. As Michael Worboys has put it, increasingly disease would be "constituted in the relations between bacteria and individual bodies." What was now likely to be left out of the equation was the broader environment in which both bacteria and bodies resided. At the same time, human health was defined much more narrowly as the simple absence of disease.[19] The corresponding role of public health was to identify the pathogenic agents and the specific pathways of contamination, and then to remove or block them. Rather than a state of equilibrium between a body and its surroundings, health was now the result of rendering a body more perfectly closed. Human bodies were clearly separable from their environments. This was a distinctly modern view.

Implicit in these new understandings of disease was a rewriting of nature. Whereas nineteenth-century medicine had acknowledged the interpenetration of body and environment and the importance of local environmental conditions, for the followers of Chapin and Hill the world was now completely external, and increasingly irrelevant, to the human body. Illness was no longer caused, for instance, by the "bad water" of a particular locality but by water (presumed to be initially benign) that had been polluted by human wastes and the germs within those wastes.

"Contaminated" (as opposed to "bad") water was the result of improper hygiene rather than a characteristic of a given environment. Bacterial pollution could theoretically be traced back to its *human* sources and then remediated. The contrasts with miasma were manifold. Not only had miasmatic disease eluded precise definition, but concerted attempts to isolate its source had only yielded confusion. Miasma's inchoate nature made it impossible to delimit and track; it lurked within landscapes, often in latent forms, even as it varied across both space and time. And although miasma might be minimized through careful environmental management, landscapes were variable and always unpredictable; consequently, they required ongoing surveillance and careful monitoring. And the most afflicted places could never be fully cured.

But in the modern view, local difference should not matter, or at least not very much. The consolidation of germ theory and the rise of modern public health provided an arena in which distinct places and unique ecologies were produced, however hopefully, as abstract space. Much like the earlier carving up of North America into a grid of potential real estate parcels, new ideas about health tried to erase the particularity of a given landscape in favor of an assumed homogeneity. Germ theory idealized the environment as an essentially pure space, a space limned from the bacteriological laboratory in which materials had been carefully arranged to render all but the pathogens under study unimportant. What mattered to health were not the idiosyncrasies of a particularly regional ecology (the direction of winds, the character of the soils) and the reactions of situated bodies (male or female; white, Indian, or Asian) but merely the presence or absence of germs. Recast as space, the landscape no longer had an agency of its own so far as health and disease were concerned. Local places receded while bacteria assumed center stage. Bodies, in turn, were viewed as separate and distinct from the spaces they occupied. Disease was contained within the pathogen; "health" was now a condition of the individual body.

Whatever the apostles of germ theory might say about the insignificance of the local environment, however, in practice it was impossible to completely discount. As Henri Lefebvre has observed, "Abstract space is not homogenous; it simply has homogeneity as its goal." Even as germ theory rhetorically circumscribed the agency of nature, public health officials could not ignore the spatialization of illness. It seemed clear that certain kinds of environments harbored and incubated disease, but the "bacteriological revolution" had surprisingly little to say about this. For the most part, early bacteriological work in California was confined to verifying the presence or absence of specific diseases in specific individuals. It offered little information on the question of how disease was spread, other than to emphasize the critical role of human-to-human transmission.[20]

But contagionism was always a somewhat idealized view. Bacteria and parasites, however localized, still moved through environments, both internal and external, that conditioned illness. And the pathogens that most concerned California's health officials—those that caused tuberculosis, typhoid, dysentery, malaria—all spent some time outside the human body. Typhoid bacteria could survive for weeks in water and sewage, depending on the temperature and chemical characteristics. Even more to the point, the mosquito discovery had not dissociated disease from the landscape so much as it had redescribed the relationship. Contemporaries recognized, much as their predecessors had, that environmental conditions were crucial to the propagation of malaria, though they explained the etiology somewhat differently. That exceptionally wet weather and flooding in the valley brought on severe outbreaks was now due to the fact that receding floodwaters left innumerable small freshwater pools on the valley floor and also filled cellars and cesspools, all of which provided ideal breeding sites for anopheles mosquitoes. Similarly, while a prolonged drought or cold weather was said to kill the disease, entomologists now pointed to the lower number of mosquitoes. Moreover, the absence of malaria in southern California, which had long been a puzzle to California physicians, could now be explained by the paucity of freshwater in the region, which limited mosquito populations. The absence of malaria in the mountains was due to the parasite's inability to survive in temperatures below sixty degrees.[21] Although professionals invoked a different set of explanations, the role of the environment remained.

Malaria provided the most obvious example of the spatialization of illness and the continuing importance of the local environment, but even typhoid—a disease that in many ways epitomizes the successes of germ theory—was also dependent on environmental conditions. The most important route of transmission was water supplies that became polluted by infected human feces. At the time, doctors puzzled over how long typhoid bacteria and other germs could survive in water or other environmental mediums. Meanwhile, rising populations and the proliferation of towns throughout California had generated something of a regional sewage crisis. In the first decade of the twentieth century, the number of incorporated towns in California went from forty to two hundred, while many unincorporated communities and farm labor camps also sprang up in response to the industrialization of agriculture. Most large cities had constructed sewage and water supply systems in the late nineteenth century, but most rural communities had neither. Country residents relied on privies, cesspools, and local surface waters to dispose of waste.[22]

By 1910 repeated outbreaks of typhoid fever had drawn the attention of the California State Board of Health. Over a five-year period (1909–14), there were 322 deaths from typhoid in the Sacramento Valley and 311 deaths in the San Joaquin. Yet even within rural California, typhoid rates varied dramatically. The regions near the Sacramento and San Joaquin Rivers were the most afflicted, and local health officers acknowledged that the environment itself was at least "a predisposing cause." The long, dry summers, followed by heavy autumn and winter rains, inevitably unleashed large quantities of sewage contaminants, which were carried downstream in floodwaters. In the upper Sacramento Valley, the river flowed rapidly through a canyon, which meant that there was little chance for contaminants to settle out. In the lower portion of the valley, the floodwaters did not rush away; instead the river's natural levees channeled excess water into natural overflow basins—a series of troughs and large sloughs that could remain filled with water for days and sometimes weeks. During peak floods, as much as one-third of the valley might be underwater. Polluted floodwaters then seeped into the ground, contaminating local wells for tens of miles; yet the region's overall aridity meant that residents could not rely on rainwater collection as an alternative. Both ground and surface waters became sources of disease.[23]

The study of hay fever, an illness strongly correlated with both place and season, also resisted the move away from environmental models of disease. Springtime outbreaks became increasingly common in the Sacramento Valley in the first decades of the century, and local doctors attributed the disease to locust and orange pollen—products of the region's agricultural development. Hired by the Southern Pacific Railroad to study the problem, the botanist Harvey Monroe Hall identified the native black walnut tree, which had been planted extensively in valley towns as an ornamental, as the likely source. And much of Hall's evidence came from his study of the local landscape and its bodily effects. As he noted, "The abundant pollen sifted down over the city [of Colusa] just at the time when the disease was most prevalent, and the disease disappeared soon after the close of the flowering period." Hall also noted that patients' symptoms improved when they left the region temporarily.[24] Although Hibbert Winslow Hill and his followers might insist on the irrelevance of the broader environment, disease in California was still linked with both climate and local ecologies.

DISCOVERING THE UNSANITARY LANDSCAPE

Not only did certain diseases retain environmental dimensions, but public health officials also retained their environmental focus. Invoking the new bacteriological theories, they declared their intention to rid environments of disease-causing germs. Whereas most nineteenth-century writers believed that the reduction of sickness in California's Central Valley would depend on both careful cultivation and the attentive management of bodies, the new public health experts argued that health would come from sanitation. It was no longer a question of adaptation and acclimatization. To avoid disease, both environments and bodies needed to be kept clean.

Sanitation was not a new concept in the twentieth century, but it achieved new importance and much more widespread application in the century's early decades. The idea of sanitation as a solution to disease had its roots in the middle of the preceding century, when medical theories themselves were far more environmentalist. The key document was Edwin Chadwick's famous *Report on the Sanitary Condition of the Labouring Population of Great Britain* (1849). In that report Chadwick had forcefully linked disease to the "filth" present in the urban environment and had recommended environmental solutions: waterworks, sewers, paved streets, and ventilated buildings. These ideas would underwrite the uban sanitation movement of the Victorian era. The advent of germ theory several decades later did not, as many have noted, radically change public health strategies.[25] Rather, many of those strategies would be refined and extended.

In California, the consolidation of sanitation as a modern public health strategy owed much to the outbreak of bubonic plague in San Francisco in 1900 and again 1907. In response to the first outbreak, public health officials drew on nineteenth-century ideas of miasma and located the source of disease in the generalized "filth" of an already pathologized Chinatown. Officials inspected, chemically disinfected, and physically destroyed much of the neighborhood, despite the protests of residents. By the time of the second outbreak, bacteriological work had uncovered the decisive role of fleas and therefore rats. Yet even though their measures had a more specific target, public health officials oversaw a sanitation campaign that was strikingly similar to, if more extreme than, that adopted a few years earlier. Dr. Rupert Blue of the U.S. Marine Health Service (the precursor of the U.S. Public Health Service) instructed his inspectors to replace or destroy wooden structures wherever feasible. He insisted that the city's sidewalks, public markets, and wharves be paved. With the cooperation of city officials, Blue radically altered the local environments of San Francisco; in fact, he hoped to make the city "one block of concrete throughout." At Blue's urging, city officials also adopted new public health regulations. They required that animal manure be contained in metal boxes, that garbage be placed only in regulation cans with tight-fitting lids, and that houses, chicken coops, and stables be constructed in accordance with new building codes. When the epidemic was finally brought under control, officials credited Blue's vigorous sanitation program.[26]

In the wake of the second antiplague campaign and the election of the Progressive governor Hiram Johnson in 1910, the California State Board of Health emerged as one of the most professional and powerful in the nation. One of the new board's first priorities was the establishment of a permanent division of sanitary engineering. In 1913 the board hired its first full-time sanitary inspector. Two years later the board established the Sanitary Engineering Bureau, with a full-time director, a biennial budget of $30,000, and principal responsibility for inspecting and regulating sewage systems and water supplies. That budget would increase rapidly in ensuing years. In 1921 the chief sanitary inspector was granted his own Division of Sanitation along with funds for two more full-time inspectors.[27] Although public health officials invoked germ theory as the scientific rationale for their new powers, the growing emphasis on sanitation underscored a much older story: the continuing dependence of bodies on their environment.

What was new in the twentieth century was the explicit concern with the dirtiness of *rural* environments. In fact, public health experts now insisted that the most difficult sanitary challenges were posed by the less populous but seemingly more "backward" countryside. As one leading reformer noted, the California "ranch" seemed to supply the essential features of a healthful environment; nonetheless, it was continually visited by malaria, typhoid fever, diarrhea, and other diseases. Nationally, much of the impetus for the new focus on rural areas came from the Rockefeller Foundation. In 1909 the foundation funded an effort to control hookworm in the rural South, using as a model American efforts in

Puerto Rico. A series of malaria demonstration projects followed in 1916. Largely as a result of these efforts, the U.S. Public Health Service established a division of rural sanitation in 1914. At the same moment, sanitation became an important facet of the agricultural extension work that was expanded through federal aid in 1916 and 1917.[28]

The particular concern with rural sanitation was somewhat ironic given that so much had been written for so many years about the benefits of the countryside to health. Since the mid-nineteenth century, American cities had been vilified as spaces of ill health, the countryside lauded as their opposite. Characterized by overcrowding, contaminated water supplies, and smoke pollution, nineteenth-century cities had much higher rates of disease and mortality than the surrounding area. These conditions generated a substantive urban environmental reform movement in the 1880s and 1890s, even as they encouraged wealthy Americans to seek out homes outside the cities. But by 1910 assumptions about the relative health of the city versus the country had been nearly reversed. Some attributed the change to material improvements in urban life: purer water supplies, piped sewage systems, and regular garbage collection. As conditions in the cities improved, Progressive reformers voiced concern that life in rural areas was not advancing as rapidly.[29]

The sudden concern with rural sanitation in California and elsewhere cannot be explained solely by the prevalence of rural disease or the improved conditions of cities, however. While death rates were decreasing more rapidly in urban areas, the countryside still had a significantly lower overall mortality.[30] So why such concern? Why did so many urban elites, including those in California, prioritize rural health and sanitation in this period? Officials pointed to the geographic distribution of disease. While death rates from disease were higher in the city, certain diseases that were rare in urban settings remained widespread in rural areas. Although tuberculosis was more common in California's major cities, malaria and typhoid were more prevalent in the valley. But health officials no longer interpreted the geographic clustering of disease in environmental or topographic terms. Public health discourses recast these and other diseases as "man-made," despite the relevant environmental factors. Both typhoid and malaria came to be labeled rural diseases and were explicitly linked to the poor state of sanitation in the countryside. The State Board of Health made a point of noting that it regarded outbreaks of typhoid fever "as a reproach to the sanitation and civilization of the community in which it was contracted."[31]

Reformers also emphasized that urban health was linked to rural

health, through automobile travel and especially through the food supply. As both people and things became increasingly mobile, diseases bred
in the countryside would ultimately be found in the city, where they
might spread even more quickly. For instance, the increasing number of
typhoid outbreaks in the 1910s that were linked to contaminated milk
prompted the regulation of dairies. Experts also expressed alarm at the
potential of the automobile to spread disease more rapidly than before.
The State Board of Health warned that "fagged city dwellers turn to the
country for rest and recuperation, but alas, only too many return with
developing cases of typhoid or malaria." Throughout the 1920s the
board would have to assure restive tourists that it was indeed safe to
travel to the Central Valley.[32] The sanitary "backwardness" of rural
America threatened the modernized spaces of urban America.

In addition, many writers emphasized that the presence of malaria
was hindering rural California's economic development—by reducing
the efficiency of workers and keeping away both potential settlers and
agricultural laborers, who had become integral to the state's farm economy. In 1910 the State Board of Health estimated that the state suffered
an annual loss of more than $2.8 million from malaria in the Central
Valley, attributing most of this loss to reduced labor productivity (estimated to affect six thousand men) and depreciated real estate values. In
some areas, property values had reportedly declined by as much as 50
percent, and in many communities local elites denied or downplayed the
presence of malaria in an effort to keep land prices up.[33]

The tension inherent in California's agricultural labor system also
motivated the state to take an interest in rural sanitation. In the early
twentieth century, the continued consolidation of landholdings and the
turn toward intensive horticultural crops contributed to a growing
demand for temporary wage labor. It was an open secret that most farm
laborers in the state faced extremely bad living and working conditions.
A pivotal moment came in 1913 when laborers at the Durst brothers'
hop ranch in Wheatland went on strike to protest their low pay and hor-
rible working conditions. Local law enforcement came to the aid of the
Dursts, and violence broke out—leaving four people dead and scores
injured. A state investigation followed, and California's Progressive leaders laid primary blame for workers' radicalism on the abysmal work
environment. According to investigators, both dysentery and typhoid
were rampant at the Durst Ranch, and subsequent inquiries revealed
similar conditions at most of the state's farm labor camps: inadequate
and unsanitary toilet facilities, overcrowded and unscreened housing,

and widespread illness. Much of the Progressive solution to labor unrest in California would hinge on the science of sanitation. Only six weeks before the Wheatland strike, the legislature had passed the Labor Camp Sanitation Act, granting the State Board of Health authority—but not funds—to regulate the conditions of farm labor camps. In the aftermath of the riot, the California Commission on Immigration and Housing (the state agency charged with overseeing migrant labor) made camp sanitation its priority, immediately hiring their own sanitary engineer, issuing a series of pamphlets on recommended sanitary practices, and initiating a campaign to encourage growers to adopt sanitary reforms.[34]

Though concerns about illness, economic loss, and labor unrest were real enough, the push for rural sanitation was also motivated by long-standing concerns about race. Although the effect of the California climate on white residents was no longer widely feared, it was recognized that disease, especially certain rural diseases, could still affect the body in profound and potentially disturbing ways. Many physicians openly equated certain rural diseases—especially hookworm and malaria—with the racial deterioration of whites. They increasingly blamed these diseases for creating an inferior breed of whites that were plagued by poverty, low intelligence, and lack of vigor. The concern with the degeneration of rural whites meshed with broader concerns about racial competition and white "race suicide" articulated by Progressive elites such as Theodore Roosevelt.[35]

Racial concerns took on particular valence in California. While white Americans had firmly established themselves as the dominant social and economic class in the region and they no longer seriously doubted their suitability for the California climate, many worried ceaselessly about their future numerical dominance. The golden state lay at the edge of the American empire; it comprised, in the words of more than one prominent western reformer, "the frontier of the white man's world." And by their very nature, frontiers threatened whiteness. The reliance of California agriculture on nonwhite labor heightened these concerns. Rather than a region of white farm families, the Central Valley relied ever more heavily on a racialized and migrant labor force. In particular, it was the immigration of Asians and, more recently, Mexicans that disturbed white Californians, including Progressive reformers. Much of the impetus to improve rural life in the region rested on the desire—in the words of Elwood Mead, a prominent agricultural reformer—"to keep rural civilization in California white." The vulnerability that whites now faced was not due to the local climate, but it was no less physical.[36]

Malaria was believed to prey particularly on whites. In an influential volume published in 1909, the historian W. H. S. Jones argued that the decline of ancient Greek civilization could be best explained by the presence of malaria and the attendant physical and mental degeneration that left the Greek population vulnerable to invasion by inferior peoples. In a speech before the Smithsonian Institution, Ronald Ross, the British scientist credited with the discovery that mosquitoes spread malaria, drew out the lessons for modern Britain and America, arguing that if left unchecked malaria could have a similarly devastating effect on the contemporary rural population. Warning against any Darwinian complacency on the part of whites, Ross told his audience that the groups most at risk from malaria were the stronger races, especially those with "fair, northern blood." The U.S. Public Health Service concurred, declaring that malaria was responsible for "inertia, loss of will power, intemperance and general mental and moral degradation" in its victims, white as well as black. California's public health establishment shared these racial concerns. In a report on health conditions in the Central Valley, one official warned his California colleagues that "no place where malaria seriously prevails is habitable by white people. They may live in such a region but they do not thrive." As part of its malaria control work, the State Board of Health undertook a psychological study of schoolchildren in one northern California district and found that malaria had likely caused "mental impairment" among the young population, with an average loss of 0.75 of a year in mental age. California elites, including the public health establishment, routinely lamented the "pallor," "sluggishness," and "weakness" of the state's rural inhabitants.[37]

Complaints about malaria in California were often explicit discussions of race and immigration. "In my own neighborhood I could name ten families who have purchased fruit farms, fought malaria for a year or two, rented to Japanese and moved away," complained one valley resident to the State Board of Health. Or, as a rancher in Penryn put it, "We could not keep any white men at work. They had malaria and would leave as soon as able to walk." Rice cultivation generated even more concern on this point. Rice farming in the Sacramento Valley was pursued almost exclusively by nonresident owners, who relied heavily on Japanese tenants to farm the land under short-term leases and cared little about providing the necessary drainage or adequate housing and less about mosquitoes. Unscreened and overcrowded shacks housed workers in the midst of flooded, insect-filled fields. An engineer's report on rice growing in northern California noted that "dead shade trees, poor roads, mosquitoes, unsanitary conditions, and Oriental tenants as neighbors make the rice farm intolerable to most Americans." William Herms, who became California's leading expert on malaria in the early twentieth century, explained the racial implications of the disease to a gathering of northern California's elites: "There are many localities in California where malaria is virtually the only blight, and these include some of our most fruitful and beautiful portions of the state, made almost unhabitable for the white man, and now having been largely turned over to the Oriental, who apparently can get along, tolerably well at least, though infected with malaria."[38]

Like his predecessor Thomas Logan, William Herms was deeply concerned about malaria in the Central Valley and its effect on the region's white farmers. Even with the advent of germ theory—a seemingly race-neutral discourse—race, environment, and disease remained intertwined in discussions of public health. But there was a crucial difference between early-twentieth-century discourse and that of the colonial period. It was no longer a question of acclimatization on the part of whites, for Herms was far more confident than Logan that the landscape itself could be reformed. The fact that malaria preyed on whites was no longer evidence that they might not be suited to the region but of the need to make the elimination of malaria a priority. Where environmental medicine had implied the need for human adaptation, germ theory implied the possibility of disease eradication. It is perhaps not coincidental that California's first state-funded malaria control effort occurred in Anderson, a town that public health investigators took note of for its "strikingly pure population of native born Americans."[39] The attempt to eradicate malaria in the Central Valley would be merely a new chapter in the long-standing effort to ensure that California would become and remain the home of those who could call themselves "white." Remaking rural landscapes would simultaneously rehabilitate rural white bodies.

Although the expansion of sanitation work in the early twentieth century built on similar practices from the preceding era, understandings diverged. Thomas Logan and his contemporaries assumed that the causes of disease had to be sought in nonhuman, as much as or more than human, factors. In Logan's opinion, the principal reason for investigating endemic and epidemic disease was "to show *how far man's agency has to do in the matter.*" But the new reformers took a different view. As one of California's leading health reformers put it, "We are very largely creatures of our environment and are subject to it in large measure, but by virtue of our intelligence we have the power to change conditions from the unlivable to the livable."[40] The new sanitation experts emphasized, indeed they often exalted, the power of humans to control their environments. In their minds, disease and health no longer represented a complex mixture of human and nonhuman agencies. Even while they acknowledged the role of the environment in some diseases, the new experts asserted that this was only a temporary condition; once landscapes had been made fully modern and aseptic, they would be essentially irrelevant. While bodies were, in some sense, still outcomes of their environment, those environments were even more fundamentally outcomes of those who inhabited them. Thus even more important than the remaking of unruly landscapes was the disciplining of ill-behaved people.

The disciplinary implications of modern hygiene movements have been most thoroughly explored in the literature on colonial public health. Scholars examining sanitary reform in places as disparate as India, the Philippines, and southern Africa have emphasized the way in which concerns about "health" legitimated colonial norms and were used to justify the increased policing of native populations. Like landscapes, bodies themselves could be placed along a continuum of modernity. Modern—and therefore healthy—bodies were those that appeared most completely closed to and least permeated by the surrounding environment. Bodily discharges (of feces, mucus, blood) had to be minimized and tightly controlled. Dirt needed to be kept at bay and quickly washed off. Insect bites had to be avoided. To the extent that disease was present among native populations, it was increasingly read as an indictment of the people themselves as much or more than of the local landscape. In colonized environments such as Panama or the Philippines, American sanitation experts labeled native practices dangerous to both themselves and others. What flowed from the new model was, in the words of the historian Gyan Prakash, "a discriminatory sanitary order."[41] The threat of disease could thus underwrite calls for outside intervention and "expert" (i.e., colonial) control.

Scholars have largely overlooked the equivalent pathologizing of the domestic rural body in the same period, however—even though the literature on rural sanitation in the early twentieth century was in constant conversation with a contemporary literature on colonial sanitation in the tropics.[42] Tropical hygiene and sanitation located the persistence of disease in the unsanitary behavior of natives, calling attention to yet another set of differences between colonizers and natives, the civilized and the uncivilized. But many of the conditions that supposedly established these differences—especially the presence and prevalence of certain diseases—

could also be found in the American countryside. And public health officials blamed those diseases on the habits of rural people. As the historian Deborah Fitzgerald has observed, early-twentieth-century reformers saw farmers as one more retrograde group—"uneducated, rural, backward, and in need of civilizing." That was nowhere more true than in issues of hygiene and sanitation. By the standards of the urban middle class (from which the public health profession drew most of its practitioners), farmers were excessively dirty. They were apt to coexist with insects. They were more likely to harbor certain germs and parasites within themselves. And most disturbingly, their wastes were always in evidence, an unfortunate and objectionable sign of the way in which certain bodies still leaked into and contaminated the surrounding environment, thereby generating more ill health. Much like native bodies in the tropics and the poor and nonwhite bodies that inhabited American cities, bodies in the rural United States were not properly separated from their environment.[43]

While all rural people were pathologized, there was still a racial hierarchy. Within the multiethnic population of California, the most frequently targeted individuals were those now deemed "foreign," like disease itself. California's new sanitary inspectors made a point of singling out Japanese farms and communities, which they routinely found "in flagrant violation of vitally important sanitary requirements." Inspectors complained of the noncooperation of Chinese residents and the ignorance of Mexicans when it came to the essentials of hygiene, and various writers reiterated their concerns. "Certain nationalities are not inclined to put much thought or effort in keeping their premises clean or in sanitary condition, or in disposing of their waste other than simply dumping it out upon the ground," wrote one inspector. "Such a class needs constant supervision."[44] William Cort, a member of the State Board of Health, assumed, despite an admitted lack of evidence, that nonwhite agricultural workers were spreading numerous diseases in rural California:

> The increase of irrigation in the agriculture of the state is bringing together large numbers of Oriental and Mexican laborers, under conditions ideal for the spread of animal parasites. In certain regions the laborers live crowded together under unsanitary conditions and there is an enormous amount of soil pollution. Human faeces is even used for fertilizer in some places. In these regions the conditions of temperature and moisture, and the habits of life of the people, are favorable for the spread of such parasites as hookworm, Strongyloides, stercoralis, the intestinal protozoa, the common intestinal roundworms Ascaris, Trichuris and Oxysuris and the fluke disease, if they can find suitable intermediate hosts.[45]

As Cort's comment suggests, the discovery of the unsanitary rural landscape was accompanied by a recasting of the inferiority of California's nonwhite residents. Disease was no longer assumed to be a natural if unfortunate condition of the landscape. Someone was to blame. And if humans rather than environments were ultimately responsible for disease, it was also true that certain human groups were consistently singled out. Although invoking bacteriology, public health practice often cast racial difference in terms that were more cultural than biological. The *habits* rather than the biology of nonwhite groups would be even more frequently cited to explain higher rates of sickness and death. Poorly disciplined people were indicted as a threat to themselves, and they were increasingly held responsible for unsanitary environments that threatened others.[46]

DIVISIONS OF LABOR: SANITARY ENGINEERING

The continuing interest in community habits and rural environments would seem to undercut any idea that twentieth-century medicine narrowed its focus to germs, laboratories, and the self-contained body. How was it that sanitation work could expand even as public health doctors were proclaiming the irrelevance of the environment? Part of the answer lies in professional divisions of labor. The coding of nature as external to human bodies did not halt the study of the environment's contribution to disease; it merely relocated it. Doctors, inspired by bacteriology, were turning their focus toward the isolated body and those bodily substances that could be analyzed in the laboratory (blood, mucus, feces, urine). Meanwhile, other professionals—notably sanitary engineers and medical entomologists—would step into the breach. Neither engineers nor entomologists concerned themselves with identifying and defining disease or treating sufferers—all that would remain the domain of doctors. But these professions would play a leading role in disease prevention. And quite unlike the new medical bacteriologists who confined themselves to the laboratory, engineers still took as their starting point the relationship between human beings and local environments. In 1918, just as Hibbert Winslow Hill was writing that health was the outcome of "attending to the individual, rather than the surroundings; improving the person rather than the premises," a leading engineer declared that "it is difficult to conceive of man apart from his environment."[47] As professional divisions of labor hardened, the environment became increasingly the domain of sanitary engineering. In fact, public health engineers would eventually name their work "environmental sanitation," and in the postwar decades new environmental problems such as air pollution would fall into their domain.

Although there were professional resentments among doctors and engineers in the health profession, there were few if any outright conflicts. For the most part, they had different things to do. If they did not share a coherent ideology, it was because they did not need to. By and large their practices did not overlap. Physicians confined themselves to the clinic and the laboratory; engineers went to the field. As Mol has observed in her study of contemporary medicine, different kinds of work may go on, and different ideologies may persist, so long as the parties are located in different places. That engineers continued to insist on the importance of the environment while bacteriologists asserted the environment's irrelevance did not necessarily present a problem to either group. "Practice isn't preceded by a principled discussion," Mol reminds us. "What seem to be logical incompatibilities are not disturbing. They don't make life more difficult: they make it easier."[48] California's nineteenth-century physicians had blended modern with ecological concepts of disease in their work; they could be both clinicians and environmental scientists. That was no longer possible or even desirable in the early decades of the twentieth century once germ theory held sway. Yet even as theories of disease hardened, practices remained diverse.

Whereas the new bacteriologists relied on the microscope, the principal tool of the sanitary engineer was the inspection. Implicit in the inspection process is a normative standard—the belief that there are correct and incorrect ways to do things. Inspections are exercises of power, the power to erase differences in the interest of another goal. Backed by the authority of the new laboratory medicine and the growing regulatory powers of the state, sanitary inspections demonstrated the power of the new public health professionals to dictate appropriate environmental conditions and to reorganize both urban and rural places. The job of the sanitary engineer was to identify and then erase those environments likely to produce and sustain disease. Put another way, sanitary inspections were the tools by which public health officials struggled—without ever fully succeeding—to create homogeneous, abstract, and uniformly modern spaces out of particular social and ecological landscapes.

In 1913, amid rising concerns over typhoid fever, the State Board of Health hired Edwin T. Ross, its first staff engineer. Upon assuming his new job, Ross immediately began a series of inspections in California's rural environments. The available evidence suggests that Ross was well intentioned and energetic. His work took him to many different places: summer resorts, canneries, ranches, schools, farm labor camps, hospitals, packinghouses, slaughterhouses, jails, lodging houses, dairies. Ross focused much of his energy on sewage and the rural water supply, which were frequently suspected of transmitting typhoid. But he was not tracking germs or administering laboratory tests. Despite their allegiance to the new bacteriological theories, for Ross and most of his engineering contemporaries bacteriological tests had limited practical usefulness in assessing local conditions. Investigators were not sure what they were looking for or even what the presence of given microbes actually meant. The most pressing question, whether a certain water supply was responsible for a given typhoid outbreak, was something that the new laboratory analysts could not even answer with certainty.[49] Instead Ross relied on his own observations and experience.

Thomas Logan had traveled across California to collect local information so that he might better understand the relationship between environment and disease; Edwin Ross went to the countryside to cite its defects. He found much to criticize. He noted that animals were typically kept near water that was also used for drinking. Slimy vegetation lined the banks of many small reservoirs, and wells were often located near privies. But Ross's concerns extended beyond the purity of water supplies. He objected to the prevalence of flies and other insects. He criticized farmers for allowing dead animals to remain lying in fields or ditches. And he showed an exacting interest in toilet facilities and garbage disposal. He wrote indignantly and often of the poor condition of privies but was even more disgusted by the fact that, while women at least used the awful privies, most farm men relieved themselves wherever convenient. He excoriated against the rural tendency to leave garbage strewn about dwellings and farmyards and was appalled by the sight and odor of animal manure. No doubt Ross agreed wholeheartedly with a colleague who wrote that "throwing this material outside the barn door and allowing it to accumulate there indefinitely is not only disgusting, but should be considered a criminal act."[50]

If healthfulness could no longer be read from an isothermal map, it could be ascertained from a sanitary engineer's inspection report. By 1913 long-standing rural practices had become the object of official scorn. The adjective that continually crops up in Ross's reports is *filthy*.[51] Sanitation work translated the diversity of California's ecologies into a progressive narrative of space. Environments were not ecologies that might be incommensurable with one another; they were more or less aberrant, more or less sanitary. And as Ross's reports imply, it was the engineered ecologies of the middle-class city that were unquestionably superior: paved roads and yards, modern plumbing, standardized buildings with designated uses, concrete canals, the absence (or complete segregation) of domestic animals, the invisibility of human and animal excrement. The objectives of public health work merged seamlessly with a preference for clean and well-engineered landscapes.

In addition to conducting routine inspections, Ross was called upon to investigate specific disease outbreaks. When a typhoid fever epidemic struck the small valley town of Hanford in 1914, the State Board of Health sent Ross and the bacteriologist Dr. W. A. Sawyer to investigate. It was Sawyer's job to confirm that typhoid was present and to identify the sick individuals by collecting and testing samples of feces. (Biological samples were now the only evidence of disease that local persons were officially asked to give. The testimony of residents and afflicted persons was no longer recorded in the board's documents.) It was left to Ross to diagnose and remedy the environmental conditions that had allowed typhoid to spread, though at the time public health officials were unsure how it had originated. They eventually traced the outbreak to a church dinner and subsequently confined a local landlady and cook as a suspected "carrier."[52]

Despite their assertion that they had localized the disease in the body of a woman, authorities embarked on a massive cleanup campaign. In response to ninety-three cases of illness, Ross hired dozens of inspectors to enforce proper sanitation measures. Upon completion, he reported that 232 garbage cans were installed; 426 toilets were screened; 535 privies were made fly-proof; 76 privies were demolished; 43 buildings, 58 basements, and 670 yards were cleaned; 301 loads of rubbish were removed; 3,965 square feet of concrete floor were installed; and 2 miles of irrigation ditch were cleaned. It is hard to know exactly what Ross hoped to accomplish by cleaning irrigation ditches, among other things, and how that might or might not have been related to the spread of typhoid that emanated from a spaghetti dinner. The reordering of local environmental arrangements was motivated by commitments other than those explicitly underwritten by germ theory. If a landscape harbored disease, that was reason enough to reorder it in a way that was pleasing to engineers. Though Ross's work at Hanford might seem, in some ways, insignificant, it was but one of tens of thousands inspections that he conducted over more than thirty years with the State Board of Health. In one two-year period, Ross reported a total 4,418 inspections and reinspections.[53] And there were many more sanitary inspectors like Ross who were simultaneously remaking landscapes and ideas of health—privy by privy, ditch by ditch.

Less than a year after hiring Ross the State Board of Health asserted that "continued and repeated inspections" would be necessary to ensure the proper sanitary condition of the California countryside.[54] Although the dominant discourse of modern public health asserted that the natural environment—soil, air, water—was merely pristine space that had little effect on heatlh, bodies and spaces remained linked in the practice of Ross and hundreds of other sanitary engineers. The presence of widespread illness had come to signify disorder rather than an inherent incongruence between the bodies of people and the landscape they occupied. Human bodies produced certain kinds of spaces: clean or dirty, organized or unruly, clean or contaminated. And these spaces in turn produced healthy or unhealthy bodies. As a result, the environments that drew Ross's attention were, to his mind, not natural in any meaningful sense; rather, they were (the poor) environments that humans themselves had created. The job of the sanitary engineer was to produce rationalized spaces out of chaotic human ecologies. In the hands of Ross and his colleagues, public health became a tool of environmental management and a rationale for reorganizing both landscapes and communities.

REINTERPRETING MALARIA: SANITARY ENTOMOLOGY

The discovery that the mosquito was the principal vector of malaria in 1897 seemed to bring one of the nineteenth century's most vexing diseases firmly into the domain of the bacteriological revolution. Yet malaria, like other vector-borne diseases, always fit somewhat uneasily into the germ paradigm. Although the malaria plasmodium could be isolated under the microscope, it nonetheless spent much of its time outside the human body—in the bodies of other animals and insects. Moreover, local environmental conditions were obviously critical to the propagation of malaria. From one perspective, the mosquito discovery did not overturn the ideas of nineteenth-century miasmatists so much as confirm them, by calling attention to the ecological features that supported anopheles: warm temperatures, moist climates, damp soils, stagnant water, and streamside vegetation. As W. F. Bynum has observed in the Indian case, "Mosquitoes simply reinforced the sense of place for malaria."[55]

Initially there was some debate over how best to combat the disease—whether through medical therapy (the administering of quinine) or through environmental control, with doctors favoring the former and engineers the latter. However, malaria was relatively quickly cast as a problem of sanitary engineering. In fact, public health officials had enthusiastically hailed the mosquito discovery as one that would "open up a new and extensive field for sanitary effort." Malaria control would become the domain of engineers and the new environmentally oriented professionals, medical entomologists.[56]

The triumph of an environmental approach to malaria in the United States had everything to do with the American campaigns against yellow fever and malaria in the tropics—Havana in 1898 and the Panama Canal Zone in 1901. These efforts had focused on eliminating mosquitoes through environmental management: draining swamps, clearing vegetation from streams and ditches, removing standing water, and poisoning larvae wherever they were found. America's tropical colonies had served as a kind of field experiment for new methods in public health, and the perceived successes of the campaigns in Cuba and Panama were instrumental in elevating the strategy of insect elimination to the position of gospel within American public health work. Moreover, the success of the Panama campaign raised expectations of the North American landscape. As one California newspaper editor wrote, "The Panama Canal Zone was cleared of mosquitoes and yellow fever. . . . Why should [such measures] not be employed here?"[57] If the disease-ridden tropics could be sanitized, then certainly so could the American South or the Central Valley.

As environmental techniques took hold, physicians' interest in malaria all but disappeared after the turn of the century—a striking contrast to their almost obsessive interest in miasma and malarial disease in the late nineteenth century. In California, few articles on malaria appeared in the state's principal medical journal, and those that did appear were likely to be written by public health officials rather than by practicing or academic physicians.[58]

The state's approach to malaria in California would be principally the work of one man, William Brodbeck Herms, a young parasitologist at the University of California and a leader in the field of medical entomology. Growing up in a malarial section of Ohio, Herms had planned to be doctor; it was the discovery of the mosquito vector that led him instead to entomology. Herms would emerge as the leading expert on mosquito control, and would write (with his California associate, the engineer Harold Farnsworth Gray) the principal textbook in the field. Within a few years of the Panama campaign, Herms had begun mosquito-control work in California, following similar prescriptions if less coercive measures. He initiated his first antimosquito campaign in 1910, in the fruit-growing community of Penryn. It was financed primarily by the local packing company in response to concerns about the labor supply. Herms traced Penryn's problem to the haphazard irrigation practices that were typical throughout the valley. In most cases irrigation channels made use of natural sloughs and old streambeds. Connecting ditches were constructed across the shortest possible distance to minimize the labor required. Farmers used the natural contours of the landscape, but their actions also altered the local hydrology—creating standing pools, poorly drained fields, and sluggish canals.[59]

Entomologists were attentive to the details of regional ecology. Nonetheless, they ultimately argued that "irregular" landscapes should be systematically reworked, and from the beginning engineers were their principal allies. Both professions endorsed general principles of environmental rationalization. Their reports on malaria control consistently contrast the unplanned landscapes they found with the orderly landscapes they created. Photographs of overgrown irrigation ditches and meandering sloughs appear alongside those of neatly engineered canals. According to Joseph LePrince, the engineer in charge of the Panama effort and Herms's mentor, streams needed to be "properly trained, that is, reduced to a minimum width of uniform cross section and freed from stones, grasses, and debris that would interfere with flow or velocity or furnish hiding-places for larvae." Despite differing ecologies, the same techniques could be applied in California. Good engineering practice required one to overwrite idiosyncrasies in favor of certain general principles. In the Central Valley, sanitation engineers advocated abandoning the naturally occurring channels and sloughs and replacing them with shorter and straighter canals that could be more easily maintained. For poorly drained lands, Herms advocated the construction of ditches (preferably lined with concrete) to draw off excess water. The valley's swamps and marshlands were to be drained or diked wherever practical. Like LePrince, Herms insisted that stream banks should be cleared of vegetation and their sides steeply cut: "A clean pond, with sharp, deeply cut banks, need not be a menace as a mosquito breeder."[60]

The ongoing development of the valley landscape vastly complicated engineers' work. Herms explained that even though anopheles mosquitoes were native to the region, recent changes to the landscape had increased the insect's numbers and range. New settlers were pouring in and altering the land to suit their purposes—leveling fields, constructing ditches, filling banks. Especially critical to the increase in malaria was the

Figure 6. Irrigated fields in the San Joaquin Valley, circa 1920. Courtesy California Historical Society, FN-30672.

spread of irrigated agriculture: between 1900 and 1910 the length of main irrigation canals in the state went from 5,100 to 12,600 miles, and this did not account for all the minor canals and laterals that now laced the landscape. Moreover, speculators often bought large swaths of land, made only the most minimal improvements, and then sold lots to prospective and inexperienced fruit farmers. And, in any case, the methods and tools were crude. Leveling a field involved hitching a metal scraper to a couple of mules and dragging it across the ground, and irrigation works were often nothing more than hand-dug trenches. Both engineers and agricultural experts noted the ensuing problems: "high spots that do not get water, low spots where the crop drowns out, ditches that break, drains that are too shallow or that do not work." In retrospect, it is not surprising that an agricultural colony begun at Los Molinos in 1912 suffered from leaking canals and rampant malaria. Settlers enticed into the project abandoned their lands and refused to pay their debts, leaving the developer nearly bankrupt. In 1918 the opening of the Anderson-Cottonwood irrigation district caused a similar epidemic in the Redding area. Leakage from new canals and laterals provided hundreds of acres of shallow seepage ideally suited to the production of *Anopheles freeborni*, and infection rates reportedly reached more than 50 percent.[61]

The rapid construction of railroad lines and highways in these decades also created microenvironments throughout the valley that favored the species. Laid out so as to minimize construction costs, transportation routes interrupted natural drainage patterns, generating innumerable seeps and ponds. Certain sections of the Southern Pacific Railroad were known to be intensely malarial, which had prompted the railroad to plant eucalyptus trees along section lines in earlier decades. As Herms and Gray would ruefully observe, "After long and sad experience, [we] are convinced that it is only by accident that either a railroad or a highway engineer ever places a culvert properly. Usually it is set too high so that a pool or swampy area forms on the upper side of the embankment; occasionally it is set too low so that water remains in the culvert itself." In northern California, mechanical gold dredging operations had also exacerbated the malaria problem. The development of sophisticated machinery at the end of the nineteenth century made it possible and profitable for mining companies to churn up immense amounts of earth to uncover relatively small amounts of gold. When they were finished the topsoil was buried beneath immense piles of rocky cobbles. For humans, these areas were rendered essentially useless, but mosquitoes found them ideal. The potholes and small lakes that developed amid the dredge mounds were all but impossible to drain and provided innumerable breeding sites for anophelines.[62]

Consequently, mosquito control work was an unending task, and Herms had neither the funds nor the labor that he wanted for the job. Legislative appropriations never matched his requests, and struggling farmers were hostile to taxing themselves for what most still saw as an unnecessary service. Thus, though Herms consistently advocated engineering methods and "permanent" control, he also endorsed the use of oil as a supplemental if inferior tool. Many valley communities eagerly seized upon the less labor-intensive alternative of oiling—and later pesticides—to kill mosquitoes, forgoing the more arduous tasks of refining their irrigation works and providing adequate drainage.[63]

Like Ross, Herms made little practical use of bacteriology in his work. The presence of malaria in a community was easily discerned, as it had been in earlier periods—by the pale, yellowed complexions and lethargy of residents—although now it could be "confirmed" by taking blood smears from suspected sufferers. Once a malaria problem was identified,

Herms relied on inspections to detect the sources of malaria in a community, and he defined "sources" in environmental rather than bacteriological terms: poorly constructed irrigation canals, hoofprints left by livestock, overgrown ponds, abandoned wells and borrow pits, natural sloughs and braided creeks, and poorly drained soils. For Herms the causes of malaria still lay within the landscape. He insisted that even the slightest infraction might produce disastrous results; the source of disease could be something as seemingly minor as a leaking hydrant near a fruit-packing house or an uncovered rain barrel. "Almost invariably it is the apparently insignificant place that breeds the Anopheles," he declared. Such detailed attention to local landscapes merely underscored his environmental orientation. As Thomas Logan had insisted on the need to collect as much environmental data as possible, Herms insisted on the overriding "importance of careful and close inspection." [64]

Unlike Logan, however, Herms no longer questioned the ability of the Central Valley to support the kind of society that he desired. What was different now was that the natural landscape itself was, in some essential way, exonerated—or at least provisionally written out of the disease equation. As Herms insisted, "Malaria is a disease not indigenous to our soil, nor to our sunshine, nor to our luscious fruit, nor to the clear, cool waters that flow from the Sierras: it is an alien that has crept in stealthily and has occupied the length and breadth of these fair valleys."[65] No longer an endemic disease, a natural condition of the Central Valley that potentially limited the uses to which the land might be put, malaria was now nothing more than an intruder that needed to be eradicated. This was a paradoxical conclusion, of course, because, as Herms himself emphasized, environmental conditions were responsible for the presence and proliferation of anopheles mosquitoes in the Central Valley. But by localizing disease in certain bodies—whether insect or human—germ theory implied that eradication of the pathogen was not only possible but also the most desirable solution.

The emphasis on disease and vector eradication was telling. Entomologists combined their understanding of ecology with a public health framework that emphasized *human* agency and engineering solutions. When mosquito control experts discussed local environments, they emphasized how those environments might be remade to conform to their own ideals. It was no longer a question of studying and adapting to the local ecology but of studying, and then efficiently annihilating, disease-bearing insects. Moreover, entomologists and engineers showed little interest in the broader effects of their control programs. Their emphasis on drainage and ditching in California had obvious implications for the valley's waterfowl, which depended on the marshes, sloughs, shallow ponds, and seepages—the very "irregularities" that engineers were bent on eliminating. Entomologists also enthusiastically and indiscriminately introduced the aggressive mosquito-eating fish *Gambusia* into the valley's rivers and streams without considering its effect on other species. Their use of oil threatened both, as well as birds, and their wholesale clearing of brush and streamside vegetation removed potential nesting sites. When wildlife officials began to raise concerns about these issues in the 1930s, however, entomologists and mosquito control engineers were dismissive, and they would remain so throughout the 1970s. While others might attend to the health of the land, their concern lay with human health, narrowly defined.[66]

Medical entomologists' somewhat instrumental interest in ecology was nowhere better exemplified than in their insistence that insect elimination should be pursued for its own sake. From the beginning of their campaigns in the tropics, Ross, LePrince, and others had argued that antimosquito work should not be limited to malaria-bearing species but should encompass "pest species" as well. This vilification of insects in the public health literature often far exceeded any empirical evidence of their role in disease transmission. Entomologists and sanitarians emphasized the possibility that any mosquito might be a vector of *some* disease, even if yet unknown. One engineer went so far as to claim that all mosquitoes, since they disturbed sleep and kept people indoors, constituted a menace to health. California's Board of Health agreed, chiding the community of Stockton for focusing solely on malaria-bearing species rather than simply eradicating their mosquitoes indiscriminately. The board's *Monthly Bulletin* attempted to blur the line between disease vectors and pest species, arguing that the "biting and buzzing [of mosquitoes] cause a loss of sleep and nervousness that are factors in ill health," regardless of whether they bore the malaria parasite. The housefly was also a target for elimination. Long suspected to be a carrier of typhoid fever, the housefly was now indicted by leading entomologists for its potential role in innumerable diseases: cholera, dysentery, enteritis, tuberculosis, anthrax, yaws, smallpox, and dermatosis, to name but a few.[67]

Even as the diseases themselves diminished, or the role of insects in their transmission was rebutted by new evidence, the entomological fraternity did not change its prescriptions. Instead entomologists argued that the role of insects in the transmission of *other* diseases would soon be discovered, so why take any chances? In their widely used textbook on mosquito control, Herms and Gray asked rhetorically, "Where can we draw the line between mosquito abatement for the protection of public health, and mosquito abatement for the promotion of comfort or the prevention of economic loss; and is it necessary or advisable to draw such a fine line?" Where indeed. Increasingly these distinctions were nullified in public health work, as any insect or animal that might possibly prove a threat to human health was targeted for extermination. In the same period the discovery that California ground squirrels could harbor bubonic plague and that some coyotes were infected with rabies generated unrelenting "control" programs against these animals that, in retrospect, are difficult to comprehend.[68]

Here was an interesting side effect of the emphasis on bacteriology. By relentlessly localizing disease in particular insect or animal bodies, germ theory made it possible to view certain species as inherently dangerous and therefore without ecological value. In their optimism that they understood the sources of disease, and in their zeal to eradicate it, California's public health professionals embraced species extirpation as their preferred strategy. In this view, disease was not the outcome of a relationship between a body and its environment but an unwelcome intruder that had to be stamped out wherever it resided.[69]

In fact, medical entomologists were often less likely to acknowledge the complexity of ecological relationships than many laypeople, who held on to older ideas about "the balance of nature." "The annihilation of a species, whether complete or relative, always brings opposition on the part of not a few persons," Herms acknowledged. "Few ideas are more firmly rooted in the mind of the average man or woman than that nature has brought forth nothing that is useless in the economy of the human family. It is time and again asserted that this or that insect, though it is known to be disease transmitting, must be good for something, otherwise it would not be in existence, and should, therefore, not be exterminated or even molested." But Herms's reply was unequivocal: "In answer to this, it should be said that parasitism is quite certainly an acquired habit."[70] According to Herms, the presence of parasites and insect vectors was not a natural condition of the environment, part of a complex ecology, but an aberration, something that called out for correction.

Modern public health thrived on a paradox: it insisted on the need for certain environmental changes while denying that the environment played an active role in the production of disease. Despite the adaptation of anopheles mosquitoes to valley environments, the role of geohydrol-

ogy in the spread of typhoid, and the importance of native and introduced plant species to the rise of hay fever, public health experts talked
as if the environment were at most a minor variable. If a landscape produced disease, it simply needed to be reordered in a way that was hostile
to insects and microbes but pleasing to engineers. And in contrast to
nineteenth-century discourses of hygiene and environmental medicine,
which had emphasized the need for local knowledge and human adaptation, modern sanitation discourse emphasized a set of uniform practices
that were applicable across all spaces. More often than not, local particularity was now studied so that it might be that much more thoroughly
erased. The goal of health hinged on the creation of appropriately "modern" landscapes that were by definition disease-free, or nearly so. Aseptic
and abstracted, modern landscapes—however hypothetical—were not
regarded as unpredictable ecologies; they were assumed to be silent backdrops for human action.

EXHORTATION AND RESISTANCE

In 1909 President Theodore Roosevelt's Commission on Country Life
had called forthrightly for the better "supervision" of public health in
rural communities.[71] To some extent this would soon be accomplished
through coercive legislation, such as laws that governed the handling of
milk, forbade spitting in public, or banned the use of common public
drinking cups. But public health inspectors and state legislators could
only do so much. To eliminate rural diseases such as typhoid and
malaria, farmers themselves needed to adopt the tenets of hygiene and
sanitation work. Thus the principal means for accomplishing health
reform was education, encouraging rural people to adopt voluntarily the
norms of bodily and environmental conduct articulated by state officials
and elite reformers.

The theories and critiques of professionals were not always rapidly or
widely accepted, however. The very vigorousness of the period's educa-
tional campaigns suggests a degree of public skepticism. "The farmer is
keen for improvements," wrote Herms, but there was more than a hint
of defensiveness in his prose. Even Herms had to admit that the typical
farmer had "no use for the germ propagandist." Although public health
officials held that germs and insects were the only causes of illness, they
still fielded numerous complaints about odors, bad air, and foul landscapes. When residents of the valley town of Gridley, the site of several
new farming colonies, filed a grievance over sewage odors, the State

Board of Health sent the engineer Charles Gilman Hyde to investigate. Hyde admitted that the town stunk, but, to no one's surprise, he did not think smells constituted a public health threat. "The odors which may be produced will not cause sickness," he wrote. Nevertheless, he duly noted the local community's unscientific perception of the problem: "Due, however, to mental reactions and the psychological effects of odors there may appear instances where sickness will be attributed to this cause."[72]

Many people held on to the belief that foul odors caused illness but were far more skeptical of the role of insects. Though most farmers spared no effort to control agricultural pests, the idea of attempting to exterminate entire species in a given area had no place in either popular or professional thought before the early twentieth century. The dominant view held that bugs were a part of nature, with a given—if annoying— role. And despite the new scientific discoveries, many people remained skeptical that insects were the source of their health problems. The mosquito theory often contradicted common knowledge, as many mosquito-infested regions had no malaria. Even a decade after the mosquito discovery, many rural physicians still believed that malaria was caused by bad air.[73]

In 1910, when Herms began his antimosquito work in Penryn, most residents still connected their malaria problem with "night air," mining slickens, and swampy landscapes. In fact, while Penryn took up the anti-mosquito campaign, four nearby towns refused. Despite Herms's personal appeals, many communities remained "either hostile or indifferent" to the idea of insect control. Even among those who acknowledged the possible role of mosquitoes, many still believed that insects acquired the malaria parasite from the local swamps and marshes. The role of flies in disease transmission was also widely doubted. One California man complained that state-sponsored fly eradication was misdirected. "I know they [flies] make a lot of dirt, spoil picture frames and such, tickle your nose in the morning if you don't get up, but they make a nice food for young poultry," he wrote; moreover, "Only a few years back they were considered a blessing, as they eat stuff that would make harm." Another farmer skeptically confronted an earnest state health official on the issue: "I have lived among the flies nearly seventy years and am still alive and well."[74] Environmental frameworks of disease often persisted even more strongly in informal and local discourse.

Reformers derided such resistance to their efforts as more evidence of rural "backwardness." As the State Board of Health complained, many country residents believed "that because they live close to nature there is little or no danger from poor sanitary surroundings," which only showed that education was "far more needed in rural communities." Colonial public health officials typically stressed that the inherent backwardness and incapacity for self-governance of native subjects required high levels of state intervention and control. In the United States, the state was necessarily more constrained, dictating greater reliance on education and exhortation: local cleanup weeks, lectures to schoolchildren, tracts aimed at farmers, and traveling exhibits.[75]

In 1909 the board targeted California's rural population for such an exhibit. With the help of the Southern Pacific Railroad, the board inaugurated a "sanitation car"—a public health exhibit housed in a railroad car that visited many of the state's rural communities. The exhibit contained scale models of sanitary and unsanitary conditions, specimens of "disease-carrying flies and mosquitoes," and graphs that illustrated the economic costs of preventable disease. Officials touted their model of the "improperly maintained farmhouse," complete with manure piles, unscreened windows, and a dilapidated and poorly constructed privy. The exhibit traveled to ninety-two towns in eleven months bringing the message that the disorderly and unhygienic space of the farm needed to be reordered and disinfected.[76]

Public health reformers worked especially hard to convince people that insects (rather than landscapes) were dangerous reservoirs of disease. The anti-insect campaigns of the period did not lack for passion and intensity. Several leading entomologists wrote tracts for popular audiences as well as texts for university students in which they self-consciously redrew the image of insects. Herms's popular textbook on medical entomology was organized around the various species of insects, detailing their life histories and then listing all the diseases that they might possibly carry. L. O. Howard, a leading figure in the field, wrote *Mosquitoes; How They Live; How They Carry Disease; How They Are Classified; How They May Be Destroyed* (1901) and followed it up with *The House Fly—Disease Carrier: An Account of Its Dangerous Activities and of the Means of Destroying It* (1911). A Stanford University professor, Rennie W. Doane, published *Insects and Disease: A Popular Account of the Way in Which Insects May Spread or Cause Some of Our Common Diseases* (1910). Herms followed with his own indictment of the fly in *The House Fly in Its Relation to Public Health* (1911), in which he described the fly as "revoltingly filthy, feeding indiscriminately on excrement, on vomit and sputum." To drive home his point, Herms emphasized that when feeding, the fly "regurgitates droplets used in liq-

Figure 7. The California State Board of Health's "Sanitation Car" in 1909, a didactic public health exhibit that traveled to more than eighty rural California towns. Reproduced from H. O. Jenkins, "A Traveling Sanitation Exhibit Directed by the State Board of Health of California in 1909," figure 27. Courtesy of the California History Room, California State Library.

uefying solid food, and extrudes droplets of excrement as well." Communities were encouraged to offer prizes to schoolchildren for bringing in the most "swatted" flies and for the best screened homes. Herms optimistically asserted that "there is no reason why [the housefly] should continue to exist, and its death knell is being sounded wherever communities care for the health of the individual."[77]

Officials focused much of their effort on women and children, whom they viewed as more amenable to their message. In his mosquito control campaigns, Herms targeted rural schoolchildren—teaching them the basics of mosquito biology, requiring them to write essays on mosquito control, and even urging them to implement control measures at home in the face of their parents' disinterest. (Herms went so far as to publish exemplary student essays in his text on mosquito control.) He also found support for his efforts in the women's clubs of rural towns, such as those in Bakersfield and Oroville, who took up the sanitary campaign in their communities. Sanitary reform fit comfortably with women's traditional role as protectors of their family's health and also offered an opportunity for women to extend their influence beyond the home.[78]

In contrast to middle-class town women and their children, adult farmers—especially men—were likely to be skeptical of both the theories and the methods of the new rural reformers. Mosquito control officials in California complained repeatedly that valley farmers showed "little cooperation" on account of their skepticism. When a federal commission surveyed farmers on the question of rural health and sanitation, the overwhelming majority replied that they found existing sanitary conditions satisfactory. Local environments were themselves a site of contestation. What constituted a proper environment was, to some extent, in the eye of the beholder. Public health experts claimed bacteriology as their support; recalcitrant farmers claimed experience. Although it is unlikely that farmers were familiar with contemporary mortality statistics, those numbers nonetheless would have confirmed their belief that the city remained a more deadly environment than the countryside.[79]

A STILL VULNERABLE GEOGRAPHY

Although sanitary prescriptions would be accepted only gradually and incompletely by many laypeople, health officials publicly insisted that any disease-ridden communities could be eventually reformed. For California's colonizers, this was a hopeful message. Germ theory helped recast the natural environment of California as neutral space with respect to health and human beings as the principal agents of disease. In fact, by the 1920s both typhoid fever and malaria were on the decline in California, and public health officials were not reticent about congratulating themselves for these successes. Ironically perhaps, the response to these diseases had relied less on laboratory discoveries than on methods of environmental control that were prefigured in nineteenth-century environmental medicine. The reduction of typhoid in California and elsewhere probably owed more to the introduction of water-supply treatment methods (filtration and chlorination) than to more accurate diagnoses or the corralling of wastes. The decline in malaria is hard to assess, even now. The disease was already in decline by the time of the mosquito discovery, and although control likely benefited some communities, other factors also contributed to the decrease, such as improved nutrition and rising standards of living. Later observers would connect the decline of malaria in some valley communities to the widespread introduction of dairy cattle (whose numbers rose rapidly after 1910), pointing out the California anopheles preferred not only cows but also rabbits, horses, and many other mammals to humans. However mixed the reasons, public health officials gladly took credit for health improvements, and they grew increasingly confident of their own methods. That confidence emerged not simply from their elite social position but from material developments for which they now claimed responsibility.[80]

But even as public health professionals congratulated themselves, the unpredictable contribution of the environment to disease remained, albeit always in the background. The history of plague is a telling example. In 1907, during San Francisco's second plague outbreak, California public health officials were confident that the elimination of rats would bring the elimination of plague from their state. However, the following year researchers identified plague in the ground squirrels of neighboring Contra Costa County, still largely a rural region. Infected squirrels were soon found in several areas of the Central Valley. Health officials assumed that native squirrels had been infected by rats and that the disease was spreading east from San Francisco. Their response was a vigorous extermination campaign. By February 1911 the State Board of Health boasted that 320,000 squirrels had been killed, and officials were again optimistic that they had eradicated the disease. However, when plague reappeared in the following decade, the University of California epidemiologist Karl Meyer began to classify the ground squirrel infection as "sylvatic plague"—an acknowledgment that wild rodents constituted a natural reservoir of the pathogen. Meyer's unconventional (and controversial) position was influenced by his training in both zoology and veterinary medicine, as well as his interest in the pioneering work of the animal ecologist Charles Elton. Moreover, it was also prescient, for in the mid-1930s the California squirrel population suffered a far more intense and widespread outbreak of plague, and by 1940 sylvatic plague had been found as far east as North Dakota. Meyer, much to the chagrin of California's public health establishment, now argued that the disease had probably always been present on the American continent and, in fact, could never be eradicated. Outbreaks were merely the result of humans settling in plague areas that had heretofore been sparsely populated. Meyer's formulation turned contemporary public health assumptions on their head: disease was not the invader; humans were.[81]

Other physicians continued to point to the regional climate as a cause for concern. Writing in 1918, Dr. William Cort, a member of the State Board of Health, warned that "on account of a semitropical climate, a close relation to the Orient and the tropics and a continually shifting population, [California] needs to be constantly on guard against the introduction and spread of diseases caused by animal parasites."[82] Here, in the space of a single sentence, Cort both localized disease in the bodies of immigrants (the "continually shifting population") and warned of the environment's role ("a semitropical climate") in disease production. Cort's comment, like Meyer's work on plague, suggests that among some medical men modern concepts of disease coexisted with an appreciation for ecological factors.

Did doctors still fear the climate's effect on health in 1918? Cort's concerns, like those of Meyer, grew out of his background in zoology as well as his training in tropical medicine. The latter had emerged as a modern medical specialty in 1898 with the publication of Patrick Manson's classic text, *Tropical Diseases: A Manual of the Diseases of Warm Climates*. Manson's work became the definitive statement of a new scientific, tropical medicine that supposedly supplanted an older medical geography of the tropics—though the continuities between the two are equally striking. Manson argued the need for tropical medicine as a specialty based on its focus on parasitic (rather than bacterial) diseases—diseases that were dependent on geographically specific vectors for their transmission. The field's claim to distinctiveness rested in part on the need to study the environmental factors that influenced disease transmission in tropical locations. Its practitioners maintained strong links to both biology and zoology, and they diligently studied the ecology of both parasites and their vectors. In other words, tropical medicine was a discursive location, much like environmental sanitation and medical entomology, in which the links between environment and health were openly acknowledged and seriously studied.[83]

That the State Board of Health would have a specialist in tropical medicine as one of its members is in itself revealing. In 1909 the secretary, William Snow—though a traditional sanitarian—argued that the medical establishment needed to recognize California's "special problems" and the corresponding importance of tropical medicine to the region, given the multiplicity of "tropical" diseases that "are now present or may at any time invade our territory." In 1917 the board created the Division of Parasitology—devoted to the study of those tropical diseases that occurred in California. William Cort headed that division, though he was but one of several California doctors with similar interests. Several physicians, entomologists, and zoologists were actively engaged in the field, and the zoology department at the University of California, Berkeley, provided them with an informal institutional base.[84]

That tropical medicine would find a home in California was yet another indication that neither the state's diseases nor its environments could be fully subsumed under a narrow bacteriological model of disease. It was a recognition of what had been there all along. Those engaged in tropical medicine discussed persistent environmental factors that were otherwise marginalized in modern medicine and the new public health. For some, California's geography and climate were still cause for concern. The mild winters and moist microclimates of the valley and foothill regions fostered the survival of many parasites and their vectors. Dr. Herbert Gunn, an expert on hookworm, warned that a "great portion of our state will afford ideal conditions for the cultivation and dissemination of this disease, and . . . there is a possibility, apparent and real, of this state becoming a hotbed of uncinariasis." Dr. Creighton Wellman went a step further, comparing California to America's tropical possessions: "California with its equable climate must be classed by the student of the geography of disease as part of the subtropics. We have no winter to break in upon the routine of the tropical parasitic affections which we may import, and they are consequently able to become endemic. Hence the matter of prevention is as important to us as it is to residents in some of our colonies."[85]

Those with an interest in tropical medicine were concerned not only with the state's existing environmental conditions but also with the changes currently under way. Much like Marshall Chipman, who warned of the health effects of mining and deforestation in the 1880s, William Herms and other entomologists worried incessantly about the spread of rice culture in the Sacramento Valley, which depended on large amounts of standing water. Introduced experimentally in 1910, rice spread quickly across the region, and so did mosquito problems. By 1916 Californians had put at least 80,000 acres into rice, and that acreage would continue to grow—to 162,000 acres in 1920 and to more than 250,000 in 1947. California farmers flooded their rice fields just prior to seeding in late spring (April or May) until sometime in September— prime mosquito season. In most cases, the fields drained into existing road ditches. The roads themselves soon became flooded, while the partially drained fields became "veritable bogs" and a breeding paradise for mosquitoes (especially the vector *A. freeborni*). By 1920 the haphazard spread of rice production had created severe public health problems in the northern part of the state. Nearby residents who were not rice farmers were calling for the prohibition of the crop, and public health officials lobbied for stronger control measures and a program of preventive quininization.[86] When California's political and business leaders began advocating for massive water supply projects in the 1920s and 1930s, medical entomologists would argue that the extension of mosquito control work was nothing less than imperative.

Obviously concerns about the environment's effect on disease remained; however, they were now typically coupled with the assertion that the ultimate source of disease in California was immigration, the state's "continually shifting population," in William Cort's phrase. Already under attack for their supposedly unhygienic habits, immigrants were also assumed to be "carriers" of disease. Although the California environment sustained and fostered certain parasites, disease only entered into sanitary spaces in the bodies of infected individuals. Health officials warned that geography made California especially vulnerable to carriers, lying as it did on the Pacific Coast, with San Francisco serving as a major embarkation point both to the tropics and to Asia. "As a result of our present intercourse with tropical countries incidental to colonial expansion," wrote one doctor in the *California State Journal of Medicine*, "bilharzias, uncinaria, filarial and a host of other parasites with euphonious names and evil designs are threatening us with an invasion."[87] The land itself was benign, but it was still vulnerable.

The origin story of malaria was rewritten in the early twentieth century along just these lines. The presence of malaria in California was now typically attributed to Italians. The reigning theory held that Italian laborers who entered the region as railroad workers in the 1850s were responsible for introducing the parasite: "These carriers, arriving in a country already the home of the anopheline mosquito, readily established foci of malaria in the great valleys and along the foothills of the western slope of the Sierras." Moreover, Italian immigrants were seen as an ongoing source of the disease in California. Thus, Herms demanded the removal of Italian railroad workers as part of his malaria control campaign in Penryn. In 1915 the State Board of Health actually declared that malaria had been "unknown" in the state before the 1850s.[88] The possibility that malaria might have been endemic to the region was no longer considered, and the presence of malaria in California in the 1830s and 1840s, prior to the gold rush migration, was officially forgotten.

Any immigrant group other than those from northern Europe was suspect, but the principal concern of California's public health establishment always lay with "nonwhite" immigrants—those from Mexico, Asia, and various "tropical" countries. "Orientals," a supposed "foci for infection," were blamed for a wide variety of parasitic diseases as well as trachoma and bubonic plague. The State Board of Health actually established a laboratory on a houseboat so that their inspectors might better target Asian populations in the Sacramento–San Joaquin delta for bacteriological scrutiny. In the late nineteenth century, doctors had blamed the presence of hookworm in California on miners from Austria, Spain, and Italy, but in the early 1900s, medical writers were more likely to attribute it to immigrants from Hawaii; by the 1920s the state's public health literature firmly connected it to recent immigrants from South Asia. In 1916 the discovery of typhus fever in Texas prompted the U.S. Public Health Service to initiate a radical disinfection and quarantine program for Mexican immigrants. The subsequent discovery of twenty-six cases in California, primarily among Mexican railroad workers, led the State Board of Health to develop its own inspection and quarantine requirements.[89] While immigrants might be hypothetically reformed, the more immediate task was to guard against them.

Persistent concerns about disease "invasion" and the possibility of new epidemics underscored the sense that California, despite almost three generations of white settlement, remained a kind of biological and environmental frontier. The concern had shifted, however, from the potential failure of white settlement from the East to that of nonwhite invasions from the West (Asia) and South (Mexico and Latin America). Many whites remained anxious about the regional geography and the possibility that a "tropical" California might reemerge from within. While new understandings of disease helped to naturalize white settlement in California, the descendants of those settlers still feared their region remained uniquely susceptible to disease. Of course, that only made the need to modernize both recalcitrant residents and rural environments that much more imperative.

In his study of colonial public health in the Philippines, Warwick Anderson has written that the extension of modernity into the colonies was predicated on a transcendence of the body, however hypothetical, that came through the adoption of proper hygiene. It is equally true that modernity, both in the United States and beyond, was predicated on the erasure of local and contingent environments and their attempted re-creation as homogeneous and controlled space. As Lefebvre has observed, "The tendency toward the destruction of nature does not flow solely from a brutal technology; it is also precipitated by the . . .wish to impose the traits and criteria of interchangeability upon places."[90] Western medicine had long resisted that erasure; however, medicine and public health became modern when they embraced, at least rhetorically, a framework that separated human health from the particularities of local environments, relegating the latter to other, less prestigious disciplines. Diseased bodies were now taken as evidence of the need for modernization rather than as indications that the local landscape needed to be studied more closely for its effects on health. On the one hand, compartmentalization and professionalization helped achieve certain results, by fostering new types of knowledge, in particular the bacteriological knowledge generated in the laboratory. On the other hand, the implications for local environments would prove far more ambiguous.

Yet attempts to separate bodies from environments were always both incomplete and temporary, and, moreover, they were not the whole story. Environmental sanitation continued to address the role of the local environment in disease production even as the work of engineers and doctors increasingly diverged. The somewhat marginal disciplines of tropical medicine and medical entomology tracked the contribution of environmental factors to the spread of disease-causing organisms, relying on both ecology and biology. And occasionally there were other suggestions that public health officials recognized that the environment was perhaps not as irrelevant as mainstream medicine and the germ propagandists maintained. In the 1930s Karl Meyer diagnosed an outbreak of "sleeping sickness" among San Joaquin Valley livestock as western equine encephalitis; the same disease would be diagnosed in humans a few years later. Combining laboratory analyses with careful epidemiological fieldwork, Meyer connected the disease with warm weather and irrigated pastures and hypothesized that mosquitoes might transmit the disease. Later work would identify the mosquito *Culex tarsalis* as the principal vector. Like *A. freeborni*, *C. tarsalis* was native to California, but the changes associated with irrigated agriculture had radically increased its numbers in the early twentieth century. At the end of World War II many feared that the state's environment would facilitate the emergence of malaria, encephalitis, and even dengue fever when infected soldiers returned from abroad. Those fears would prompt the legislature to quickly approve $700,000 in emergency funds for mosquito control work, as well as massive use of the new chemical DDT.[91]

Perhaps most revealing of the limits of the modern model of health in California, however, was the disease "valley fever" (now known as *Coccidioidomycosis*), named for the San Joaquin Valley, where it is endemic. Typically, valley fever resembles a cold or flu, and patients recover relatively quickly, but in severer cases the disease erodes the body's extremities. The disease was first diagnosed in the state in 1894; the pathogen was identified a few years later. Yet it would be decades before physicians could account for its transmission. Contemporary science locates the source of disease in a soil fungus that enters the human body via airborne spores. The fungus has precise environmental requirements and is highly localized, restricted to certain arid and semiarid regions of the Americas—although the spores can travel long distances in the atmosphere. It becomes more easily airborne, and infection rates increase, during periods of drought. In the late twentieth century, local outbreaks were correlated with the construction of new subdivisions on undeveloped land and with archaeological excavations. In other words, valley fever is, in every sense, an environmental disease. And at the turn of the century, locals had a remarkably similar understanding. They classed the illness with the valley's multitude of fevers. (It was misdiagnosed as malaria as late as the 1930s.) Moreover, its prevalence in the valley suggests a modern medical reason for long-standing fears of winds and newly cultivated land. Yet early-twentieth-century physicians struggled to combine the fever's apparent geographic distinctiveness with their own ideas of disease, attributing it variously to poor hygiene, insect or animal vectors, or immigrants from the tropics. Although the fungus had been consistently identified in the soil, until the 1940s researchers maintained that the source was more likely infected rodents, such as pocket mice or kangaroo rats.[92] The idea that the land itself might be the source of disease had become difficult to imagine. Modern models of disease both furthered and constrained medical thinking.

Valley fever is only the most obvious example of medical bacteriology's failure to materially separate disease, much less health, from the environment; yet the field was far more successful in achieving a discursive separation, which would, nonetheless, have significant material effects. It is important to recognize that state intervention in the arena of public health began when germ theory enjoyed its greatest dominance. Early-twentieth-century public health strategies were built on a historically contingent model of bodies, environments, and disease, but those same strategies would shape perceptions and practices for decades to come. Even as understandings of disease became more complex, practices would still be shaped by that earlier history.

In fact, much of medicine and public health in the 1930s and 1940s continued to embrace a "modern" view of both disease and the body, and a correspondingly narrow view of health. The dominant medical specialties turned away from the study of the environment; doctors perceived bodies as closed or nearly so; and public health officials insisted that human communities were the source of their own health problems.

In California considerable effort would go toward ensuring the provision and maintenance of clean water supplies, proper sewage and waste disposal, and improving community hygiene. Long-standing concerns about the health of migrant workers as well as their potential to transmit disease would become even larger issues during the 1930s, as thousands of poor migrants from the Southwest poured into California, many from states with high rates of malaria infection. In the 1940s a severe epidemic of diarrhea and enteritis among migrant families in the San Joaquin Valley would prompt increased attention to sanitation in the fields and labor camps of the Central Valley. At the same time, the importation of agricultural laborers from Mexico under a series of binational agreements renewed concerns about alien carriers of disease and led to additional measures to screen, vaccinate, and disinfect incoming immigrants.[93] These efforts would all be justified, however vaguely, by references to germ theory and "modern" science.

Emboldened by the declining rates of diseases such as typhoid and malaria, health officials believed that their ongoing reconstruction of local ecologies was rapidly rendering the North American environment irrelevant to disease. In the decade after World War II, however, environmental changes would create new health problems, and changing ideas of the body would make the relationship between disease and the local environment newly visible. These developments would also challenge the overly simple equation that linked modernity, and modern landscapes, with health.

4

Modern Landscapes
and Ecological Bodies

Only yesterday mankind lived in fear of the scourges of smallpox, cholera, and plague that once swept nations before them. Now our major concern is no longer with the disease organisms that once were omnipresent. . . . Today we are concerned with a different kind of hazard that lurks in our environment—a hazard we ourselves have introduced into our world as our modern way of life has evolved.

Rachel Carson, *Silent Spring*, 1962

Oh, lots of times you get dizzy while working, but you don't usually dare to take the day off or you'll get fired. One time I felt so bad I went to the doc. He said it was just flu. You hear that all the time: just flu. I knew some fellows over by Taft that it happened to. They all tooken sick, and went to Dr. J——. They were so sure they had flu they never even bothered to tell him what they'd been doing. They'd been mixing parathion for a crop dusting outfit over there. Funny they'd all come down with flu at once, don't you think?

Every year we go up north to pick cherries and peaches. About three years ago, a whole bunch of us got parathion poisoning in the peaches near Modesto. I heard 85 people got poisoned. Maybe some of them died. I don't know for sure. But I know one lady, Mrs. C——, was really in bad shape. And ever since then, whenever she picks in an orchard where they've been using parathion—oranges or anything else—she gets real sick, and has to be under oxygen.

Statements of California farmworkers to the
California Department of Public Health, 1969

On July 8, 1949, several crews of farmworkers arrived at a pear orchard in the Sacramento Valley near the town of Marysville. They had come to harvest. The day was hot and humid with little wind. Shortly after lunch a few of the men fell ill. By midafternoon, at least a dozen pickers were too sick to work. Some laid down in the fields where they had been working; others walked off the job. Soon the sick men were dripping with sweat, and most were also retching violently. Others found their arms and legs twitching uncontrollably. All were taken to the hospital.[1]

A few months earlier, Herbert Abrams, chief of California's Bureau of Adult Health, had given a talk before the Entomological Club of Southern California. Abrams told his listeners that America's rural environment had undergone a fundamental transformation in the twentieth century, and, as he saw it, the industrialization of farming was generating new public health problems. He was especially concerned about a new class of chemical pesticides known as organophosphates, or OPs. These compounds were being rapidly introduced into the California environment, despite the fact that many were known to be highly toxic. The most common OP, parathion, was a cheap, brown liquid, only a few drops of which could kill a human being. Abrams told his listeners that he was already aware of one death, though the chemicals had been in very limited use.[2] Seven months later the pear-picker incident at Marysville bore out his warning. The sick workers were diagnosed with acute parathion poisoning.

The pear-picker poisonings signaled the arrival of a new kind of illness in California's Central Valley, one that would necessitate a reconsideration of the environment's role in disease. These workers had not swallowed or been in direct contact with parathion. The orchard they entered had been sprayed twelve days earlier. Their exposure was merely to lingering pesticide residues. The new and unfamiliar reactions of bodies to their environment—violent tremors, uncontrolled vomiting, profuse sweating—demanded interpretation. In 1949 the dominant assumptions of public health still ascribed disease to individuals. Modern frameworks of health developed and propagated since the early decades of the twentieth century held that impure and undisciplined bodies were themselves the source of a polluted environment, and migrant farmworkers were the most pathologized group in the state. The concern with the diseases harbored by migrant laborers had reached something of a fever pitch in the 1930s, when tens of thousands of migrants from the Southwest arrived in California. Although health officials pointed proudly to several decades of sanitation work that had made California and its Central Valley into a modern and orderly space, the health of farmworkers remained a seemingly intractable problem. Health officials openly worried that migrants carried and transmitted malaria, tuberculosis, and various diarrheal and viral diseases.[3] Given this history, many people interpreted outbreaks of twitching limbs and massive sweating as symptomatic of the weak and disease-prone bodies of the workers themselves.

But the frameworks of hygiene and sanitation would ultimately fail to contain the problem. Diseases like malaria, plague, valley fever, and typhoid revealed that even infectious disease could not be separated from its environmental context. Now the rise of new types of disease—often chronic and noninfectious—would further challenge the modern model. As both medical experts and laypersons struggled to come to terms with a series of new health threats—pesticides, air pollution, radioactive fall-out, and chemically polluted water supplies—they were forced to acknowledge the critical role of the modern environment in the production of disease. The environmental changes of the postwar period were ushering in new types of illness, in the Central Valley and elsewhere in the United States. One result was that modern frameworks of the human body and the promises of sanitary engineering competed and intermingled with other frameworks, at once old and new, that constructed the body as permeable and the landscape as dangerous.[4] Sanitation had sought to render the environment passive, but new kinds of illness pointed even more strongly to the environment's active role in shaping health. The problems posed by pesticides in the Central Valley would become central to rethinking—or perhaps rediscovering—the relationship between bodies and environments in the mid-twentieth century. Although "environmental medicine" was a concept supposedly relegated to the past, "environmental health" would now emerge as a focus of public health work.

THE POSTWAR LANDSCAPE

The process of agricultural modernization was already well under way in the valley in the 1920s and proceeded even more rapidly in the decades after World War II. Modern technology brought changes to California's agricultural landscape that even the casual observer could not fail to notice. The beautifully engineered irrigation canals of the Central Valley

Project now wove smoothly through the landscape even in the driest months and years, bringing water from the northern Sacramento Valley into the drier San Joaquin. At the same time, new irrigation technologies and improved machinery enabled farmers to level and irrigate fields that once had been devoted to pasture or ignored altogether. Farming spread into drier regions and onto the hillsides, and much of the southern valley took on a greener appearance. The west side of the San Joaquin Valley, long ignored because of the absence of water, became one of the region's most productive areas. By 1950 horticultural products (fruits, nuts, vegetables) had replaced traditional field crops in many areas and accounted for 63 percent of California's agricultural production.[5]

Meanwhile, developments in mechanization and hybridization changed the way that farmers farmed. University of California scientists had developed specialized harvesting machines for the region's fruit and nut crops as well as for cotton plants. Hundreds of new plant varieties were brought into existence and into the ecology of the valley: tomatoes with tough skins that could withstand mechanical harvesting, graperoot stocks that were resistant to the devastating phylloxera pest, higher-yielding and whiter strains of cotton, larger and juicier nectarines, frost-resistant avocados.[6]

Dramatic changes in the land were accompanied by social changes. The higher input costs, along with the bias of federal crop subsidies toward larger growers, pushed many small farmers off the land. Although California had historically had many large farms, it was only in the postwar period that the number of farms began to decrease. Chronic agricultural surpluses and falling profits during the 1950s forced out many farmers and encouraged greater concentration in California agriculture. In the same period, California emerged as the nation's leading state in farm income, supplanting Iowa. In 1962 agriculture would become the state's leading industry, with California farmers producing record harvests.[7]

As fewer owners controlled more acreage, the demand for wage labor increased. Unlike most modernized agriculture, which is characterized by high levels of machine mediation, fruit and vegetable production has never been fully amenable to mechanization but remains, even today, a relatively labor-intensive enterprise. Human hands are still required to prune trees and to pick the most sensitive and difficult crops such as peaches, strawberries, lettuce, and cauliflower. Migrant labor has been a staple of California agriculture since the late nineteenth century, but agricultural expansion coupled with the intense specialization in fruits and

Figure 8. The California Aqueduct under construction in the San Joaquin Valley. Massive state-funded irrigation projects allowed for the continued expansion and industrialization of valley agriculture in the postwar decades. Courtesy California Historical Society, FN-36329.

vegetables created a growing demand for transient, seasonal labor. As the number of farmworkers declined nationally in the postwar period, their numbers increased in California. Between 1940 and 1982 the number of farmworkers in the state rose by more than 200 percent. In 1990 nearly 80 percent of all farm work in the state was performed by hired laborers. Although some of these laborers were permanent residents, most moved in search of seasonal work at some point during the year. Farmworkers were overwhelmingly migrants, constituted by the flows of a global labor market that recruited the most exploitable workers to pick the state's most profitable agricultural products.[8]

Since the 1920s an increasing number of California's farmworkers have come from Mexico. Early on, the reliance on Mexican laborers had met considerable resistance from both small farmers and committed nativists—and concerns about the importation of disease had been a key argument for curtailing Mexican immigration. The labor shortages produced by World War II encouraged large growers to lobby for looser immigration restrictions, and the federal government responded by agreeing to sponsor the migration of Mexican agricultural workers into the southwestern states. For their part, growers drew on arguments about Mexican bodies, claiming that they were uniquely suited to labor in the hot and difficult environment of the valley. Intended as a temporary measure, the bracero program was discontinued when the war ended, only to be reinstated in 1950 after intense lobbying by agricultural interests. In 1959 California reported 206,000 seasonally employed farmworkers, of whom 83,000 were braceros. By definition, braceros had no permanent homes in the United States but moved from labor camp to labor camp. Like most California farmworkers, the braceros did not work directly for a grower but for a labor contractor who supplied workers to multiple growers. The workforce on a postwar California farm was always changing, variously composed of migrant residents, braceros, and illegal immigrants. They were perhaps the most socially invisible and politically powerless group in postwar California.[9]

The reports of the California Department of Agriculture from these decades evoke the widespread confidence of agricultural experts in modernization. Written in the straightforward prose of bureaucrats, the reports themselves are mundane documents. Field inspectors and department heads catalog for their superiors how they have spent their time— the meetings they attended, the complaints they answered, the inspections they conducted. They also offer a running commentary on the state of farming in their respective regions, noting the effects of cold weather on sensitive crops, the successes of new seed varieties and harvesting techniques, the spread of weeds, the latest insect infestations, and the labor situation. Almost incidentally, these reports reveal a regional landscape that was awash in chemicals.

The department's records make clear that the modernization of California's agriculture depended not only on mechanization, irrigation technology, advances in plant hybridization, and the recruitment of a vast labor supply—although it required all of those—but also on massive amounts of the new agricultural chemicals. Inspection reports are filled with references to chemicals: the number of new pesticides registered for use; the commencement of spraying to counteract the Mexican fruit fly, the beet leafhopper, and countless other pests; the confiscation of fruit that had been found to have unacceptable levels of DDT residue; warnings issued to pest control operators for leaving empty pesticide containers laying about; reprimands to pilots who sprayed pesticides without the proper licenses; the calculation of crop losses resulting from the accidental drift of chemical defoliants; the damage to fruit trees subjected to excessive spraying.[10] In the 1950s and 1960s California farmers applied pesticides on more acres and in larger quantities per acre than elsewhere in the country. Estimates from the period typically put the state's share of pesticide use at 20 percent of the national total, but no one knew the actual amounts applied. By 1955, 7.1 million acres, or two-thirds of California's cropland, was being treated with chemicals. By the mid-1960s more than sixteen thousand pesticides had been registered in California, and farmers increasingly relied on multiple applications of multiple chemicals.[11] Although not as visible as mechanical harvesters or irrigation projects, agricultural chemicals were also changing the ecology of the Central Valley in profound ways.

Discussions of the turn toward organic chemicals in the postwar period almost invariably focus on DDT. Used effectively to combat malaria by killing mosquitoes during the war, DDT was hailed as a kind of wonder chemical. It was quickly employed throughout the United States (including California) in both mosquito control and agriculture. Other chemicals in the same class as DDT—chlorinated organic hydrocarbons—also came quickly into use, including substances like DDE, endrin, aldrin, dieldrin, and toxaphene. But the chlorinated organic hydrocarbons were but one of several new groups of agricultural chemicals. In addition, there were chemical herbicides and defoliants, such as 2,4-dicholorphenoxyacetic acid (2,4-D), a hormone-based weed killer, which became ubiquitous in field crops such as rice and barley. (It would later gain notoriety for its use as a defoliant in the Vietnam War.) Also used widely in postwar agriculture were the acutely toxic organophosphates.[12]

As with DDT and 2,4-D, chemical manufacturers aggressively promoted OP pesticides for use in agriculture. Farmers were receptive. Often the turn to these chemicals was prompted by the resistance of insects to DDT—something that was already in evidence by 1950. In 1949 California farmers were applying parathion and tetra-ethyl phosphate (TEPP). In 1951 they added demeton (systox) and EPN; in 1953, malathion and chlorthion; in 1956, dipterex and the extremely toxic metacide. In 1958 the introduction of thimet into cotton production was

accompanied by several serious poisonings. That year also saw the introduction of tetram, disulfuton (di-syston), mevinphos (phosdrin), azinphosmethyl (guthion), and carbophenothion (trithion). In 1964 the
highly toxic carbamate pesticides made their first appearance in
California agriculture. Growers liked these new pesticides for several
reasons. They typically decayed more rapidly than organochlorine compounds (e.g., DDT) and thus posed fewer residue problems. They were
also effective against some of California's most difficult pests—such as
the red scale—and they offered an alternative as pest resistance to DDT
increased over the course of the 1950s. Moreover, the most popular
organophosphate—parathion—was also very cheap; it was developed in
Germany in the 1930s and had never been patented.[13]

It is, however, impossible to track the historical introduction of chemicals in a detailed way. Complete records of pesticide applications in
California have been maintained only since the early 1990s. Previously,
county agricultural commissioners kept records of the number of acres
treated and the types of restricted pesticides in use but little or no information on quantities. Consequently, chemicals slipped into the valley's
landscape without leaving much of an archive.[14]

NEW KINDS OF ILLNESS

Those who pursued and advocated agricultural modernization in the
postwar years were thinking primarily about efficiency and profits, not
about ecology or health. The development of modern sanitation and
mosquito control had supposedly severed the link between disease and
the environment. Thus, the modernization of agriculture proceeded
from the assumption that nature was merely another set of inputs—
water, land, soil—into the agricultural process, inputs that could be
managed to increase production. The environment was considered a
critical economic factor but not a medical one. It quickly became apparent, however, that the modern agricultural landscape was producing
more than good yields and record harvests; it was also producing disease. In some areas, the expansion of irrigation was creating hordes of
disease-bearing mosquitoes, sparking outbreaks of malaria and
encephalitis. But there were also new illnesses that seemed to be connected to the new pesticides.[15]

Some valley residents questioned the broader effects of the new agricultural chemicals from the moment of their introduction. Some farmers
pointed to the damage that the chemicals did to "nontarget" crops. For

instance, when the herbicide 2,4-D came into use in 1948, it generated widespread opposition among beekeepers, hay growers, vineyard owners, and some citrus ranchers who urged that it be more tightly restricted or even banned. Livestock owners throughout the valley complained that their animals were being poisoned routinely by the new chemicals. Often farmers or their insurers went to court to try to recoup property damages, and inspectors at the Department of Agriculture spent much of their time investigating complaints of pesticide and herbicide damage and compensating affected growers. By the late 1950s pesticides were implicated in several fish kills in the valley, as well as in the decline of certain bird populations.[16]

Early concerns about the effects of pesticides went beyond property damage and wildlife.[17] As they watched vines wither and cows sicken, valley residents were already questioning in the 1950s how these chemicals might be affecting human health. Some asserted that agricultural spraying had made them sick—such as Mrs. L. F. Cannon who complained of illness after a pilot mistakenly applied parathion to an orchard surrounding her house. A few rural residents were already calling for a ban on parathion and other highly toxic chemicals; others were complaining about the felt effects of agricultural chemicals, seeking answers and reassurance from health authorities. One man wrote to the Department of Health in 1957 complaining about an airplane that had sprayed unknown pesticides in the area: "I was made very sick—vomited and had loose (no control) bowel movements—pounding at throat and tired ache behind ears. The desire to vomit stayed with me a full week though I never did vomit again. I am not used to this locality—came here in March of this year—I do not know the farm practices—and am much concerned over the possibility of another exposure." Another resident wrote to the editor of a valley newspaper stating his intention to move his family out of the area because of his concern for their well-being in the face of so many hazardous substances.[18]

Most individuals did not take the time to write down their complaints, and even so, the written complaints of individuals such as Mrs. Cannon are only rarely retained in government archives. The record of illnesses that we have is surely only a fraction of those that occurred. However, in response to the pear-picker poisoning of 1949, the Bureau of Adult Health began to publish statistics on the occurrence of occupational disease attributed to agricultural chemicals. The basis for their statistics came from reports of illness that were required under California's workman's compensation law. Throughout the 1950s reports of pesticide-induced occupational disease increased in the state. By 1963 the California Department of Public Health (CDPH) was reporting that agriculture had the highest rate of occupational disease in the state, more than 50 percent higher than any other industry, even though these statistics did not take into account the fact that pesticide-related illness was vastly underreported.[19]

In that same year, ninety peach pickers became ill near Modesto in the northern San Joaquin Valley; one died. The poison in question was again parathion. Although the evidence indicated that the chemical had been applied in accordance with existing regulations and manufacturer's instructions, public health officials declared that the leaves and branches of peach trees throughout the central part of the state were laden with dangerous levels of residue that made harvesting a potential death sentence. Moreover, when California officials arrived to investigate the poisonings, they tested not only those who had been made obviously ill but also those workers who had no apparent symptoms. Blood tests of thirty-three people who were working in the affected orchards but showed no obvious signs of illness revealed that twenty-five of them had subclinical symptoms of poisoning. Then investigators went to local motels and labor camps looking for workers who had picked in other fields, where no illnesses had been reported. They tested forty-five workers and found that sixteen of them (35%) had been seriously affected by OPs.[20] Though the samples were quite small, the results were disturbing. They suggested that the incidence of pesticide effects among field workers was high even if not obviously visible.

The peach-picker incident was the most well publicized, but it was just one of several similar incidents that followed. According to CDPH statistics, thirty-seven workers became seriously ill in five instances in 1966. In 1967 peach pickers in the northern San Joaquin Valley were again poisoned, this time after exposure to azinphosmethyl ethion. The following year a crew of orange harvesters fell ill in the San Joaquin Valley. The mass poisonings recorded by the CDPH are disturbing in themselves; but the records of the Department of Agriculture indicate poisoning episodes of which health officials remained unaware, and health officials themselves complained that the vast majority of poisonings went unreported. In 1968, when the CDPH finally conducted a study that attempted to capture the overall incidence of poisoning, researchers found that 85 percent of those interviewed reported at least one symptom that might be pesticide related, and 20 percent reported five or more symptoms.[21]

The dangerousness of OP compounds had never been in question. Since the late 1940s scientists have recognized that their toxicity resides in their ability to inhibit the action of cholinesterase, an enzyme crucial to the normal functioning of the nervous system. Although the first symptoms are mild—runny nose, blurred vision—they can quickly progress to muscle twitches, tremors, convulsions, severe abdominal cramps, vomiting, bronchial spasms, and irregular heartbeat. In the severest cases the respiratory and heart muscles stop working, and death quickly follows. Relatively little is known about the chronic effects of these compounds, though mounting evidence suggests that they may produce delayed neurological problems. They are also linked to several types of cancer. More recently still, organophosphates have been implicated, with some controversy, in Gulf War syndrome.[22]

Farmworkers themselves, though they had no knowledge of toxicology, recognized the dangers of the new chemicals. Although there are few records of workers' own experience of pesticide exposure, in the late 1950s Henry P. Anderson—a young public health investigator from the University of California—undertook an investigation of bracero laborers with funding from the National Institutes of Health. What is interesting is that the report itself, and the interests of Henry Anderson, does not revolve around, or even include, questions of pesticides. In Anderson's report, the risks posed by pesticides warrant only the briefest footnote. Anderson asks no questions on the subject, and though he is clearly an advocate of farmworkers, his focus is classically modern: hygiene, sanitation, and farmworkers' lack of access to modern medical care. Yet as he interviewed braceros in spring 1958, the concern about pesticide-related illness appears in the statements of the braceros themselves in response to *other* questions:

> While I was here last September, I got sick as a consequence of my work. I was picking tomatoes near Oxnard. My fingernails became infected as a result of poison that was on the tomato plants. Some of my fingernails fell off. It was very painful to work.

> While I was working for —— Farms, I got sick. My mouth puffed up and swelled. I think it was because of the poison they put on the plants. It hurt a lot.

> I got sick here. My eyes hurt very much. I don't know what caused it, but it may have been something they sprayed on the trees.[23]

In the late 1950s farm laborers experienced agricultural modernity in many ways, not the least of which was through the reactions of their bodies—reactions that they could describe but not control. Workers complained of severe rashes, eye problems, temporary paralysis, nausea, dizziness, and shortness of breath. These were new kinds of health problems that were not easily captured by public health discourses of hygiene and contagion. Regardless of what public health officials might think, workers still read their bodies as a kind of instrument whose limits and illnesses measured the health of the land. Farmworkers located disease not in their own bodies or in their own communities but in a landscape that they found foreign and physically threatening, and one over which they felt they had little or no control. Workers referred to the chemicals simply as "la medicina" and gradually came to associate certain illnesses with certain locales and crops, an assessment that occupational health researchers would eventually confirm. "My daughter gets swollen hands and feet, and welts, when picking tomatoes," reported one farmworker to the CDPH. "My husband gets very sick at the stomach when picking the lemons and valencia oranges. I get eye irritation with all the jobs around here: plums, grapes, peaches, tomatoes." Despite the repeated assertion that the advances in medical care and sanitation had made the United States healthier than nonindustrialized countries such as Mexico, for workers the calculus of health was not so clear. As one stated, "I do not believe that the braceros who have been to the United States return to Mexico in better health. From my own experience, I do not believe this." "Our contracts have not expired, but we are leaving," another worker told Anderson. "Two men in our area died recently from heat exhaustion. Others have become sick. We want to get out of the country while we are still alive."[24]

Despite the common insistence that Mexican workers were ignorant, in the 1950s they understood the dangers of pesticides better than many public health officials and far better than most of the public. Workers' understandings emerged from everyday experience in the fields, an experience that public health experts lacked and about which they had not bothered to ask. As one bracero explained to Anderson, "I have talked to many braceros from my village who have worked in the United States. Many of them are in worse health when they return to Chorinzio than they were when they left. The reason for this, I believe, is that they have to breathe in too many chemicals that have been sprayed on the plants where they work."[25] Workers' own interpretations of illness turned the racist notion of "fit" between Mexican bodies and the California environment on its head. In their epistemology, the modern environment, rather than Mexican bodies, was the site of pathology.

INTERPRETING ILLNESS

Organophosphate poisoning was a new kind of disease. When there had been no organophosphates in the environment, there had been no organophosphate poisoning. If disease can be thought of as the outcome of certain kinds of practices, then in this case it was also the outcome of certain places. Material changes to the landscape were producing effects in human bodies, and public health officials, farmers, and workers all had to reckon with the results. But developments, no matter how novel, interact with existing frameworks of understanding.

The dominant framework of public health in the 1950s still lodged disease principally in bodies, ascribing it to individuals and their communities. Twentieth-century medicine had hammered its lessons of contagion, hygiene, and racial susceptibility into many people's minds. Physicians and laypersons alike were trained to locate the cause of disease inside the body. Thus it is not surprising that pesticide poisoning among nonwhite laborers would frequently be blamed on the laborers themselves. Labor bosses and growers had an obvious interest in downplaying illness in order to keep workers on the job. The diagnoses of physicians who were employed by growers, or who were members of the same rural communities, were likely to be shaped by their own social and economic concerns. But most doctors were also unfamiliar with pesticide-related illness. They knew little or nothing about the chemicals in use, the symptoms they caused, or the multiple ways in which they might enter the body.[26] Misdiagnoses were common. Because many of the symptoms of organic phosphate poisoning (headaches, nausea, vomiting, cramps) could be associated with viral or bacterial infection, farmers as well as medical professionals often interpreted sickness among workers as evidence of their inherent susceptibility. When only a few pickers were affected, heat stroke was commonly suspected. Frequently, worker complaints of illnesses were dismissed as "hysteria." However, when large groups of pickers came down with such symptoms, they were usually interpreted as evidence of more typical illnesses, such as food poisoning, common diarrhea, or the flu. As one doctor from Tulare County later testified before Congress, "They [the workers] joke about the field boss calling them all sickly, saying, 'You people are always passing the Hong Kong flu to one another on the way to work.'" This construction of illness rendered it not only viral but also a completely foreign agent—an alien disease residing in the bodies of alien workers—with no connection either to pesticides or to the local environment.[27]

Because so many of California's farmworkers were ethnically Mexican, discussions of susceptibility to pesticide poisoning invoked and easily intertwined with racialized discourses that had raged around the body of the Mexican immigrant for decades. The importation of Mexican laborers under the bracero agreement, like the influx of Mexican workers in the 1910s, had rekindled long-standing fears about the threats to (white) public health posed by certain racial and ethnic groups. And although public health officials rejected biological racism, they now portrayed Mexican cultural differences as so important and so deeply rooted as to be a kind of racial proxy. Those in the medical establishment, as well as both political supporters and opponents of farmworkers, drew on racialized discourses of hygiene that asserted that farmworkers' lack of education and cultural background made them more susceptible to all kinds of disease, whether infectious diarrhea or parathion poisoning. These discourses of sanitation and hygiene obscured the emergence of pesticide poisoning as a new disease. Instead health officials repeatedly critiqued the "unhygienic" habits of Mexican workers, such as drinking from a common cup, as well as their reliance on folk remedies and traditional healers. That such workers suffered more illnesses than other groups in California was merely evidence of their lack of modernity. As the president of the California Medical Association told state legislators investigating pesticide concerns, "People just must be more careful about their personal hygiene if they are going to avoid any difficulty." Adopting a similar analysis if a more sympathetic tone, a state health official insisted that this group required much greater supervision in the use of agricultural chemicals. Either way, much of the problem of pesticide poisoning seemingly lay with the workers themselves. Once again, illness was conflated with notions of racial difference.[28]

It is possible to ascribe the dismissal of pesticide poisoning by growers and community physicians to their self-interest and even their duplicity. However, even those who advocated on behalf of farmworkers overlooked the problem of pesticides for nearly two decades. Instead, when discussing health, liberal advocates adopted a sociological analysis, arguing that workers' illnesses were rooted in their poverty and exclusion from American society. This, in fact, had been the approach of Henry Anderson when he interviewed bracero laborers in the late 1950s. And to some extent, it would be the approach of public health officials in the 1960s. The California Department of Public Health issued a report on farmworker health in 1960 based on a study by Bruce Jessup, a physician at Stanford University with a strong interest in the relationship between poverty and health. Predictably, the report emphasized the prevalence of bacteriological disease among workers and called for better sanitation and more access to health insurance and medical care. Although it coincided with a report of a special governor's committee that addressed pesticide risks to consumers, there was almost no overlap in the two documents. The committee mentioned the risk chemicals posed to workers only in passing even as it dwelled on the much smaller risk that pesticide residues on fruits and vegetables posed to consumers.[29] Despite their radically different political implications, what the sociological discourse of exclusion shared with the discourse of infection was their belief that disease was a problem rooted in human communities. They both overlooked the role of the environment and thus continued to exonerate a landscape that was becoming ever more infused with chemicals.

But not everyone ignored the environmental dimensions of the problem. Environmental approaches to disease had never been completely erased in the fields of medicine and public health. Rather they had been marginalized and, to some extent, rewritten along more modern lines. In California the small group of experts who were trying to draw attention to pesticide poisoning recognized the the problem was, at least in part, environmental. These were the CDPH's specialists in occupational health. One was Dr. Irma West, who began working on pesticide issues in California in the early 1950s and subsequently emerged as a national expert on farmworker exposures. Another was Dr. Thomas Milby, who came to work for the CDPH in 1961 and soon after found himself investigating the peach-picker poisonings. Although they shared an office with their colleagues in bacteriology and sanitation, occupational health physicians (in California and elsewhere) had a training and orientation that was somewhat different. Occupational health derived its professional identity from its focus on the work environment. Emerging as a progressive response to nineteenth-century industrialization, industrial hygiene (what would later be known as occupational health) studied the diseases peculiar to certain trades. Since its inception, industrial hygiene had highlighted the link between the health of bodies and the condition of their environment. It focused not merely on the body of the worker but also on the work environment: the modern factory. And like sanitation, it owed much to nineteenth-century theories of environmental disease and the assumption that the environment shaped both the physical and the moral qualities of individuals. Not incidentally, occupational health was considered a marginal and low-status subspecialty in the immediate postwar decades.[30]

Moreover, by the 1950s occupational health physicians took a much narrower view of the "environment" than their nineteenth-century predecessors had. Following the institutionalization of germ theory and the rise of laboratory medicine, occupational health researchers had moved out of the factory. In the 1920s toxicity tests on animals became their principal methodology, and their research focus became the "maximum concentration level" (later renamed the "threshold limit value," or TLV)—the highest concentration at which a chemical failed to produce harmful effects in laboratory animals or, alternatively, the lowest concentration at which it generated such effects. By 1945 researchers had established maximum concentration levels for 136 compounds.[31]

The toxicology laboratory afforded certain intellectual possibilities while foreclosing others. Ideas, no matter how abstract, always contend with a larger, and very material, world. We may like to believe that we can think what we want, but even our thoughts are products of the environments we occupy. In a climate-controlled laboratory, surrounded by graduated cylinders, precision balances, and carefully bred rats, it was possible to think about precisely measuring the amount of a given chemical that must be present before it caused consistent biological effects in a group of nearly identical animals. In the laboratory the environment is deliberately "controlled" to the point that it becomes invisible. This was a forthright attempt to overcome the problems of local particularity that made occupational health something less than a true science in the eyes of its leading twentieth-century practitioners. When researchers moved into the laboratory, they abandoned their concern with the broader work environment in favor of a focus on certain chemicals. Or, put another way, in toxicology research the environment was reduced to a set of discrete chemicals. In fact, the environment of the laboratory is carefully constructed so that agency can be ascribed solely to the chemical under study. Other factors are purposefully eliminated.[32] As bacteriology had collapsed the agency of nature into the agency of a specific pathogen, so modern toxicology had collapsed it into the agency of a specific chemical.

In contrast, researchers typically viewed "field" research as messy, inexact, and often unreliable. The same divisions exist today. Depending on your point of view, the beauty of the laboratory is that the environment can be kept at bay. The "field" never conforms to the idealized space of the laboratory. Criticisms of field research typically focus on investigators' inability to fully control or measure the relevant variables. Inconclusive results are cast as a failure of experimental design or implementation. In this model, reliable (read "real") knowledge can only be generated in the most artificial and contrived space. But it is worth questioning the way in which local environmental particularity is glossed as a problem of the field versus the laboratory. The supposed difficulties and problems of field research emerged only after the laboratory became the preferred site of knowledge production in modern science.[33] By way of contrast, nineteenth-century medical topographers embraced environmental complexity. For them, the appeal of California lay in "the sheer variety of its landscapes and climate," a circumstance that was likely to produce immense amounts of new knowledge.

Although trained in toxicology, California's occupational health officials found themselves in fields and orchards, not laboratories. And they quickly noticed that poisonings could not be predicted using the standard techniques of industrial toxicology. Mass poisonings were not necessarily correlated with obvious variables such as the quantity of pesticide applied or the elapsed time between application and harvest. Instead poisonings seemed to occur somewhat randomly. A field that had been sprayed with parathion only a few days earlier was harvested without incident, while another that had been sprayed more than a week or even a month before left several workers vomiting. The peach-picker incident was a case in point. Toxicity depended not only on the chemical in question but also on the particular environment in which it was applied.[34]

Although occupational health experts had long acknowledged the importance of environment to health, they had also assumed that the work environment was fixed, relatively predictable, and ultimately amenable to their control. A questionable assumption even for the most stable factory environments, it bore no relationship to the reality of agriculture. In the fields of California the local environment repeatedly escaped the descriptions of delocalized laboratory science on which occupational health had come to depend. Threshold limit values determined in the laboratory had little relevance to the safety of any given orange grove or vegetable field. As one investigator would later lament, in practice the amount of toxic residue on a given crop depended on "the vicissitudes of environmental factors."[35] The unpredictability of the natural world continually frustrated those who sought to describe and manage the agricultural environment. Place mattered in multiple ways.

Some of the earliest concerns about agricultural chemicals stemmed from the fact that they could not be spatially contained. Unlike factories or laboratory chambers, farmers' fields had no walls. Pesticides intended for a given field routinely migrated downwind. The use of airplanes to apply chemicals made the problem significantly worse. Not only were pesticides applied with even less control, but planes often sprayed the wrong field or dusted adjacent homes, gardens, or roads. Even if applications reached their intended target with little "drift," subsequent winds could whip up contaminated dust and rains could wash chemicals into irrigation ditches and streams, where they moved downslope.[36] However neatly demarcated, the space of the industrialized field could not be bounded or controlled; it expanded and contracted with the movement of air, water, and soil.

Yet even within the space of a given field, environmental variation affected pesticide toxicity. When CDPH officials investigated the outbreak of poisoning among peach pickers in 1963, sampling for parathion residues in the orchard suggested that levels were not high enough to produce acute effects in workers, which led investigators to hypothesize that a degradation product was responsible. Later work would establish the role of the oxygen analogs of organic phosphates. Created under conditions of intense sunlight, these compounds have a toxicity ranging from two to hundreds of times the toxicity of their predecessor chemicals. In the case of the peach pickers, the deadly parathion had been converted into a much deadlier chemical form, paraoxon, which had a toxicity roughly three and a half times greater. Moreover, research suggested that smog could increase the production of oxons.[37] The California climate was capable of producing both superior fruit and extremely toxic substances, but the production of toxicity, like the yield of peaches, varied from year to year and from field to field. Once organophosphates were introduced into the environment, their toxicity was affected by a multitude of factors that experts often could not recognize much less control.

The recurring yet unpredictable illnesses of California's agricultural workers underscored the fact that toxicity could not be adequately understood in the laboratory. Instead researchers had to return to actual environments. Spurred by the growing public and political concern over pesticides in the late 1960s, federal authorities would direct research money to the problem, and public health researchers from the University of California's School of Public Health would conduct a barrage of field experiments over the next two decades. As university researchers pursued the issue of worker exposure and pesticide toxicity, they identified a multitude of significant environmental "variables." Important climatic factors included not only temperature and rainfall but also wind velocity, incident radiation, and humidity. Also relevant was the type of soil: soils with high clay content were likely to bind the pesticides and slow their dissipation; soil moisture, on the other hand, could increase the dispersion rate. Perhaps it is not surprising, then, that studies of pesticide levels in different orchards located in the same area often yielded dramatically different results, even when application rates had been the same. A study of parathion decay conducted in the 1970s revealed that pesticide residues in the same California field could vary by as much as ninety-fold, depending on the time of year the pesticides were applied.[38] Much like the nineteenth-century medical topographers who tried to understand malarial fever, late-twentieth-century occupational health researchers found themselves attending to local environmental details.

Seemingly every environmental factor that researchers thought to consider had some effect on toxicity. The type of crop mattered: citrus, peaches, and grapes were among the most likely to generate toxic exposures because of their abundant foliage and the location of the fruit deep within the branches. But when researchers looked beyond the basic crop type, they found that even the variety could influence toxicity: a Minneola tangelo was not the same as an Orlando tangelo, nor was a Temple orange the same as a Washington navel. As one pair of investigators concluded, pesticides were so sensitive to environmental conditions that the level of residues would also be highly variable even when they had been applied to the same crops in precisely the same way. To understand the onset of illness, investigators found themselves asking numerous questions about the environment. What was the weather? What varieties were involved? Had it rained? Had it been windy or calm? Was it foggy or sunny? Was the field located in a valley? On a hill? On a north slope or a southern one? It was precisely this environmental "sensitivity" that made mass pesticide poisoning events unpredictable. Toxicity was not simply a quality of a given chemical; rather it was a complex relationship between a chemical and the environment in which it was applied.[39]

Two leading researchers, William Popendorf and John Leffingwell, concluded in 1982, after almost three decades of work on the issue, that all organophosphate pesticides were "environmentally sensitive" and that their "decay variability" was "at the root of the sporadic history of pesticide residue poisoning reports."[40] The gridded fields, the engineered irrigation canals, the neat rows of cotton, lettuce, and orange trees that lined the Central Valley all suggested a landscape that was highly ordered and eminently under human control. But control is always an illusion that is more or less sustained by both human and nonhuman elements. Mass outbreaks of illness among workers revealed that the modern farm was actually a chaotic, unpredictable ecology. And this ecology did not stop at the tip of a branch; it extended into the interior of the human body.

The modern body, whether that of a public health official or Mexican farmworker, was supposedly cosmopolitan, able to move safely from one environment to another so long as the rules of hygiene and sanitation were followed. But instead movement now seemed to obscure a deeper relationship between bodies and their previous environments, between sick workers and modern orchards. Industrial hygienists had relied historically on monitoring workers' bodies for signs of illness and limiting harmful exposures—through modifying work processes, limiting the time that workers were exposed to hazardous substances, providing protective equipment, or some combination of these. But unlike factory workers, the bodies in question here were constantly moving—a necessary condition of California's industrialized agriculture. Individual workers found themselves laboring in many environments over the course of a year. Migrants passed from field to field, county to county, state to state, often not knowing where they would be the following day and unable to recall all the places they had already been.[41] What this meant for occupational health specialists was that they could not identify the boundaries of the workplace. In contrast to the assumption of bounded and stable spaces, the space that migrant farmworkers occupied was always discontinuous. Confronted with these realities, occupational health professionals could not even begin to calculate worker exposures. Throughout the 1950s and 1960s, public health officials in California would complain that mobile bodies could not be adequately monitored or studied.[42]

Those bodies that could be located further complicated the issue of toxicity. Industrial toxicologists recognized in theory that the response of individual bodies to particular compounds might vary considerably. But in practice they assumed the inherent similarity of all bodies, that a given exposure would generate a predictable effect (that was the basis of TLVs). In reality, workers' responses to pesticide exposure differed immensely, and attempts to quantify the effects of particular exposures seemed only to yield more variables that required quantification. The absorption of pesticides by a given individual varied with work rate, work style, personal habits, and the type of clothing worn. Most disturbing was the recognition that the more toxic organophosphates had cumulative effects, which rendered exposed workers subsequently more susceptible to systemic poisoning.[43] This was the reverse of nineteenth-century acclimatization. Residence in the valley or another polluted environment rendered individuals only more susceptible to that same environment. The longer one stayed, the sicker one got. Individual bodies were not merely different from one another, they each had a distinct history, and these histories mattered critically to the production of disease.

Moreover, certain pesticide combinations were discovered to have dangerous synergistic effects. Malathion, for instance, one of the OP pesticides least toxic to human beings in isolation, can become highly toxic in the presence of certain other chemicals. And while in a typical factory the universe of chemicals was known, in the case of crop workers no one could say what chemicals workers had been exposed to over the course of a week or a season. By the mid-1960s more than sixteen thousand different chemicals were registered for use with the California Department of Agriculture, which itself advised farmers to undertake multiple applications of multiple chemicals. Farmers themselves constantly adapted their chemical applications in the face of new pest and virus outbreaks, which were always unpredictable. In such a landscape, field workers had no way of knowing their previous exposures; yet a person already exposed to certain chemicals was at much greater risk of becoming acutely ill in the presence of organophosphates. Bodies themselves became sensitized and still more vulnerable to the modernized landscape. Occupational health professionals had no way to predict the outcome, or even to determine what they termed a worker's "exposure history."

The idea of "exposure history" asserts, against the normalizing tendencies of modern toxicology, the relevance of an individual's experience. Toxicity, in other words, is not simply the result of the interaction among a given chemical, its environment, and a standardized body. What also matters is the contingent history of the particular body in question. Had she been exposed to parathion before? How much and for how long? Where else had she worked, and under what conditions? What other chemicals had she come into contact with? The quality of "toxicity" emerged out of the relationship among a particular chemical, the surrounding environment, and a particular body with its own history of exposures and injuries. It defied any easy notion of "agency" or disease causation.[44]

The human body, like the natural environment, was unpredictable and resistant to quantification. Given the fact that toxic effects varied among individuals, occupational health experts came to believe that the best method for uncovering OP poisoning was a blood test for the enzyme cholinesterase, which had been suggested in 1950. Yet as investigators seized upon cholinesterase as a kind of litmus test for exposure in the early 1960s, they came to realize that even "normal" cholinesterase levels varied widely among individuals. Everyone agreed that OP exposure depressed cholinesterase, but there was no agreement on how much depression was significant. In a few cases, doctors even found that blood levels of cholinesterase could be normal, despite the presence of severe symptoms. Moreover, levels of cholinesterase in the blood were only an approximation, and not always a good one, for levels in the brain, which most researchers felt was the real variable of interest.[45]

Biomedical science, like any discipline, is shaped by its past—for better and for worse. Models of occupational health were derived from the turn-of-the-century factory, in particular, from the manufacture of leaded materials. In the factory environment, lead is most readily absorbed by the body through inhalation. Accordingly, the traditional focus of industrial hygiene had been on airborne contaminants; that, in turn, determined the direction of early research on pesticide exposure. Techniques existed for measuring air contamination and human respiration, so these were the variables measured. By the late 1950s investigators had generated a considerable amount of quantitative data on air exposures during harvesting, and the relatively low level of airborne pesticide residues reassured growers, manufacturers, and many in public health that organophosphates did not pose a serious public health problem.[46] But the reality of the fields was more complicated. Poisonings continued.

In fact, skin exposures were the more significant, yet still unquantified, pathway. Skin exposures were not only difficult to quantify, however; they were also more difficult to conceptualize. In the prototypical exposure model that gradually emerged, investigators focused on four critical variables: the amount of pesticide applied, the amount remaining in the orchard when workers began harvesting, the amount of pesticide that entered the worker's body, and the response of the body to different amounts of exposure. In such a model, the connection between body and environment is clearly indicated by the line that connects "environmental residue" to "worker dose." But bodies and environments were nonetheless confined to their own distinct boxes, which are punctuated only by a slender arrow. The iconography is revealing. When investigators spoke of "exposure pathways," they implied that such pathways were narrow routes of entry that could be regulated, tracked, or even blocked.[47] The surface of the body was assumed to be well defined, a boundary between the individual and the outside world that was breached only in specific instances, and exposure itself was assumed to be finite and discrete rather than an ongoing process that involved multiple chemicals on trees, in air, in water, in food.

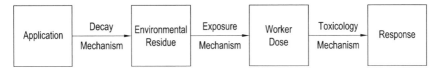

Figure 9. Conceptual model relating pesticide application to acute poisoning among farmworkers. While they took into account environmental factors, occupational health specialists still envisioned bodies and environments as distinct and separable entities. Redrawn from William J. Popendorf, "Exploring Citrus Harvesters' Exposure to Pesticide Contaminated Foliar Dust," *American Industrial Hygiene Association Journal* 41 (September 1980): 652–59, figure 1, 653. Reprinted with permission of the American Industrial Hygiene Association.

But crucial to understanding the prevalence of pesticide poisoning in California were the ways in which workers' bodies became completely intermixed with their environment. Observations made in the field revealed the limits of the language of "environmental residue" and "exposure pathway." Thomas Milby, the official who had investigated the peach-picker poisonings and subsequently became director of the CDPH, described the process of picking fruit in this way: "[Fruits] do not grow at the tips of branches. In order to cut a cluster of grapes, or pick an orange or a peach, it is . . . necessary to penetrate a leaf curtain. In the course of only an hour or two, the worker is drenched with whatever liquids may be clinging to these leaves: he has inhaled quantities of whatever dusts the picking process has rendered airborne; he has gotten these substances up his shirt sleeves, down the front of his shirt, down the back of his neck; he has gotten them in his eyes and ears; he has gotten them in his mouth and throat."[48]

It was the lack of mechanical mediation and the resulting bodily engagement—the mixing of bodies, leaves, sweat, and dust—that rendered fruit picking unlike the modern assembly line and especially dangerous in the presence of organophosphates like parathion. Organophosphates are absorbed rapidly through the skin and are typically even more damaging when a person sweats or wears contaminated clothing. Both the hot environment of the Central Valley and the intricacy of orchard work made it impracticable for workers to wear gloves, masks, or other protective equipment. And even these items offered only the illusion of an impermeable boundary. In practice, bodies could not be held separate from the environment in which they labored for eight, ten, or twelve hours at a time. Moreover, the effort to develop threshold limit values for the new chemicals assumed that exposures were limited to an eight-hour day. These values did not take into account the fact that a worker might also live in a contaminated environment, wear the same contaminated clothing for days, eat contaminated fruit, and wash with contaminated water. Crews of workers were frequently sprayed directly by overflying planes, leaving them drenched in chemicals. When no water was provided in the fields (a not uncommon occurrence), workers drank the contaminated runoff of irrigation canals. Sometimes entire crews were forced to spend the night in recently sprayed orchards. Here, given the social and environmental realities of farm labor, the notion of the body as a discrete entity penetrated only by narrow "pathways" began to break down. Migrant bodies and toxic chemicals both challenged public health experts' control of space.[49]

Thus, while researchers acknowledged that the skin could be penetrated and that exposure pathways were multiple, they nonetheless sought to reestablish and fortify the boundary between the body and its environment.[50] So researchers intensified their focus on the body. They struggled to quantify pathways. Recognizing, as Thomas Milby did, that workers were literally drenched in pesticides, occupational health researchers tried to break the drenched body into component parts and to measure how much parathion was likely to enter through a head, an arm, or a chest. They compared exposures of experienced versus inexperienced workers, of those who rolled up their sleeves and those who did not, of those picking lemons and those picking oranges. They tried to determine what parts of a body were most vulnerable to penetration, to enable the design of appropriate protective clothing or different work routines.[51]

The occupational health emphasis on discrete boundaries marks the modernist desire to see bodies as both cosmopolitan and separate, or at least separable, from their environment. Much as Hibbert Winslow Hill had argued in the 1910s, the body's skin was presumed to be a kind of boundary or frontier, something that separated an individual from the environment, rather than a space of contact, connection, and transfer.[52] That boundary was critical to the continued modernization of the landscape, for it underwrote the assumption that people could be relocated at will and that ongoing environmental changes would not be registered in their bodies. Modern ideas of the body acknowledged the environment's ability to affect health but only in limited ways. Those ideas had made possible certain kinds of medical knowledge, but in the case of pesticides they were overly optimistic. They implied that the impact of chemicals was discernible and ultimately controllable, that the relevant "exposure

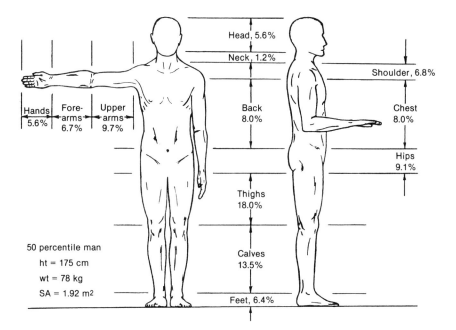

Figure 10. Despite numerous local and individual variables, occupational health researchers tried to develop models that could predict the occurrence of pesticide poisoning. This diagram, illustrating a "human dermal surface area model," shows which parts of a farmworker's body typically receive the heaviest dose of pesticides under field conditions. Reproduced from William J. Popendorf and John T. Leffingwell, "Regulating OP Pesticide Residues for Farmworker Protection," *Residue Reviews* 82 (1982): 125–202, figure 5, 157. Reprinted with permission of Springer Science and Business Media.

pathways" could be identified and blocked. At heart, occupational medicine was always an effort to contain the dangers of industrialization, not to challenge it. The goal of its practitioners lay in engineering compatibility between fragile human bodies and the industrial work environment, and modern models of the body underwrote that project. But while those in occupational health struggled to work within the modern model, other developments were profoundly challenging it.

ENVIRONMENTAL HEALTH AND ECOLOGICAL BODIES

To some, the emergence of a landscape that produced disease because of its modernization might seem to undermine modern concepts of health and disease. Sanitation and modernity were no longer a guarantee of health; in fact, the opposite was true: the most modern, the most technological, seemingly the most controlled spaces could also be the most physically threatening. That at least was the experience of farmworkers in the Central Valley. But these developments, while undeniable, were subject to different interpretations. What emerged in the immediate post-war decades were not only new forms of disease but also multiple discourses about the relationship between environment and health. These discourses shared the recognition that the environment was a critical factor in the production of disease. They differed, however, in how they conceptualized the human body and in how they construed the sources of illness. Was disease produced by discrete agents, or was it the outcome of a body in interaction with a complex environment? Could illness be ascribed to specific chemicals, or was it better understood as the outcome of a multiplicity of factors? Could disease be controlled by shoring up the body and improving individual habits, or would it require fundamental changes in the work environment?

When occupational health professionals recognized the problem of pesticide poisoning, they sought to fit the problem within their existing practices and understandings. Drawing on both toxicology and modern models of the body, they assumed that chemicals were the ultimate source of illness, that the offending substances could be identified and controlled (through sampling orchards, better spray equipment, more thorough training of applicators), that the boundaries of the body could be fortified (through protective clothing, improved equipment, more frequent washing). Bodies and environments might overlap, but they remained essentially distinct. And at first glance, the practices of occupational health seemed as if they might be capable of containing pesticide-related illness, if only they could be implemented. Perhaps it might have worked. But pesticides were not the only new health problem, nor was the Central Valley the only landscape generating new health concerns.

In the early 1940s residents of Los Angeles began to complain about "smog" and the eye and throat irritation that it caused. Because of its undeniable effects on the body, air pollution immediately raised concerns and questions about the broader environment. Ordinary people were quick to infer correlations between air quality and illness, often invoking remnants of an older discourse about health and environment. Those concerns grew more pronounced in late 1948 when an unusually severe and persistent smog killed nineteen people in the industrial town of Donora, Pennsylvania, over the course of several days. Residents of Donora pushed an otherwise resistant Public Health Service to conduct a broad health-effects study, which revealed that nearly half of the local population felt they were physically affected by smog. In fall 1954 several days of choking smog in Los Angeles made air pollution a central issue in California's gubernatorial election. A CDPH survey conducted in 1956 revealed that smog was now a significant factor in people's decisions to relocate in or away from Los Angeles. Local newspaper accounts of frustrated residents selling their homes and moving away seemed to support these conclusions.[53] Much like their nineteenth-century predecessors, at least some mid-twentieth-century Californians were still seeking a healthy landscape in which to live, and Los Angeles no longer fit the bill.

Americans in the 1950s confronted not only the problem of smog but also that of nuclear fallout. When atomic tests conducted by the United States in Nevada and the Marshall Islands produced dangerous and unexpected levels of radiation over large areas, fallout emerged as both a national and an intensely local concern. The revelation that strontium-90, a component of fallout with a twenty-eight-year half-life, was widely distributed in the country's food and milk supply created a pervasive sense of anxiety. By the middle of the decade, the health effects of fallout were a common subject of newspaper commentary, best-selling novels, and popular films as well as an issue in the 1956 presidential election. In California a fallout episode that followed a Soviet test in 1958 generated widespread concern about contamination of the state's food and drinking water and numerous inquiries to the health department.[54]

Pesticide concerns, moreover, were not confined to farmworkers in the Central Valley. In fact, the problem of pesticide residues in food first attracted the attention of politicians in the United States in the 1920s. The issue received sustained attention in the 1950s when Rep. James Delaney of New York held a series of hearings on the subject, focusing on the risks of eating foods that contained small quantities of DDT and other organochlorine chemicals. Those hearings continued for several years, evoking testimony from Food and Drug Administration officials, consumer groups, agricultural representatives, and public health experts. In 1954 Rep. Arthur Miller of Nebraska successfully sponsored a bill that required the U.S. Food and Drug Administration to set tolerance levels for allowable pesticide residue levels to ensure that any chemicals remaining on food products posed no health risks to the consumer. At the conclusion of the hearings four years later, Congress passed the Delaney amendment to the Food, Drug, and Cosmetic Act, which established "zero tolerance" for suspected chemical carcinogens in the food supply. Then, in 1959, the secretary of the Department of Health, Education and Welfare revealed that much of the nation's cranberry crop contained residues of aminotri-azole, a chemical that was known to cause thyroid cancer in laboratory rats. The incident drew national attention to the dangers of pesticide residues in foods, and sales of cranberries plummeted.[55]

All these developments made the 1950s a critical and confusing decade for the public health profession. Faced with new problems for which they had no answers and no protocols, health experts began to rethink their approach to health and disease. The formulations of early-twentieth-century authorities such as Hill now seemed quaintly naive. In 1948 the newly formed World Health Organization pointedly critiqued the narrow definition of health as the simple absence of disease, declar-ing that health should be understood more broadly as "a state of com-plete physical, mental, and social well-being." California public health officials would frequently reiterate this more expansive definition in the coming decades. Moreover, those same officials recognized that the prob-lems posed by pesticides, air pollution, and fallout indicated that the causes of illness were neither discrete nor easily traceable, nor were they ultimately located in the bodies of others or confined to the space of the factory. Instead disease might once again be a quality of certain land-scapes. In a statement that Thomas Logan would have been comfortable with, a leading California official wrote that "the objectives of public health have to do with man's successful adjustment to, or his judicious manipulation of, the particular environment of which he is a part." In 1960 the U.S. surgeon general testified on these emerging health concerns before a congressional committee, which in turn appointed the Committee on Environmental Health Problems. That committee, chaired by Paul Gross, a chemistry professor at Duke University, issued its report in 1962. The Gross Report strongly criticized the existing "categorical separation" of the Public Health Service's environmentally oriented pro-grams (water pollution, air sanitation, occupational health, etc.) and stressed the need to consider environmental threats to health, whether chemical or biological, in an integrative way.[56]

Most professionals portrayed these developments as radically new, as the outcome of the country's postwar technological development and what they saw as a related shift from infectious to chronic disease. Some, however, acknowledged continuities with the past. "Now we've come full circle," declared Malcolm Merrill, head of the California Depart-ment of Pubic Health, in 1962. "The health status of an individual, a community, or a nation is a continuous interplay between personal health and a healthful environment." Sanitary engineers, in particular, had never abandoned their environmental focus; however, Frank Stead, chief of the CDPH's Division of Sanitary Engineering in the immediate postwar decades, retrospectively critiqued the focus of early-twentieth-century sanitation work. He pointed out that modern public health had focused too narrowly on the presence or absence of specific contaminants, and only at or near the point of human contact. Stead insisted that professionals needed to broaden their focus and consider the "environmental system as a whole." Writing specifically of the problems posed by pesticides, he countered the assumption that the cause of illness lay solely in individual chemicals, maintaining that California's entire agricultural landscape was now a "chemical reaction platform."[57]

Even as health officials like Stead insisted on the need to consider the "environment" in a broad and integrated fashion, however, even as they delineated the multiple ways in which the environmental changes might ramify in the human body, they remained under the influence of germ theory and their own professional history. While they argued that health needed to be understood more broadly, they returned to familiar conceptualizations. In practice, they continued to focus on pathways of infection and to look for specific etiologic agents. When farmworkers collapsed in the fields, what they wanted to know was, which chemicals caused illness, and how did they enter bodies? This was an old story, except that the traditional part of the bacteria was now occupied by chemicals such as DDT and parathion. What was new was the dawning recognition that human beings, even experts, were not fully in control of the actions that they set in motion. As Irma West, an occupational health specialist with the CDPH, observed, the problem with pesticides lay in "their ability to do much more than is expected or desired of them."[58]

Of course, the controversies over air pollution, radioactive fallout, and pesticides were not confined to experts. As much as public health experts were responding to events, they were also responding to concerns voiced by ordinary individuals. California officials found themselves besieged with complaints in the 1950s, first regarding smog and then fallout and pesticides. As the decade wore on, public complaints commanded the attention of political leaders, who in turn looked to experts to provide both information and solutions.

In the 1950s ordinary people did not draw on the technical language of toxicology to make their complaints. Instead, like their nineteenth-century predecessors, they drew principally on experience. Many suspected that chemicals and pollution might be responsible for their health problems, and some sought out medical practitioners who were sympathetic to their environmentalist hypotheses. At the level of the everyday, people still believed that health was a subject available to common sense as well as medical diagnosis. In 1958 one resident complained to the CDPH after a fallout episode that he coughed up blood after being outside during the worst period of exposure. "I believe that had I known the radiation was of that intensity I would have stayed indoors as some protection," he wrote, and he went on to assert the relevance of his self-diagnosis given the inadequacy of modern medicine: "Perhaps all the above is coincidental, but since we know little concerning radiation on individuals my guess is as good as any." Yet, for the most part, public health professionals in the 1950s derided lay evidence and dismissed public concerns over pesticides.[59] That, however, would become increasingly difficult to do.

In early 1962, as the Gross committee was publishing its report on environmental health for the U.S. Public Health Service, Rachel Carson published a series of articles in the *New Yorker* that focused on the health and environmental effects of pesticides. She would publish *Silent Spring* that fall to both instant acclaim and vitriolic attack. Many scholars date the emergence of modern environmentalism, and the public's enthusiasm for "ecology," to that moment. While popularly remembered as a book about the effect of pesticides on wildlife, *Silent Spring*'s central concern is human health. Or, perhaps more accurately, it is the unacknowledged connections between health and the environment. Certainly Carson described the effects of pesticides on birds, animals, and fish to great effect. Yet rhetorically these early chapters merely laid the groundwork for the core of her book, which described how these chemicals might affect human beings. One chapter is devoted to the effects of pesticides on cells and human genetics, while the book's longest chapter—ominously titled "One in Four"—analyzes pesticides as potential causes of cancer. As Carson explained to her editor, her principal intention was to set new threats to human health "within the general framework of disturbances of the basic ecology of all things."[60]

Silent Spring cast issues of health and environment in new terms for many people. Or perhaps these were not new understandings but existing understandings to which Carson gave new sanction and powerful articulation.[61] That was what made the book so subversive. At the same moment that a professional discourse of "environmental health" was emerging in places like that occupied by the Gross committee, Carson put forth a popular discourse of ecological health that would challenge the models promulgated by contemporary medical experts.

Carson herself had no training in medicine. Her academic background was in biology and zoology, first as a graduate student at Johns Hopkins University and later as an employee of the U.S. Fish and Wildlife Service. Both of her previous books had been about the sea. Ultimately, Carson's contribution was not that she linked human health with the environment—after all, medical entomologists, specialists in tropical medicine, experts in occupational health, and even farmworkers themselves had continued to do just that. But Carson successfully articulated new ways to specify that linkage. She put forth a popular discourse of ecology that linked the quality of soil, water, and air to animal and human physiology in ways that many laypeople and some experts found compelling. As a writer, she willingly crossed the professional divide that had come to separate the study of human bodies from the study of the nonhuman environment. What she described in nontechnical terms for her readers was the way in which contaminants cycled through environments and then entered both animal and human bodies, through air, water, soil, and complicated food chains—how DDT applied to a lake in northern California (to control gnats) was taken up by plankton, passed from plankton to plant-eating fish, from plant-eating fish to carnivorous fish, from carnivorous fish to grebes and gulls. She was convinced that once released into the environment, pesticides would find their way not only into animal but also human tissues, through one means or another. For Carson the marine biologist, human beings did not exist in a blank space that was traversed by chemical and bacteriological contaminants; rather, they existed in particular ecologies all of whose parts were interlinked.[62] A disruption in the landscape would lead, eventually, to a disruption in the health of those who occupied the landscape.

Carson's point of view emerged from her training in zoology, her knowledge of ecology, and her growing understanding of the causes of human cancer. In all these ways, *Silent Spring* was a product of mid-twentieth-century science. However, Carson's way of framing the issue of human health had more than a little in common with that of California's nineteenth-century doctors. Of course, she was not writing of miasma, nor did she doubt the role of mosquitoes in the production of malaria. However, her view of health owed much to her strong sense of corporeal connection to the landscape, and her ecological perspective made her far more willing than most contemporary physicians to postulate the intermixing of bodies and environments. Carson was particularly disturbed by the recent recognition that chemicals could cross through the placenta—long thought to be an inviolable barrier between the fetus and the outside world.[63] In *Silent Spring* she portrayed human bodies as infinitely permeable and therefore as products of the landscapes they inhabited. In her view, diseases could not be understood solely as a quality confined to an individual body; rather it was the outcome of a body in interaction with a series of environments. And she was decidedly pessimistic about the ability of experts to protect and reelaborate the assumed boundaries between human beings and the larger world. The presence of disease signaled the need for people to pay much closer attention to their environment.

There is a well-established tradition in scholarship of reading popular concerns about both disease and environment as incarnations of a long-standing diatribe against modernity. "Disease incidence," as the historian Charles Rosenberg has observed, often becomes "an argument for social reform, an indictment of a pathogenic society."[64] From this perspective, the discourse of ecology and health that emerged in the wake of *Silent Spring* can be read as a reformulation of the nineteenth-century attack on cities as spaces that bred both immorality and ill health. In these accounts, whether the object of reform is the turn-of-the-century industrial city or late-twentieth-century agriculture, concerns about health expressed archetypal anxieties about modernity and modern life. Stated this way, the problem is essentially a cultural one: it is an argument about how society should, or should not, be ordered.

To read pesticides as a cultural issue is an important insight, but to stop there ignores a material context that also requires explanation: dying birds and fish, trembling and paralyzed farmworkers, rising cancer rates in the industrialized countries. Regardless of the position one took on modernity, the mid-twentieth century witnessed material developments that called for both interpretation and response. And even those who tried to contain these problems within resolutely modern frameworks often found those same frameworks frustratingly inadequate.

Moreover, the argument about modernity embedded in discussions of ecological health was far more fundamental than its critics have typically recognized. To the extent that modernity is predicated on a human-nature divide, the position of Carson and many of her followers was less a moral indictment than a direct challenge. Purveyors of modernity have assumed and relied on clear boundaries between culture and nature, bodies and environments, humans and animals. But the movement of chemicals from the laboratory to the soil, from rivers to bloodstreams, from fish into breast milk all suggested that those boundaries were more fictional than real. By following the complex path of chemicals, Rachel Carson had undercut a fundamental modern assumption.[65]

Carson's book also challenged contemporary assumptions about scientific knowledge and evidence. Carson had made herself intimately familiar with the literature on modern toxicology. But while she used toxicological evidence, she more often critiqued its limitations. Her arguments emerged primarily from ecological observations and medical case histories—the field and the clinic rather than the laboratory. She drew on the work of ecologists and field biologists, whose work was grounded in the careful observation of particular landscapes, and she marshaled medical histories as important forms of evidence. Much like Thomas Logan, she cultivated a network of local observers who wrote to her of their own experiences.[66] Her contribution was to amalgamate dozens of local stories and to insist that these local observations constituted an important source of knowledge that should not be dismissed simply because they emerged outside the laboratory and were therefore labeled "anecdotal." She emphasized that the evidence of experience—however incomplete, however constructed—was critical to gauging the effects of these new substances, especially since laboratory tests could not assess long-term and cumulative effects. Moreover, controlled laboratory spaces and inbred rats could never account for the complexity of actual bodies and actual environments.

Carson's insistence on the relevance of local experience stands in sharp contrast to the erasure of environmental particularity in so much of the concurrent medical and public health literature. *Silent Spring* spread the activity of knowing health widely, recruiting the experiences of wildlife experts and housewives, the materials of the toxicology laboratory and the clinical physician, the natural histories of grebes in a northern California lake, and the experience of farmworkers in the Central Valley.[67] All of these entities produced potentially useful knowledge about chemicals, environments, and health.

The response of medical professionals to Carson's book was mixed. On the one hand, she had relied on the work and ideas of several physicians who had investigated the health effects of pesticides, especially their link to chronic disease. These included Morton Biskind, an M.D. in Connecticut who connected many of his own illnesses to dry-cleaning chemicals and who had since sought to document chemically related illness in his patients; and Malcolm Hargraves, a physician at the Mayo Clinic interested in the potential link between petroleum hydrocarbons and blood disorders such as leukemia. Carson's most important source, however, was William Hueper, a controversial researcher at the National Cancer Institute who had devoted his career to studying the relationship between industrial chemicals and chronic disease in factory workers. What these physicians shared was not only their interest in the health effects of chemicals but also an emphasis on clinical medicine and case histories—a kind of local knowledge of its own. Their work was routinely marginalized by their laboratory-oriented peers for just that reason.[68]

The mainstream of the medical profession took a more cautious stance. Most physicians who responded publicly to *Silent Spring* hewed closely to the modernist view. They acknowledged the evidence of laboratory toxicology while dismissing personal and clinical accounts of illness and disability as unscientific, or at least unreliable. Moreover, they insisted on a narrow definition of causality, locating the source of disease in isolated chemicals. Consequently, an organization such as the American Medical Association (which Carson considered a direct antagonist) could maintain that the harmful effects of chemicals were unproven and that popular fears about the chemical permeation of human bodies were vastly overdrawn.[69]

Despite this medical conservatism, the history of modern environmentalism indicates that Carson's ecological language resonated strongly in 1960s America. The book immediately found fertile ground and a wide audience. Moreover, on its publication, dozens of individuals wrote to Carson, often recounting their personal stories with chemicals, as if reading *Silent Spring* had given new meaning to long-standing illnesses and persistent suspicions. Many more wrote to local newspapers and magazines, or conversed in garden clubs and informal community networks.[70] In this way the evidence of local people, though it lacked the imprimatur of the laboratory or the channels of professional science, began to travel. Connections were made. Local experience in chemically laden environments became something more than local, even if not quite scientific. And the interpretation of certain kinds of disease suddenly became open to debate. Carson had drawn on local places and local experiences to formulate a broad discourse about ecology and human health; that discourse, in turn, flowed back to affect understandings of local environments. That would be true in many places, including the Central Valley where the problem of poisoned farmworkers—understood for more than a decade as a problem of occupational health—began to take shape as a much different kind of problem. It would subsequently emerge as an environmental problem.

REMAKING DISEASE

When the individual experience of illness had been deemed irrelevant and laboratory evidence was either equivocal or unavailable, pesticide poisoning could be said not to exist. Disease is at once material and cultural, and its emergence depends on the existence of certain material conditions as well as a relevant language and set of practices. Over the course of the late 1940s and the 1950s, the tools of occupational health—clinical observation and cholinesterase testing—had made pesticide poisoning into a medically recognized disease, but it was known only as a somewhat narrow problem of occupational health, a problem that concerned only farmworkers and a few professionals. An occasional paper attested to the issue in obscure journals of industrial hygiene and entomology. But no newspaper articles appeared, no congressional hearings were convened, no advisory committees were formed, no legislation was passed.[71]

This changed in the late 1960s. At that point, the problem of farmworker exposures emerged into public consciousness and political debates, and pesticide poisoning became an unequivocally real disease. It began to garner public research funds and prompt some efforts at regulation. And while California occupational health specialists would be there to cite the data they had gathered over two decades, the reasons for this new visibility for the most part lay outside their own profession. It lay instead with California's farmworkers and the new environmentalism inspired by Carson.

In 1962 the National Farm Workers Association (later the United Farm Workers Organizing Committee, or UFW) began organizing farm laborers in the southern part of the Central Valley, and over the next several years the union would repeatedly call attention to the appalling working conditions that characterized so much of California's agriculture. Yet, in terms of health, the focus of union leaders initially followed that of the state health department; they stressed the need for sanitation and access to medical care. While seeking ways to improve the health of farmworkers, the union understood their diseases as the outcome of exclusion and marginalization. Until 1967 the UFW appeared either unaware or uninterested in pesticides as an issue. In that year the union opened its own clinic in Delano to provide medical care to striking workers and soon discovered that many workers had symptoms of pesticide poisoning. The following summer, an incident of mass poisoning occurred in an area where the UFW was organizing: nineteen orange harvesters in Lindsay fell ill in a field that had been heavily sprayed with parathion more than a month before. In the ensuing months, leaders of the union effort educated themselves about pesticides and their effects and even initiated legal action against the Kern County agricultural commissioner in an effort to make local authorities reveal the amounts and types of pesticides that were being applied. By the beginning of 1969, the UFW had made worker health and safety its primary issue, granting it precedence over long-standing demands for collective bargaining.[72]

In 1969, as the issue began to appear repeatedly in the popular media, pesticide poisoning symbolized more than the oppressive conditions suffered by farmworkers. It was also one among several indications that human health was under threat from the modern environment. The evidence of farmworker illnesses had come full circle. Rachel Carson had used that evidence as one story among many to make her case about ecology and health. Drawing on the reports of California's occupational health specialists, she argued that farmworker illnesses were only the most obvious evidence that human health was at risk in the modern landscape. But once the discourse of ecological bodies had taken hold, farmworker illnesses could be read as a synecdoche for the problem of environmental decline. Acute disease could now symbolize chronic environmental problems. As one farmworker hypothesized, "We have now more cancer deaths. . . . Is it really caused by smoking—or does it have something to do with the breathing of all kinds of sprays of different chemicals?"[73]

The rise of ecological thinking not only recast pesticide poisoning as an environmental issue; it also challenged the rules of evidence. Modern biomedical science largely rejected experiential accounts of illness because they could not be verified by instruments in the laboratory. When a typhoid outbreak occurred, the state sought bacteriological facts that could attest to the presence of disease and dispatched a health official to collect stool samples. In contrast, reports of pesticide poisoning typically relied on clinical accounts and personal testimony. Consequently, the existence of pesticide poisoning could always be doubted, especially when physicians relied on workers' subjective retelling of their illnesses. Health officials warned of "psychosomatic symptoms" among field crews and had consistently cautioned that the prevalence of illness might be overstated if investigators relied on worker interviews rather than laboratory measurements of cholinesterase.[74] But in the wake left by *Silent Spring*, local experience in the landscape again became an important way of registering environmental pathology.

Like the people who wrote to Carson, farmworkers began to narrate publicly their own experiences with pesticides, and many outsiders were now willing to take those experiences seriously. Researchers acknowledged that poisoning was vastly underreported, and they tried to establish a more accurate estimate of its prevalence by conducting interviews with hundreds of workers, often hiring farmworkers themselves to do the questioning. One study found that the actual rate of serious illness exceeded the reported rate by a factor of three hundred. By 1970 even the California Department of Public Health was taking a new interest in the experiential testimony of workers. In a federally funded report on pesticides, the CDPH presented twenty pages of workers' accounts alongside laboratory results and the clinical reports of physicians—although the authors still felt compelled to add a disclaimer for the presence of unorthodox and "anecdotal" material. In 1969 as the U.S. Congress considered legislation to establish the Occupational Health and Safety Administration, the UFW brought several farmworkers to speak about their own experiences in the fields—chronic headaches, persistent skin conditions, recurring seizures, nosebleeds, nausea, vision problems, breathing difficulties, and the problem of "rubber legs."[75]

In 1969 workers' own words indicated that they continued to read the presence of pesticides through the reactions of their bodies, even though doctors and growers had so often dismissed their concerns. As one white farmworker put it, " After you've been in this game long enough, you learn a few things. You see the labels on the sacks; one thing and another. *Just from experience, we know what the sprays do.* If you stay home for a couple of weeks, because it's raining or something, and you start to feel better—well, you don't have to be a genius to figure out what's going on." And while bodily reactions lacked the precision of laboratory instruments, bodies could also gauge change over time. Workers claimed that their experience indicated that pesticide use was heavier in California than in other locations where they had worked, or that it had grown more intense over time. According to a fifty-eight-year-old man, "Close to twenty years ago . . . the pesticides they was using wasn't so poisonous to humans as the ones they're using now." A Mexican American women told the same interviewer, "This year, we had much irritation of the legs, back and face while thinning sugar beets. This did not happen in other years. They must be using something new and bad."[76]

Farmworkers and environmental activists did not reject the laboratory knowledge of modern medicine, but they put that knowledge into a framework that coalesced with, rather than rejected, their experience in the fields. Union leaders were quick to mobilize the data of biomedicine to verify farmworkers' accounts—pointing out, for instance, that blood tests done on children in Tulare County indicated that nearly half of those tested had abnormally low cholinesterase levels as well as abnormally high levels of organochlorine pesticides, such as DDE and DDT.[77] Cholinesterase tests were clearly a product of the laboratory, and they emerged out of a modern notion of the body as a discrete and bounded entity. But for farmworkers and their advocates, cholinesterase tests corroborated ecological understandings, both old and new, that emphasized the inherent porosity of the human frame and its ongoing dependence on the broader environment. Knowledge, once produced, can be put to work in many ways. For a public that was growing increasingly sensitive to environmental concerns and familiar with a popular discourse of ecology and health, the scientific evidence of cholinesterase measurements corroborated the experience of workers and helped to materialize a link between the local environment and farmworker health. Pesticide poisoning could no longer be read as another affliction of disease-prone laborers. Now it symbolized the ways in which the modern environment could permeate the body.

But the UFW's most potent strategy was its insistence on the permeability of all bodies and its invocation of an ecological model of health. Union supporters repeatedly pointed out that the polluted agricultural landscape posed risks not only to farmworkers but also to everyone who passed through it or consumed its products. In other words, porosity was not a quality circumscribed by class or race. In testimony before the U.S. Senate, one farmworker advocate asked, "What [do] a young mother of two children and the wife of a farmer in Lubbock, Texas, . . . a mother of four children in suburban Westchester County, New York, and lastly a farm worker in Coachella, California, have in common with each other?" The answer was that all three had been seriously disabled by pesticides. The UFW newspaper *El Malcriado* warned that unsuspecting drivers on the roads of the Central Valley were at risk from the chemicals that wafted through the air. And in its promulgation of an ongoing boycott against California's nonunion grape growers, the UFW explicitly linked farmworker illnesses with consumer concerns about pesticide residues as shared concerns about environmental health, invoking Rachel Carson as a point of reference. Union materials argued that consumers could not protect themselves without demanding the control of pesticides in agriculture, that the same chemicals posed risks to both those who labored in the fields and those who ate California's produce—though this claim skirted the fact that the substances of most concern to consumers (DDT and the organochlorine chemicals) were not typically those of greatest concern to farmworkers (OPs).[78] Popular notions of ecology and environmental health constructed the bodies of both middle-class consumers and farm laborers as enmeshed in the modernized environment, as open, porous, and increasingly at risk.

In this way, the UFW made farmworker illnesses something more than self-referential. Its pesticide campaign reworked long-standing discourses about race and disease in ways that foreshadowed the environmental justice movement of the 1980s. For the union, bodily impurity still signaled disease, but that impurity was not necessarily the result of contagion and poor hygiene; it could also be the result of a polluted environment that could not be held at bay. Nor did discussions of sanitation disappear, but their context changed. Now farmworkers argued that sanitation improvements were needed to protect not only against the spread of infectious disease that resided in specific bodies but also against a contaminated landscape. Disease was not an indictment of the allegedly infected and vulnerable bodies of nonwhite laborers; instead it was made to signify the pathology of the modern landscape.

And people other than farmworkers did begin to fear for their health. At least a few residents who were not farmworkers reportedly moved away from the valley to escape the pesticides. Some spoke directly of their personal anxiety. Lee Mizrahi, a doctor at a valley clinic who had conducted the blood tests on farmworkers' children, admitted that he had initially been skeptical about the seriousness of the pesticide issue. However, by the time he testified before the Senate in 1969, his view had changed, and his concerns extended beyond his farmworker patients: "A lot of us who are not agricultural workers are also concerned about the presence of so much pesticides in and around our homes, our children, our food. I am concerned about what appears to me to be a steady invisible trend toward the ultimate total pollution of our rural environment. I like where I live and I want to stay here, well and practicing medicine."[79]

On the other hand, some local residents were not convinced. And some were openly dismissive. Understandings of the body were still shaped if not determined by class, gender, and race. Growers and their representatives repeatedly told the press that they were not worried about pesticides, even though they and their children also lived in the valley. Even some farmworkers dismissed the danger. "I been sprayed off and on ever since 1957," a white farmworker told an interviewer in

1969. "I been sprayed head to toe with parathion. It don't make you sick—if you handle it right. But you got to handle it right. Main thing is not to breathe the dust when you dump a sack into the mixing tank. Take a deep breath and jump back fast. . . . 'Course I knew one guy got sicker 'n a dog. He'd been working with it all morning, must have had it on his hands, and when he went to eat his sandwiches at noon—hoowee! But it didn't kill him." Men, both growers and laborers, often adopted this rhetoric of masculinity and downplayed their physical vulnerability. Moreover, there was a tendency in the advocacy literature to point to white women and children as the innocent victims of errant pesticide sprays. Even the UFW consistently used assumptions about the leakiness and porosity of the female body to make its own political point. Much of the union's consumer campaign focused on the fact that pesticides were known to pass into breast milk, and breast-feeding mothers boycotting the pesticide-laden grapes at Safeway became a powerful image for the union cause.[80]

Yet those gendered differences, while still in place, were becoming less predictable. Some white male workers publicly voiced their concerns about the bodily effects of pesticides. As one put it, "Money ain't everything. You got to think of your health. . . . There's no protection from nothing at all on these spray jobs. . . . One guy I worked with many years, spraying on all different farms around this area. He died before his time, and I just know it had to be from the sprays." Concerns about environmental health could cross lines of class and race as well as urban and rural boundaries—albeit somewhat unevenly—generating a shared sense of physical vulnerability and calls for alternatives to chemical-intensive agriculture. Practically, this meant that both consumer and environmental groups endorsed the UFW boycott.[81] Farmworkers and environmentalists now wielded disease to demand the reorganization of the local environment and the larger society.

Only limited change was in the offing, however. After years of discussion, some regulatory action was taken in the early 1970s. At the urging of public health researchers and farmworker advocates, California adopted the first protective laws in 1971. These took the form of "worker reentry intervals," mandatory waiting periods between the application of pesticides and the harvesting of crops. Having sprayed a crop or orchard, the owner was required to post a notice of the type and date of spray and to allow a specified time to elapse before sending workers into the field. Waiting periods were tailored to specific crops and to specific chemicals; however, rules were adopted only for the most lethal pesticide-crop combinations. With the longest interval set initially at thirty days, everyone recognized that this regulation would not protect against all instances of poisoning. At the federal level, regulations that were markedly weaker (the longest interval was set at only two days) would only be adopted in the mid-1970s.[82] The inadequacy of these laws reflected the opposition and undeniable power of the agriculture industry, which had lobbied hard against any new restrictions. But physicians, scientists, and government officials all justified these standards by referencing a distinctly modern concept of disease. Biomedical frameworks that had emerged in a different context were applied, however awkwardly, to the new problems. Worker reentry intervals still focused on the agency of individual chemicals, ignored the larger environment, and assumed that exposure occurred through a limited number of pathways that could be measured with some accuracy. Even as ecological ideas were sweeping the country, modern concepts of the body remained in place and could be effectively mobilized to limit efforts at environmental reform.

But worker reentry intervals challenged as well as underwrote the status quo. Regardless of how weakly they were formulated, the new regulations recognized, however implicitly, the abnormality of the modern agricultural environment and the porosity of the human body. The orchard, long a symbol of health in California's promotional literature, had become a landscape that could produce disease and sometimes even death. It was also a landscape that escaped modern techniques of sanitation and environmental management and which had to be avoided, at least for a time. As nineteenth-century doctors had advised settlers to avoid the valley's miasmatic lowlands in the autumn months, now public health officials restricted farmworkers from certain fields for days or weeks at a time. That the creation of such a pathological space was now openly acknowledged, and even sanctioned by law, suggested that a substantial change in thinking had taken place. Now it was modern environments, not just (unmodern) people, that might require quarantine.

By the end of the 1960s it was clear that the century-long struggle to normalize the Central Valley landscape had either ended or entered a decidedly new phase. In the 1920s public health experts had been certain that modernity was sanitary modernity, and they had set about to manage the landscape accordingly. Before that time, the region's natural characteristics—high temperatures, swampy ground, thick fogs—had appeared to threaten health. Four decades later the processes of capitalist modernization yielded a new kind of abnormality. A landscape once understood to be naturally dangerous, then benign, was now understood to be unnaturally dangerous, at least in places, and, once again, in need of reform. Modernity, it seemed, was transforming both environments and bodies in unpredictable and undesired ways.[83] Instead of separating bodies from their environments, instead of representing the triumph of human agency over nature, technological and scientific "progress" was revealing new kinds of interdependence. Farmworkers, consumers, and public health officials all acknowledged that human health was linked to environmental conditions, that bodies were porous, often hopelessly so—even if they envisioned those connections in somewhat different ways. This recognition was the result of many factors: concern about fallout and air pollution, the publication of *Silent Spring*, and, not least, the experience of farmworkers and occupational health officials in the chemically laden landscape of the Central Valley. Bodies themselves were once again important barometers of place, while modern environmental movements were the outcome of particular kinds of experience in actual postwar landscapes.

Ideas about the environment were intimately tied to understandings of health and disease, and both were changing in the postwar decades. The corollary of a porous and vulnerable body was a complex and active environment, an environment that had to be taken into account. Vomiting, muscle spasms, and heart palpitations were only the most obvious signs that human beings were objects of environmental change and not simply its agents. As Irma West of the CDPH had opined, chemicals, though placed intentionally in the environment, consistently did more than was expected or desired of them—much more. Moreover, studies of pesticide poisoning revealed that the cause of disease could not be neatly localized in a given individual or a given chemical; rather, it was distributed in complex ways among different chemicals, plants, soils, air, clothing, and individual bodies. Landscapes, however modern, were not all the same, nor were they merely passive backdrops to human action. Even as public health investigators drew on modern frameworks of disease, their work in California agriculture was undermining those same frameworks.

Like modern environmental movements, the ideas and intentions of California's public health officials emerged out of complex and contingent experiences that occurred in specific places. Knowledge itself has a geography, often a complicated one. The toxicology laboratory was one site for the production of ideas about "environmental health"; the

Central Valley was another. Within the valley, a set of factors—some local, some not—came together to produce not only record harvests but also disease: a migrant labor system, a dry climate, low profit margins, a set of specific crops, a variety of introduced pests, the easy availability of highly toxic chemicals.

However, the investments already made in the processes of agricultural modernization were far too great to allow for a radical rethinking of the underlying assumptions, at least on the part of growers and government. For those invested in the modernizing project, including those in occupational and public health, the task now became how to contain and explain an admittedly abnormal and unhealthful landscape, an approach that was implicit in the reliance on "reentry intervals" as a regulatory strategy. Although the migration of pesticides from the branches of trees into the bodies of workers demonstrated that the boundary between human beings and their environment was often more illusory than real, those in public health were still intent on drawing boundaries. For the purposes of regulation, toxicity was still presumed to be confined to a singular piece of fruit or a particular field—a space within a space that could be walled off, in some sense, from the rest of the landscape. Though no walls separated one field from another, the boundaries of contaminated fields were to be marked by signs that gave the name of the pesticides that had been sprayed, the date of their spraying, and the time at which the field would again be safe to enter. Toxicity, in this somewhat optimistic view, was contained, temporary, and subject to quantification. Like infected bodies, toxic spaces could be quarantined until they posed no threat. Beyond those boundaries, disease was not associated with the environment. Most of the landscape remained benign; public health experts could still claim to be in control; and modernized agriculture could continue with the guidance of modern public health. At least that was the hope.

5

Contesting the Space of Disease

[We are] trying to answer the question, yes or no, is this a
safe place to live?

> Dr. Richard Whitfield of the Kern County
> Health Department, testifying before the
> California Senate Committee on Toxics and
> Public Safety, 1985

Communities expect epidemiologists to be able to reach some
conclusion about exposure and disease, but the constraints
of available information can make it difficult to distinguish
between competing hypotheses of major economic and public
health consequence.

> Peggy Reynolds et al., "The Four County Study
> of Childhood Cancer: Clusters in Context," 1996

The science of epidemiology needs a real kick in the rear end.

> Connie Rosales, mother of cancer sufferer, 1988

By the early 1970s the ecological challenge of Rachel Carson and the
United Farm Workers had been met by a reassertion of existing public health strategies. Even as the concept of environmental quarantine was being promulgated, however, the problem of a toxic landscape
was exceeding its solution. Chemicals were not stationary, and property
lines and the posting of metal signs did not stop pesticides from moving
into soil, water, air, and human bodies. In 1979 the discovery of a highly
toxic pesticide in the Central Valley's groundwater raised new fears
about the regional landscape. Further monitoring would soon reveal that
the valley environment was far more contaminated than previously suspected and that chemicals could move farther and persist longer than

anyone had predicted. What was once thought to be relatively contained was found to be diffuse. Over the course of the 1980s, it became obvious that pesticides were not confined to farmers' fields. Consequently, groundwater contamination, and then toxic air pollution, would become a focus of the state's health and environmental agencies and an ongoing concern of valley residents.

During this period, residents of McFarland, a small town in the southern half of the valley, realized that a startling number of children in their community had been diagnosed with various forms of cancer. When they began to suspect that their contaminated environment was the source of their children's illnesses, McFarland, an otherwise unremarkable place, became one critical site for rethinking how health might or might not be related to the environment in the late twentieth century. Many people, both within and outside the valley, immediately connected the health problems in McFarland—"the West's best-known cancer cluster"—to the region's history of chemical pollution. For them, McFarland's plight symbolized the devastating and unavoidable effects of the modern agricultural environment on human health. At the same time, it pointed to the unfulfilled and illusory promises of sanitary modernity.[1]

McFarland was but one of many localities in which chemicals and illness coincided and where modern understandings of health and disease would be publicly challenged. The 1980s were the toxic decade, in the Central Valley and beyond. Ushered in by the disasters of Love Canal and Three Mile Island, the decade saw hundreds of American communities struggle with environmental contamination and its implications for public health. Stories about hazardous waste and groundwater pollution—and their possible health effects—were the staple of daily newspapers, the subject of television specials, the focus of congressional hearings, and at issue in thousands of lawsuits nationwide. Contemporary films and novels articulated the pervasive toxic fears and their often ambiguous sources. Terms such as *PCBs* and *dioxin* entered the national lexicon, and fears about the environmental causes of birth defects and cancer drove many self-described homemakers to become environmental activists.[2]

In California, officials found themselves caught up in a series of environmental and public health controversies that they could not contain. Since the early twentieth century, public health experts had claimed authority based on their ability to isolate the causes of disease and to block their entry into human bodies. But in the case of McFarland, experts could neither explain nor control the migration of chemicals into air, water, and food. After years of investigation, California's health officials would insist that the McFarland environment was "safe," yet their findings would fail to allay the concerns of most local residents. Instead residents now openly questioned the methods and assumptions of public health and refused to accept expert assurances that the local environment was irrelevant to disease. Whatever view one might ultimately take of its cause, the McFarland cancer cluster and the ensuing controversy revealed that modernist ideas of place and health in the Central Valley were no longer widely shared, and perhaps never had been.

UNCOVERING POLLUTION IN THE POSTWAR DECADES

Unlike the changes wrought by large dams, the environmental effects of chemicals are often subtle. At least initially, many of the changes chemicals induce occur at a level far below that of human experience. The poisoning of farmworkers had made chemical effects visible. But the resulting environmental focus remained relatively narrow: the concern lay with exposure to either the sprays themselves or sprayed crops. Regulators did little to assess the presence of chemicals in the broader environment; they assumed that the ongoing use of chemicals had no relevance to most people's health.

Despite official disinterest, accumulating evidence suggested that pesticides were widely distributed in the environment. The first sign that agricultural chemicals were entering the Central Valley's water supplies had emerged in the 1950s, even before Carson had begun writing *Silent Spring*. Then in the early 1960s several fish kills in the San Joaquin River made it clear that agricultural practices were affecting water quality. In 1963 water engineers finally tested agricultural runoff in the Central Valley for salts, fertilizer products, and various pesticides and herbicides; they found them all but were quick to dismiss their significance. The following year, residents of the San Francisco Bay area became aware of a proposal to build an immense drain that would siphon agricultural runoff from the valley and deposit it in San Francisco Bay. This was yet one more component of the massive reengineering of the valley's waterscape that had been under way since the late 1930s. A coalition of local interests subsequently delayed construction of the so-called master drain and forced the U.S. Public Health Service to assess the health and environmental effects of the project on downstream waters. When it finally appeared in 1967, that study identified thirty-five pesticides in the Central Valley's agricultural drain water.[3]

Studies of chemical contamination in water supplies were the exception however. Through the end of the 1960s, the only monitoring done by most public drinking water suppliers was a test for bacteriological contaminants. Only in the early 1970s did the problem of chemicals in drinking water receive any regulatory attention. In 1972, while Congress was debating new drinking water legislation, the Environmental Protection Agency (EPA) announced that tests of the New Orleans water supply had revealed the presence of thirty-six organic chemicals, including three known and several suspected carcinogens. Two years later the Environmental Defense Fund (EDF) issued a widely publicized study showing that people in the New Orleans area who relied on surface water had higher cancer rates than those who relied on groundwater. The implication that EDF and others drew was that chemically contaminated surface water was raising cancer rates. These developments helped ensure passage of the Safe Drinking Water Act later that year. Among other things, this legislation directed the EPA to ascertain the scope of organic chemical pollution and to set water quality standards for any chemicals that posed health risks.[4]

The idea of water quality standards was not new in 1974. Sixty years earlier, as Hibbert Winslow Hill was pronouncing the irrelevance of the environment to health, the U.S. Public Health Service had adopted the first drinking water standards, a maximum allowable level of the bacteria *E. coli* that was intended to control the spread of typhoid in interstate commerce.[5] These standards recognized the relationship between the environment and health, but, true to the precepts of germ theory, they cast that relationship in narrow terms. Disease was believed to be contained in the specific bacteria, and the water supply was assumed to be the principal route of exposure. Although the consumption of water and food opened bodies up to environmental influences, bodies were otherwise envisioned as impermeable.

Regulators are seldom innovators. They typically build on the structures and protocols that are already in place. Thus, the federal regulations enacted in the wake of the Safe Drinking Water Act extended the approach forged in the early twentieth century. The relationship between water supplies and health was encapsulated in a set of numbers, the "maximum contaminant levels," or MCLs. Like the threshold limit values adopted by occupational health researchers and the residue tolerances adopted for pesticide-treated food, MCLs assumed that there was a level of exposure below which no adverse health effects would occur in any place, at any time. Moreover, the adoption of national standards assumed at the outset that local environmental conditions and individual differences were unimportant. So long as concentrations of certain chemicals remained below their MCLs, chemical contamination was officially deemed irrelevant to anyone's health.[6]

More than a decade after *Silent Spring,* amid growing recognition of the pervasiveness of chemical contamination and the complexity of local ecologies, regulatory strategies still relied on a modern concept of the body. Maximum contaminant levels assumed single etiologic agents, clearly defined pathways of exposure, and an external environment that could be effectively controlled and managed. In fact, none of these was true. The relationship between organic chemicals in water and chronic disease (such as cancer) defied any easy etiologic description. The existence of these contaminants in multiple places—not only in water but also in air, soil, and food—defied any attempt to isolate pathways of exposure. And the inherent complexities of the environment defied attempts at monitoring as well as management.

Moreover, the reliance on MCLs assumed that all the relevant contaminants were known and could be identified in the laboratory; yet analytic techniques did not exist for most of the organic chemicals already in use. And even where reliable techniques did exist, they were often prohibitively expensive to employ. Making MCLs even more impracticable was the fact that little data existed on the toxicology or health effects of most chemicals on which to base a standard. When the EPA finalized its drinking water regulations in 1978, it adopted an MCL for just one category of organic chemicals. Pesticides were completely ignored.[7]

Experts were at least debating the health significance of pesticides in surface water at this time, but few expressed concern that these chemicals might also contaminate underground water supplies. Though aware of the potential for chemicals to move into groundwater but lacking dramatic evidence comparable to fish kills, most experts assumed somewhat wishfully that chemicals placed in the ground stayed put. Studies that addressed the fate of pesticides in soil and water were comparatively few; scientists and farmers alike assumed that the overlying soils protected groundwater from chemical contamination, acting as a kind of natural filter. Despite the extraordinary amount of pesticides used in the San Joaquin Valley, the region's groundwater was assumed to be relatively invulnerable to contamination—because its flat topography and dry climate generated little runoff. Consequently, groundwater supplies were rarely monitored. The state of California conducted just one inquiry on the issue before 1970. In a study remarkable only for its poor design, investigators applied two pesticides (DDT and lindane) to a small test field and hastily concluded that agricultural chemicals typically degraded in place.[8]

It would take another decade, and the discovery of widespread groundwater pollution, to reveal the pervasiveness of chemicals in the valley's environment. In 1979 the California Water Quality Control Board tested some wells at the Occidental Chemical plant, located in the northern San Joaquin Valley town of Lathrop. They found excessive concentrations of 1,2-dibromo-3-choloropropane (DBCP), a compound handled at the plant and used by farmers to protect the roots of crops from small worms. DBCP had emerged as a chemical of major public health concern two years earlier, when male workers at the Occidental plant learned that it had made them sterile. DBCP's ability to cause reproductive and other abnormalities in laboratory rats had been known to chemical companies since the late 1950s; the first published accounts of its toxicity had appeared in 1961. Yet no restrictions were placed on its use. It was only when workers began to ask questions, and to link the deficiencies of their own bodies to the pesticides they handled, that the existing scientific literature was rediscovered. Within two months of that discovery, the EPA suspended the use of DBCP on nineteen food crops. Two years later both the state of California and the EPA suspended DBCP's registration.[9]

When water quality officials discovered DBCP in groundwater, they realized that the chemical was capable of moving out of the soil and into the water below. Although this was a disturbing and unexpected discovery, everyone involved still assumed the problem was localized. However, two months later regulators found DBCP in two wells in a nearby county. The state legislature then directed the California Department of Food and Agriculture (CDFA) to test groundwater throughout the valley. Those tests revealed that 90 out of 262 wells tested contained DBCP in measurable amounts. The water supplies of several valley communities were seriously contaminated.[10]

Given the region's dependence on groundwater and DBCP's extremely high toxicity, the scope of the potential problem was enormous. DBCP had been applied in high quantities on thousands of acres throughout the region for more than two decades. Almost simultaneously, environmental officials in New York discovered the insecticide aldicarb in wells on Long Island. Like DBCP, aldicarb had been used heavily in regional agriculture and had been thought to pose little environmental risk. Yet both

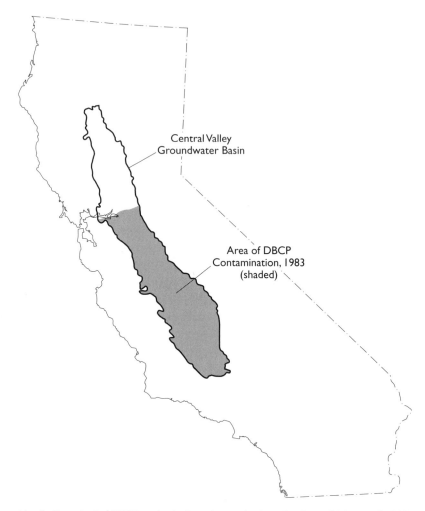

Map 2. The extent of DBCP contamination of groundwater in the Central Valley as of 1983. Redrawn from U.S. Geological Survey, *Ground Water Atlas of the United States,* vol. 1, *California, Nevada* (Reston, VA: USGS, 1995).

of these compounds were highly toxic and known to cause cancer in laboratory animals. Their presence in groundwater prompted the closing of scores of wells and indicated that the regular use of pesticides in agriculture, not just their disposal, might be polluting large quantities of the country's drinking water supply.[11]

The DBCP and aldicarb discoveries were but two of hundreds of cases of groundwater contamination that were uncovered in the late 1970s and early 1980s. As more and more chemicals were found in groundwater in more and more places, toxic contamination emerged as a nationwide problem. In 1978 Love Canal, a working-class suburb in upstate New York, became a front-page news story. Residents had found chemicals oozing into their basements, and they pointed ominously to apparently high rates of illness and miscarriage in their community. In 1980 Congress held hearings on the scope and implications of groundwater contamination nationwide, while the plight of Love Canal residents prompted the first "Superfund" legislation to facilitate the federal cleanup of hazardous waste sites.[12]

In California the DBCP discovery starkly revealed the inadequacy of existing water quality regulations, which barely addressed groundwater and ignored many dangerous compounds then in widespread use. It also quickly drew the attention of state political leaders, who immediately called for more monitoring. More monitoring identified more contamination. In 1983 California investigators found another carcinogenic soil fumigant, ethylene dibromide (EDB), in valley groundwater—also at disturbingly high levels. Aldicarb was also found in many locations. The following year the state legislature passed a bill that, for the first time, required community drinking water suppliers to periodically test their water for the presence of common pesticides. By April 1985, 57 pesticides had been found in nearly three thousand wells across twenty-eight counties—most of which were in the Central Valley. Though California appeared to have the most widespread groundwater contamination, the problem was national and even transnational. Wherever pesticides were manufactured, used, or disposed of, some were likely to turn up in groundwater. As one EPA official put it, "The more we look, the more we find." By 1985, even with only limited attempts at monitoring, the EPA had found 17 pesticides in the groundwater of twenty-three states. Ten years later, more than 140 pesticides had been found in forty-three states.[13]

Not only had existing regulations overlooked the presence of organic chemicals, they had also failed to account for the complexity of the broader environment. Groundwater had been ignored in existing water quality standards in part because it was assumed to be relatively invulnerable. But, in fact, the susceptibility of groundwater varies tremendously. The fate of DBCP in one environment may be quite different from its fate in another. The likelihood of pesticides moving into groundwater, like the potential for pesticide-induced illness in farmworkers, was contingent on local environmental factors: the amount of rainfall and agricultural runoff, the ability of particular types of soil to absorb water, the geologic conditions of the aquifer. Once again, there were countless local variables to be measured.[14] The only way to know what was happening was to return to the "field," to the specific ecologies in which all these chemicals had been used.

Environmental conditions in the Central Valley are difficult to understand much less to control. Contamination and toxicity remain spatialized in complex ways. One region of the valley differs from another. Field research would eventually reveal that the aquifer underlying the eastern side of the San Joaquin Valley was extremely vulnerable to pesticide contamination (a conclusion reached twenty years after a similar set of experts had concluded just the opposite). In part, that vulnerability lay in the soil. Most of the sediments in the eastern half of valley had eroded from the Sierra Nevada. As the mountains weathered, sand and rocks settled at their base. Consequently, the soil in this area is derived from granite; the grains are coarse, and water moves through the intervening spaces with relative ease. Moreover, groundwater in the area is often shallow, from twenty to two hundred feet below the surface. Whenever water enters the ground, it is likely to pull substances out of the soil and carry them down to the underlying aquifer. This means that when pesticides are applied to crops, they can move quickly and easily into the water below. In contrast, environmental conditions in the western half of the valley make groundwater contamination less likely. There the soil comes from the Coast Ranges. The sediments are finer and are more likely to attract and attach chemicals. Even though groundwater is shallow, it is more difficult for organic chemicals to travel downward, and once in the groundwater they move more slowly. Although contamination has been found throughout the region's aquifers, wells on the east, and most populated, side of the valley are far more likely to be polluted and to have substantially higher pollutant concentrations.[15]

Predicting the fate of chemicals in the valley environment was made more difficult by the region's history. Humans have exploited and altered these aquifers in significant ways. Although groundwater was first used in the San Joaquin Valley in the 1880s, pumping rates increased dramatically after World War II. It was the accessibility of groundwater that allowed for the rapid intensification of agriculture in such an arid climate in both the pre- and postwar decades. The increased reliance on well irrigation meant that in any year large quantities of water moved up through the wells. Groundwater was replenished not from the flow of streams (most of which were completely diverted for agricultural purposes) but from the excess irrigation water that soaked into the soil; and that water was likely to carry significant quantities of chemicals and nutrients. Pumping has even altered the direction of groundwater flow in the area. Water that used to flow generally east, toward the trough of the valley, now flows west, toward some of the most powerful pumps. Water that once flowed toward the soil surface now flows downward. Even deep groundwater, which is typically protected somewhat by overlying soils, has been affected. Thousands of wells now tap the deeper aquifers; more are drilled whenever there is a drought. Often poorly constructed by cash-strapped farmers, irrigation wells have cut through natural barriers of soil and clay and have become conduits for contamination. Inside a well, the polluted groundwater in a shallow aquifer mixes with the uncontaminated water from below. Like the farms above, the underground hydrology is a complex product of nature and human history. Consequently, regional groundwater is now more easily and more thoroughly contaminated. Groundwater hydrology in the region is also more complicated and variable than it would otherwise be. Where a pollutant ends up depends not only on the geologic characteristics of the aquifer but also on which combination of wells is being used at different times.[16]

The discovery of pesticides in groundwater and the difficulties it presented undercut the assumption that the relationship between environment and health was one that could be managed. Certainly humans had applied the chemicals, but they had never intended, or even considered, that years later those chemicals would turn up in local drinking water. Once widespread contamination was discovered, however, agencies and politicians were anxious to demonstrate that they were in fact protecting public health. Caught in a defensive posture, they turned to familiar strategies. Immediately after the DBCP revelations, both the state and federal governments quickly issued an MCL for DBCP in drinking water.[17]

Chemicals moved through the valley's air as well as its water, but air is even more difficult to test and regulate. Although researchers and community members had been concerned with pesticide "drift" since the mid-1940s, significant research on airborne toxics would not come for another forty years. Even in the late twentieth century, it remained difficult to assess air quality in a quantitative fashion. However, field tests revealed that as much as 90 percent of an applied pesticide could volatilize from the surface of soils and plants within a matter of days. Valley residents already knew through experience that ambient air could harbor pesticides; this had been the basis of complaints that dated back to the early 1950s, and such concerns had not abated. Although the CDFA did not publicly track or report the number of illnesses caused by airborne pesticides, it acknowledged more than one hundred illnesses resulting from "coincidental exposures" in 1986 alone. In one of several incidents in the 1980s, residents of the town of Ceres were evacuated after an atmospheric inversion trapped airborne pesticides (methyl bromide and chloropicrin, a component of tear gas) and generated toxic smog. People living up to two miles from the target field became ill.[18]

Again the natural conditions of the valley exacerbated the problem. Researchers were surprised to find that the valley's dense tule fog could both transport pesticides over long distances and significantly concentrate them. Sampling revealed that wintertime fog routinely contained organophosphorus compounds (e.g., parathion) that were sprayed on fruit trees during the colder months. The measured concentrations were extremely high—in some cases, several thousand times higher than what would be expected in rainwater. Not only were the pesticides present in fog, but so were their highly toxic degradation products (oxons). Moreover, contaminated fog likely poses a substantially greater health threat than does polluted air, because the inhaled particles are not quickly exhaled but linger inside the lungs. For some, the discovery of pesticide-laden fog might have recalled the much earlier concern with miasma. Once again, the valley's climate and environment threatened health—only this miasma had taken on a modern, chemical form.[19]

California officials adopted legislation to address toxic air contaminants in 1984, and their approach was the same as it had been for drinking water: the state would rely on animal toxicity studies to develop maximum allowable contaminant levels for specific chemicals. The responsibility for adopting air quality standards, however, was given to the CDFA—an agency that had shown little enthusiasm for regulating or monitoring pesticides in the past. Fifteen years after the Toxic Air Contaminants Act, the CDFA had adopted an air quality standard for just one pesticide.[20]

While no one was watching, pesticides had moved into air, soil, streams, and groundwater, and also into human and animal bodies.[21] Although the monitoring of pesticides in water and air only began in the 1980s, there was no reason to think that similar conditions had not existed for at least the past three decades. The effect of this contamination on human beings was unclear. Were these substances affecting the health of persons other than farmworkers in the immediate postwar period? The invisibility of any link between human illness and a pesticide-ridden environment might have been because there was no such material link in that place at that time, that—as many in agriculture argued—pesticide concentrations were too low to matter. Or it might have been because there was little information that could immediately reveal or even suggest such a link. Archives, including environmental archives, are only created for a purpose. And the reasons for archiving data on pesticide concentrations in water, air, and soil has been to regulate these substances. Where no desire or political will existed to regulate pesticides, there was no reason to monitor them. And so long as they were not monitored, it was that much easier to assert their irrelevance. Any environmental history of the valley stumbles upon the limits, sometimes deliberate, of the written record.

In the absence of environmental monitoring data that could be correlated with existing standards or animal toxicology studies, sick bodies provided the only evidence of danger. In the 1950s and 1960s it had been the undeniable symptoms of farmworkers that had forced the recognition of acute pesticide poisoning. But when the health concern was cancer or any other chronic condition, the latency period between toxic exposures and identifiable symptoms could be months, years, or sometimes even decades. This temporal gap, combined with the mobility of modern individuals, meant that the symptoms of bodies could not be easily connected to particular places. Thus modern assumptions about the separation of bodies and (polluted) environments could not be directly or easily challenged. Even if there were a relationship between health and place, it would have been extremely difficult to see. More difficult questions about the relationship between environment and health in the valley would soon become unavoidable, however. Those questions would find their focal point in the otherwise unremarkable town of McFarland.

THE McFARLAND CANCER CLUSTER: POLITICS AND PUBLIC HEALTH IN THE 1980S

There was little to distinguish McFarland from its neighboring agricultural towns in the postwar decades. One of several valley communities established shortly after the turn of the twentieth century, McFarland owes its existence to a couple of ambitious land speculators, James B. McFarland and William F. Laird. After buying a large parcel of land from the Southern Pacific Railroad, McFarland and his partner subdivided it into smallholdings and invited struggling farmers and businessmen from southern California to try their luck in the valley, where land was cheaper and the growing season longer.[22] Several decades later the agrarian hopes of McFarland's early settlers had given way to the realities of industrial agriculture.

Agriculture in the valley continued to become more specialized in the postwar decades. As urbanization forced farming out of southern California, the completion of major irrigation projects in the 1950s and 1960s allowed production in the valley to expand and intensify. The area around McFarland witnessed the rapid expansion of cotton and grape acreage, typically on very large farms, as well the planting of new fruit and nut orchards. Along with this expansion of farming came the increased use of fertilizers and pesticides. In the 1990s the counties surrounding McFarland—Fresno, Tulare, and Kern—accounted for 50 percent of all pesticides used in the state (and California continued to have by far the highest overall usage of any state). This reflected the fact that cotton and grapes—the two most prevalent crops in the region—were responsible for the majority of the region's pesticide use.[23] As the size of farms grew and the demand for labor increased in the decades after World War II, McFarland became one of the valley's farmworker towns. In 1980 more than 70 percent of the population was Latino, and most were employed directly or indirectly in farm labor. Like other farmworker towns in the region, McFarland suffered from seasonal unemployment and endemic poverty. In 1992 it was ranked one of California's ten poorest cities.[24]

McFarland was a town that few people knew or cared about until the 1980s. That two children living on the same street received diagnoses of cancer in 1983 was hardly something that would attract public attention. A few months later, however, a third child who lived a block away was similarly diagnosed. Although the types of cancer differed, the parents of these children began to wonder whether the illnesses were somehow connected. Anguished discussions among the affected mothers marked the discovery of the McFarland cancer cluster. Epidemiologists working for the state of California would eventually confirm thirteen cases of childhood cancer in a town of only six thousand people. Although only a small number of people were directly affected, the rate of childhood cancer in the community was more than three times what would be typically expected. Most of the cancer cases occurred in a single neighborhood of small stucco homes built in 1980 with financing from the Farmworkers Mortgage and Housing Authority (FMHA).[25]

Connie Rosales, the mother of one of the sick children, contacted the Kern County Health Department for answers. Local public health officials responded slowly. They initially assumed that there was no cluster. As residents' concern and frustration grew in the face of the local government's inaction, residents looked elsewhere. Rosales contacted the local branch of the Mexican-American Political Association (MAPA) for help. Eventually her letter arrived at the office of state senator Art Torres, a leading Latino politician in the state who was also known as an aspiring gubernatorial candidate. In May 1985 Torres announced that his committee, the Senate Committee on Toxics and Public Safety Management, would hold hearings in McFarland. Those hearings drew the attention of newspapers and television and elicited anguished testimony from residents, including Rosales. With political scrutiny now cast on Kern County, the board of supervisors declared a public health emergency. At that point, the state agreed to provide the local health department with funds to initiate a detailed study of the stricken neighborhood.[26]

Residents immediately suspected that the environment was the source of the illnesses. The widely publicized problem of groundwater contamination, in particular the valley's DBCP problem, had sensitized many residents to the health threats of chemicals. Having been alerted to the problem of nitrates in the water from notices on her monthly bills (which warned against giving the water to infants), Rosales wondered whether the same contaminants might be responsible for the cancers. Deferring to these concerns, health officials conducted nitrate tests on the local water supply. When levels were found to be within established public health limits, officials began to test for other substances. First they tested drinking water for several pesticides (including DBCP), toxic metals, and ionizing radiation; then they tested the soil around the homes and schools of cancer victims. They followed the soils tests with air monitoring for carbon monoxide (a product of the freeway that runs through the middle of the town). They also tested the inside of victims' homes for radiation, asbestos, formaldehyde, and other carcinogenic chemicals. Kern County investigators certainly found contamination in McFarland, but it was not exceptional contamination; in fact, all soil and groundwater contaminants fell within established public health limits. Officials declared the environment "safe," though none went so far as to say that it was healthy.[27]

Residents were not satisfied. In their minds (and many others') the existence of so much disease in one place pointed to the local environment as the likely cause. As in other struggles over toxic contamination, those least willing to yield to the assessments offered by government officials were women. Initially the leading activists in McFarland were mothers of cancer victims. Women typically assume responsibility for the well-being of their families, and since the late nineteenth century this obligation often has pushed them into the public arena to fight for safer environments. Like that of the "municipal housekeepers" of the late 1800s and the antimosquito crusaders of the 1910s, the activism of the McFarland mothers emerged from their desire to protect their families' health. Women, moreover, often have access to different kinds of information—that gathered through observation and conversations with other women. They, more frequently than men, are at home and in the community and therefore usually more aware of the everyday hazards their children encounter as well as subtle changes in their physical well-being. As mothers in McFarland became more aware of the prevalence of sickness and the rumored and known environmental dangers, they began to develop their own hypotheses about the cancers and to press health officials for better answers.

It was not only their social positioning that enabled women to speak out against perceived environmental health threats but also the different ways in which gendered bodies have been envisioned and experienced. As in earlier periods, the permeability of the late-twentieth-century body remained strongly connected to gender. The construction of the female body as especially vulnerable to occupational hazards had underwritten protective legislation for women, as well as sex discrimination in the workplace, for decades. Although these distinctions were strongly challenged in the 1960s by feminists (who insisted on the strength and resilience of female bodies) and environmentalists (who insisted on the permeability of *all* bodies), other developments reemphasized the significance of bodily difference. In particular, new concerns about the vulnerability of the fetus helped reinscribe the belief that women's bodies were more porous than those of men. Ideas about pregnancy and fetal health were transformed by the discovery that viruses and drugs (such as thalidomide) could cross the placenta. By the early 1980s prescriptive prenatal advice emphasized that pregnant women should avoid drugs, alcohol, and toxic chemicals. Not surprisingly, the women in McFarland worried about the pesticides they had encountered when they were pregnant, as well as the contaminated environments in which their children played. From their perspective, the spatialization of disease was related to the uses of the local landscape: the heavy application of pesticides, the contamination of groundwater, the ongoing problems with air quality. If they were to believe that the environment was not the cause, the mothers insisted, public health officials had to provide another explanation.[28]

Although the McFarland mothers challenged the disinterest of public health officials, they did not at first challenge modernist conceptions of the issue. They cast their hypotheses in terms of specific contaminants, and their principal concern was the town's water supply. To the extent that McFarland's activist residents shared the modernist framing of the problem, it may have been because that was the only way in which they saw the problem. Or it may have been because that was the only kind of argument that could gain traction within the existing regulatory system.

As media and political interest in McFarland intensified, public health officials struggled to put together a coherent response. As community members asked more questions, the limits of the existing regulatory framework became obvious. Until the crisis, there had been no monitoring of pesticides in the local environment. And though the surrounding environment in McFarland met "existing standards," public health officials acknowledged that the relationship between environment and health might not be fully encapsulated by those standards. Water quality standards existed for only a handful of pesticides, and there had not yet been any attempt to assess and regulate pesticides in air. The lack of knowledge generated more anxiety. Concerns about McFarland's environment were so intense that in 1985 the FMHA placed a moratorium on building in the community.[29] As one health official acknowledged, "The variety of chemicals being introduced into that environment gives you any number of possible sources of environmentally induced damage."[30] Soon afterward, the California Department of Health Services (CDHS) confirmed another childhood cancer cluster in the nearby town of Fowler.

In the 1950s the investigation of pesticide-poisoning among farmworkers had fallen to specialists in occupational health, who drew on their background in toxicology and factory environments to craft a response. Under mounting pressure to do something about the McFarland situation, health officials now turned to epidemiology. In contrast to bacteriology, which focuses on disease in individuals, epidemiology studies the occurrence of disease in populations. By looking at the spatial patterns of disease occurrence, epidemiologists try to uncover common causes of illness. The discipline's roots are typically dated to John Snow's famous investigation of a cholera outbreak in mid-nineteenth-century London. By mapping where the sick individuals lived, Snow traced the source of cholera to a single contaminated well.[31]

By World War II, epidemiology, much like occupational health, had become a somewhat marginal subfield of biomedicine, eclipsed by germ theory and the successes of laboratory medicine. Then, in the 1950s, the field attained new prominence when epidemiologic studies provided insights into the causes of lung cancer and heart disease—the two most prevalent forms of chronic illness in the United States. The crucial difference between postwar epidemiology and bacteriology lay in how these specialties described disease causation. Instead of emphasizing a single agent responsible for disease, epidemiologists embraced the idea that causes were multiple. And instead of trying to specify the precise mechanisms of disease etiology, they attributed disease to the interaction among a series of variables, or "risk factors," whose actions were only incompletely understood. The dominant metaphor within the field was the "web of causation."[32]

In California epidemiology had attracted some attention in the 1930s as a tool for studying infectious disease, notably in Karl Meyer's studies of sylvatic plague and western equine encephalitis. However, epidemiology would only become institutionalized in the California Department of Public Health in the 1950s. The reason lay not with infectious diseases such as plague but with the new problem of smog. As air quality in Los Angeles emerged as a major political issue, the CDPH found itself tasked with determining whether and to what extent community-wide pollution threatened public health. These early air pollution epidemiologists acknowledged that they were less concerned with identifying the precise cause of disease than with preventing illness, and they often invoked nineteenth-century models. As the state's leading smog researchers declared, "Just as epidemiology led to methods for control of water-borne disease before bacteriology permitted the accurate diagnosis of them, so it is hoped that epidemiology may assist in the control of the health effects of air pollution even before it is possible to diagnose a disease caused by air pollution."[33]

The turn to epidemiology emerged from the desire to know whether disease was in fact spatialized, the result of living in certain modern landscapes. In the case of air pollution, many laypersons already assumed that it was. Those same concerns would soon be extended to chemical pollution. A critical moment came in 1975 when the National Cancer Institute published the *Atlas of Cancer Mortality for U.S. Counties*. Each page of the *Atlas* presented a national map with individual counties color-coded to illustrate the local mortality rate of specific cancers. Dark red sections, indicating a "significantly high" mortality rate, stood out against the olive green color that represented average rates. Aggregations of red sections seemed to signal ominous cancer hot spots. The *Atlas*, much like Thomas Logan's isothermal maps, visually linked disease and place.[34]

In the 1970s the national media seized on the implied link between environmental pollution and cancer. Articles on the subject proliferated in publications ranging from *Newsweek* to *Reader's Digest, Scientific American* to *Vogue.* Several popular books took up the issue, most notably Samuel Epstein's *Politics of Cancer,* which unequivocally attributed rising cancer rates to the postwar rise of the petroleum, chemical, and pharmaceutical industries, as well as to the increasing use of pesticides and food additives.[35] Many in the research community, while considerably more circumspect than Epstein, shared the concern with environmental carcinogens. As the cancer researcher John Higginson put it in 1976, rising cancer rates "should be considered due to the entry of new agents into the environment or increased exposure to old ones, until proved to be due to better diagnosis or other forms of artefact." At the same moment, the catastrophes of Love Canal and Three Mile Island made the public acutely aware of their vulnerability to environmental contamination. And following the passage of the Superfund legislation in 1980, the EPA began identifying toxic hot spots across the country. [36]

At that point, California's public health officials expanded their environmental epidemiology work from air pollution to chemical contamination. After the discovery of DBCP in Central Valley groundwater, the state legislature demanded an assessment of health effects. The CDHS responded with an epidemiologic study that examined the association between groundwater contamination and cancer rates in the Fresno area. At the same time, officials discovered that groundwater in northern California's Silicon Valley was polluted with a carcinogenic solvent (1,1-tricholoroethane) that emanated from leaking underground storage tanks at the Fairchild Semiconductor facility. Again the agency undertook an epidemiologic study in response to public and political concern. Both of these studies suggested a link between contamination and the clustering of disease. And while the Fresno DBCP study initially attracted little attention, the Fairchild studies were widely covered by the press. Whatever its limitations, epidemiology provided a scientific language for talking about the relationship between the modern environment and the prevalence of disease and offered a means for evaluating the effects of chemicals "in the field."[37]

After the initial environmental tests in McFarland had revealed no obvious contamination concerns, public health officials bowed to public pressure and began a small epidemiologic study. With advice from the state Department of Health Services, the Kern County Health Department began a case-control study of ten affected individuals and twenty unaffected individuals (controls) selected from the same community. The goal was to identify whether the exposures or lifestyles of sick individuals in the community were significantly different from the exposures and lifestyles of healthy individuals. Researchers proceeded to render complex lives and environments into a set of rote questions whose answers could be statistically tabulated: Where have you lived and worked? What do you normally eat? Do you smoke, and how much? What illnesses have you and your relatives suffered from? What are your hobbies? Have you been exposed to pesticide sprays? To paint, lacquer, or varnish? To hair dyes?[38]

The case-control analysis, completed in October 1986, was inconclusive. Nor did it stem mounting public concern. In fact, the controversy intensified later that year when the UFW produced a video about McFarland to support their call for a renewed grape boycott. Featuring Connie Rosales and several other McFarland parents, *The Wrath of Grapes* asserted that the cancers were a direct result of growers', in particular grape growers', intensive use of pesticides. The involvement of the UFW angered the local public health establishment, along with farmers and political leaders who were openly hostile to the union. The politics grew only more charged when Thomas Lazar, a medical anthropologist hired by Kern County to work on the cluster investigation, quit his job and went to work for the UFW. Lazar claimed that county health officials were covering up the extent of the cancer problem and told the press that existing data suggested childhood cancer rates were high not only in McFarland but also in several other county communities. Although health officials insisted that Lazar's charges were baseless, they would eventually confirm the existence of clusters in two more San Joaquin Valley communities (Rosamond and Earlimart). That fall, the McFarland and Fairchild controversies contributed to the overwhelming passage of Proposition 65—the Safe Drinking Water and Toxic Exposure Act—which forbade any business to knowingly pollute drinking water with toxic chemicals (defined as those suspected of causing cancer or birth defects) and which required that the public be notified of any exposures to known or suspected carcinogens.[39]

Amid these developments, the CDHS agreed to expand their environmental studies in McFarland by gathering data on pesticide use in the area and testing for electromagnetic and radio-frequency radiation (a product of a Voice of America transmission tower located just north of town). Lazar's charges also rekindled the interest of state politicians in McFarland. Senator Torres held another hearing in October 1987, which elicited a barrage of testimony, both angry and emotional, from residents who felt that the state was not doing enough to protect them. After the 1987 hearing, Torres argued that the entire San Joaquin Valley should be declared a toxic waste site under the state Superfund law. Bakersfield's state assemblyman, Trice Harvey (a Republican), asked the governor to declare McFarland a disaster area so that emergency funds could be made available for a broader investigation. Soon afterward, officials announced that low concentrations of DBCP had been found in one of the town's wells, intensifying the sense of panic. Then, in spring 1988, presidential candidate and civil rights activist Jesse Jackson made McFarland the centerpiece of his campaign in the California primary. Jackson visited affected families and held a news conference. Accompanying Jackson were prominent political figures, Hollywood celebrities, and the national media. Soon after, the governor announced the formation of a "blue-ribbon advisory panel" to oversee the work at McFarland.[40]

As Jackson's visit made clear, environment and health concerns in McFarland were intertwined with issues of race and class. The historical experience of Mexican American communities in the West, racialized divisions of political power, and the influence of agribusiness in California all fostered a deep distrust of government in McFarland's Latino community. Jackson insisted that the environmental problems at stake in McFarland could not be separated from the issue of Latinos' political disenfranchisement in the valley. Marion Moses, a physician and UFW activist who had worked on pesticide issues since the late 1960s, told a reporter that "if this were happening in a white, middle-class office building, something [more] would have been done." Connie Rosales contrasted the response to McFarland with that to Love Canal, noting that residents of the latter community, who were overwhelmingly white, had received both political attention and government funds even though no one there had actually died. Many others voiced similar sentiments to reporters. "We just don't count for much," remarked a field worker and father of one cancer victim.[41] Adam Salinas, a resident of Earlimart, suggested that "maybe they [government officials] don't pay attention to poor people because they're poor."[42] These charges of racism and classism were fueled by the fact that, locally, political power was divided starkly along racial lines. Although almost 90 percent of the population was Latino, the white minority controlled the local government. As late as 1988, McFarland had only one Latino elected official. Important institutions—the city council, the water company, and the county health department—were dominated by whites, and less than 10 percent of the Latino residents were registered to vote.[43]

The influence of race and class was undeniable, and this was the fact that the press coverage tended to highlight. However, those divisions were not all-encompassing. There were at least a few Latinos who thought the activists had overstated the problems, and there were many middle-class whites who voiced their concern. There were also white farmers in the region who feared the long-term health effects of pesticides and who argued for less chemical-intensive methods of farming. Among farmers and middle-class whites, it was often the experience of cancer in their own families that pushed them toward this ecological view. The knowledge that DBCP was linked to both sterility and cancer had reportedly spurred many farmers to avoid it, despite the lack of a good replacement.[44]

The stance of the state toward the community was more predictable, though hardly singular. The "official" reaction to the cluster often varied. Most obviously, the political spotlight and upcoming election motivated an otherwise disinterested governor to make available some funds to study the problem. Within the bureaucracy itself, attitudes toward the cluster differed starkly among agencies. The slow response of the county health department was a result of both its sensitivity to local politics and its unpreparedness. Like most health departments in the early 1980s, the Kern County Public Health Department was staffed with sanitationists and headed by an M.D.; it had no expertise in epidemiology. The most critical division, however, existed between the Department of Health Services and the far more powerful Department of Food and Agriculture. Not surprisingly, those working for the health department were more sympathetic to residents' concerns and to the possibility that pesticides might be causing chronic health problems, though they were dismissive of lay hypothesizing and frustrated by residents' skepticism about their methods and expertise. Individuals at the CDFA, on the other hand, were critical of any explicit focus on pesticides.[45]

In January 1988 the CDHS released a preliminary report that suggested a tentative relationship between childhood cancer and parental exposure to certain pesticides. Although officials would soon back away from this conclusion citing a lack of evidence, these findings intensified the political spotlight on McFarland. Moreover, in the process of conducting that study, researchers had identified the third cluster (in Rosamond). In light of these findings, the advisory panel debated whether to suggest more intensive monitoring for pesticides or, alternatively, more epidemiological work. Although community members had lobbied for more environmental monitoring, the blue-ribbon panel opted for epidemiology. Given the opposition of agricultural interests and the

CDFA to pesticide monitoring, epidemiology was the less controversial option. Based on the panel's recommendation, the CDPH agreed to conduct what it would call "the four-county study"—a multicommunity study of cancer incidence in the four southern counties of the San Joaquin Valley. The goal was to compare the frequency and distribution of childhood cancer in the state's most pesticide-laden communities to that in other, hypothetically less contaminated regions. The hypothesis underlying their effort was that California's agricultural communities, subject as they are to high levels of pesticide exposure, might show an elevated cancer rate. Put another way, the question was, could modern epidemiology confirm that the San Joaquin Valley was a pathological space compared to other locations in California? The discovery of a fourth cancer cluster in fall 1989 (in Earlimart) elevated the concern that the valley was indeed such a pathological space.[46]

The decision to conduct the study was itself significant. It demonstrated that residents had attracted the attention of the state health establishment and that they had succeeded in changing the question that experts were asking. In 1984 the question had been, what individual exposures might have contributed to the cancers? In 1989 the question was far more radical: was the pesticide-laden environment of the valley increasing the risk of cancer among the entire population? Three of the four cancer clusters occurred in communities located in the valley's agricultural bottomlands, which were surrounded by fields of cotton, grapes, and citrus. All lay above the polluted aquifer on the valley's eastern side. The study's hypothesis was that the state's rural regions were more subject to certain diseases than its cities—not because they were insufficiently modern, but because of the very nature of their modernity. As one local health official described it, the San Joaquin valley was the site of a "grand experiment" on the human effects of pesticides. The four-county study asked whether modern agriculture had respatialized disease, creating a new form of miasma—an ill-defined pollution that permeated local landscapes and entered bodies in multiple and unknown ways.[47]

The study continued for two years. When the results were announced, they revealed far more about the limits of epidemiologic knowledge than about the material relationship between environment and disease. The scope of the study had been constrained by the fact that reliable cancer data were available only since 1987—the year that a state tumor registry was implemented in response to the McFarland crisis. The study itself was an elaborate statistical analysis that compared rates of specific cancers to those that might be expected based on national averages and those predicted by standard statistical distributions. In September 1991 the CDHS presented its conclusions at a public meeting while the UFW held a protest against pesticides outside. Lynn Goldman, then the state's lead epidemiologist on McFarland, summarized the years of work on the cancer cluster and concluded by stating that nothing in either the environment or the lifestyles of McFarland's residents could explain the cancers. As Goldman and others would write in a technical article, there was no "strong force of morbidity" operating in the region. Goldman emphasized that the aggregation of so many cases might, after all, be simply random, or, alternatively, there might be some "force of morbidity" that was not strong enough to detect with existing statistical techniques. In their report the investigators discussed the statistical challenges presented by data sets with small numbers and large variances—an implicit acknowledgment that statistical epidemiology often obscured, even when it did not disprove, what seemed otherwise evident at the level of experience. In the absence of conclusive evidence that the cancers were linked to environmental causes, the professional assumption was that they were not, which was also the answer that most of the state's political leaders had wanted to hear. The cluster, while undeniably "real," had no cause that modern epidemiology could identify, and, they insisted, it likely had no cause at all.[48] The panel disbanded itself and advised the state to formally end its investigation. "Science," they insisted, had nothing more to offer on the subject, and thus it was time to move on.

Some people, especially business and agriculture representatives, took the failure of epidemiology to link environmental contamination to the cancers as evidence of the normalcy, or at least the irrelevance, of McFarland's environment. The pathology of cancer could still be assumed to lie within the body and, in this case, within the already pathologized bodies of those who were ethnically Mexican.[49] Although ecological views had challenged this racializing logic, they had by no means supplanted it. In this view, disease and health suggested nothing about environmental change or the region's chemical-intensive agriculture. Although the agriculture industry did not invent modern understandings of health, modernist models of the body remained powerful in part because they continued to serve certain groups well. The notion of an impervious body and a passive environment continued to underwrite the project of industrial agriculture in the San Joaquin Valley.

Many residents, however, were deeply frustrated by the inconclusive results. They took the results of the study as evidence of the inadequacy of epidemiology and the modern regulatory system. Community members had hoped for a definitive scientific explanation. When the state's epidemiologists could not provide it, some community members refused to yield to an ambiguous expert knowledge that exonerated the local landscape while failing to answer their most pressing question: What had caused the cancers? Of the experience in McFarland, one activist commented cynically, "We've learned over and over again that the studies produce statistics to be analyzed away; that the tests produce numbers to be classified into safe levels or standards; and that experts can find ways to explain away anything." What seemed evident at the level of experience became invisible in the languages of modern public health and biomedicine. Connie Rosales put her frustration still more bluntly when she told a reporter that "the science of epidemiology needs a real kick in the rear end." The more officials insisted that the environment was "safe," the less people believed them. Despite the assurances of experts, many families chose to move away from McFarland; others lamented their inability to do so.[50]

Among those who were disillusioned by the failure of epidemiologists to connect the environment to health, there were several responses. Some maintained that health officials were engaged in a cover-up and were simply unwilling to confront powerful agricultural interests on the issue of pesticides. In this view, the dynamics of power combined with institutionalized racism to shape the scientific findings. Others tried to critique modern public health on its own grounds, arguing that the environmental investigation had not gone far enough. Several community leaders became reasonably well versed in the languages of environmental science and epidemiology, and they prodded officials to test for additional chemicals, to use lower detection limits, and to include more cases in their epidemiologic analysis.[51]

Others simply disbelieved the results, although they could not articulate precisely why. Most laypersons remained wedded to the validity of their own experience in the landscape and, like nineteenth-century physicians, continued to meld modernist with environmental understandings of disease. Regardless of what epidemiology might suggest, residents still *felt* that the landscape was toxic, that their bodies were instruments that measured things that epidemiology and toxicology apparently did not. They insisted that a lifetime of experience in the Central Valley provided them with knowledge that experts in laboratories or distant offices lacked or chose to ignore. After all, it was community members rather than public health experts who first identified the cluster. Lois Gibbs, leader of the community group at Love Canal and subsequently a prominent antitoxics activist, defended this approach. Local people, she insisted, "know when something is wrong. They can see dead vegetation, smell chemical odors, taste the foulness of their drinking water, and observe an increase in disease."[52]

Residents of McFarland echoed Gibbs's view. Many homes sat within ten feet of cotton fields, and residents complained of pesticide spraying and especially the use of defoliants—chemicals that cause plants to drop their leaves. Defoliants were sprayed extensively around McFarland each fall prior to the cotton harvest. Community members told health officials about floodwaters that smelled of chemicals and petroleum, and complained that their drinking water was often discolored and tasted bitter. They reported smelling pesticides in the air throughout the year. They pointed to an unusually large number of sick animals in the area. Many in the town were farmworkers and had experienced firsthand the chemicals that were applied to the surrounding landscape; many had been ill in the fields or had watched as coworkers fell ill. One field worker, Jaime Caudillo, said as much, telling a reporter, "If we get sick from it, the children must absorb some of it." In ways that echoed nineteenth-century understandings, McFarland's residents drew analogies between the landscape and their bodies, reading the reactions of plants and animals as evidence that they, too, inhabited a pathological landscape. As another resident put it, "It seems so obvious to me. The pesticides kill bugs, so why not us?" From this perspective, cancer was part of a continuum of pesticide-induced diseases that workers in the region had been familiar with since the 1950s: rashes, "rubber legs," nausea, seizures, and even death. At least one farmworker had made this connection more than a decade earlier when she suggested that acute pesticide poisoning might also be linked to rising rates of cancer and other chronic diseases.[53]

This is not to say that the residents of McFarland were better epidemiologists than those who worked for the state health department but merely that they understood at some level that the abstractions of modern biomedicine, along with its narrow definition of causality, might be deeply problematic. While residents shared modernist assumptions about disease, they also held open other possibilities.

RESURRECTING MODERN BODIES

Why linger over the problems of one seemingly insignificant community? McFarland was but one small town, in one valley, in one state. But if knowledge is local, then localities (other than the locality of the laboratory) are critical. And much of the knowledge about chemicals and environmental health in the late twentieth century emerged from particular places, from the reactions of individual bodies that were located in specific environments. Knowledge has a complex geography. Even in the late twentieth century, knowledge about bodies and environments could still be produced in multiple places—in toxicological laboratories and within the computers of professional epidemiologists, but also in the valley's fields and orchards and on the streets and playgrounds of McFarland. Moreover, understandings of disease and health cannot be separated from the social and political contexts in which they arise. In McFarland, this context included farmworkers' experience of pesticides across more than three decades, a history of racializing logic within the supposedly scientific discourses of biomedicine, and the political marginality of California's ethnically Mexican communities.

Moreover, although the issues at play in McFarland were local, they resonated elsewhere. The attention of the national media, as well as state and national political figures, indicated that McFarland's troubles had broader relevance. The town's plight was symptomatic of a growing concern with the health effects of modernity across the valley, the country, and even the industrialized world. Reports of cancer clusters surged over the course of the 1980s; one study found that state health departments received roughly fifteen hundred requests for cluster investigations in 1989 alone.[54] Although only a fraction of these requests became major public health controversies, whenever controversies emerged they inevitably pitted the residents of polluted communities against local governments and public health officials. Increasingly, laypeople refused to accept the assurances of experts about the safety or irrelevance of polluted environments. Taken together, these controversies signaled that modernity could no longer be automatically associated with health; in fact, many felt the opposite was true.

Among professionals, the rising skepticism about expert knowledge in the late 1980s was attested by the proliferation of studies on "risk perception"—an area of sociological research whose aim was to explain why laypeople did not always accept professional assessments of technological and environmental risks. Social scientists now analyzed why the public often resisted the reassurances of experts regarding the safety of things such as pesticides and nuclear power. Why did laypeople become so concerned about pesticides when they still smoked and drove cars, activities that—statistically—were far more likely to kill them? In a particularly modern irony, a whole new field of expert, managerial knowledge had developed to explain why those types of knowledges were no longer authoritative.[55]

The emergence of such different perspectives on disease is often explained as a problem of rationality, or, more to the point, irrationality. Experts emphasize the inability of laypersons to fully comprehend the relevant science. In an effort to produce knowledge that is supposedly "universal," modern science relies on a scale of analysis that is typically either far below or far above that of lived experience. It emerges from the laboratory, or from the statistical analysis of large populations. Science reduces individual experience and history to certain forms of quantitative representation: numbers representing pollutant levels and probability, graphs and diagrams that show the presence or absence of imagined relationships. Because they cannot understand the abstract representations that are scientists' stock-in-trade, or because they cannot suppress their emotional response to illness, the argument goes, laypersons often misinterpret or ignore the objective evidence.[56] On the other side, activists frequently assert that government scientists are guilty of cover-ups, or at least are unwilling to pursue aggressively the relevant questions because they do not want to confront powerful interests.

Yet the issue is not simply that laypersons cannot read the scientific evidence or that government scientists always serve power but that there exist different types of experience, different forms of knowledge, and different understandings of disease and health—though these are seldom made explicit.[57] In the case of McFarland, modern and ecological concepts of health were both in play—and where an individual stood on the cancer cluster had much to do with the model that he or she relied on.

Epidemiology itself is something of a paradox in this respect. Epidemiology had initially been applied to environmental health issues because, in contrast to toxicology, its multicausal approach could take into account the complexity of actual human bodies and their environments. Epidemiology, at least in certain forms, is explicitly ecological in its approach. But as residents of McFarland gradually realized, epidemiology (much like occupational health) was ecological in a somewhat superficial sense. In many ways modern epidemiology invoked much of the logic if not the methods of modern bacteriology. Ultimately, epidemiologists still envisioned bodies as idealized entities that were distinct from their environmental context. And while acknowledging the multiple possible causes of disease, most epidemiological studies set out to identify discrete "risk factors." For instance, in a case-control study investigators look for substances that sick persons were exposed to but which the controls were not. The goal of such a study is to link a specific exposure to a specific outbreak of disease: those who drank from a given well contracted cholera (or cancer); those whose water came from elsewhere did not. In this model, disease is still highly localized—in a given infectious agent, a certain well, an individual body. Routes of transmission are narrow and traceable and are subject to verification in the laboratory. Thus, while epidemiologists acknowledged that the causes of disease were multiple and could interact in complex ways, in practice they sought to verify independent, or nearly independent, cause-and-effect relationships. Chemical contaminants were conceptualized as akin to microbes, as singular "agents" that could induce disease once they entered the body. Moreover, only those exposures that could be quantitatively measured and linked to cancer through toxicological studies were deemed legitimate risk factors. As in occupational health research, bodies and environments were presumed to interpenetrate one another but only in limited ways that were subject to both measurement and control.[58]

On the whole, epidemiological studies of cancer in the postwar period have tended to validate those risk factors that can be directly ascribed to individuals (e.g., smoking cigarettes); other, more social factors that might contribute to disease routinely fall out of the analyses because they cannot be adequately quantified, because they are viewed as something less than truly causal, or because they are subordinated to other individual factors. Even when it takes into account ostensibly social or environmental factors, epidemiological research often treats them as independent qualities of individuals.[59] For example, in the DBCP/cancer study that CDHS epidemiologists conducted in 1982, investigators found a positive association between three types of cancer and rising DBCP levels. But they also discounted their study as inconclusive; the primary reasons were race and class. As they explained, the census tracts with higher rates of stomach cancer were both poor and predominantly Mexican American, and Mexican American ethnicity was suspected to be a risk factor in stomach cancers.[60] Implicit here was the possibility that Mexican American populations in the Central Valley might have a higher cancer rate because of their lifestyles (e.g., their typical diet) or because they were genetically predisposed to stomach cancer.

But the relationship among "race," environmental contamination, and the incidence of cancer is likely far more complex. Given how long and how deeply race and disease have been intertwined, it is impossible to fully disentangle the relationship between them. In some cases, "race" may indeed serve as a crude approximation for specific populations that share certain biological or "lifestyle" characteristics as the epidemiologic approach assumes. What the DBCP and other such studies typically failed to acknowledge, however, was that "race" was also entangled with a person's access to medical care, exposure to various toxic substances, and presence in certain kinds of environments. Histories of workplace and residential segregation have located different kinds of bodies in different places. As the history of pesticide poisoning among farmworkers makes clear, nonwhites and the working class have routinely found themselves exposed to dangerous environments that more privileged individuals have been able to avoid. Segregation has combined with racist policies to ensure that race and class are unavoidably intertwined with environmental quality at the level of everyday life. Nonwhites do have higher incidences of certain cancers; they are also far more likely to work in more hazardous jobs and live in more polluted communities. Latino farmworkers are among the most disadvantaged in this respect.[61]

Race, class, and environmental contamination are thus not independent "variables" for which researchers can statistically control. To the contrary, histories of segregation and the differing quality of local environments, along with any relevant shared genetic traits, have combined to help produce "race" as a meaningful category in epidemiologic research. Material differences in the environments that people occupy have helped to produce physical differences in specific bodies (which also happened to be racialized): depressed cholinesterase, sterility, chronic respiratory problems, perhaps even cancer. In the case of pesticides, it is not hard to see how environmental and social history may be critical factors in the production of disease. Because people do not live their lives in laboratories, the predisposition to disease is never solely a property of the individual; it is also the outcome of a person's relationship to a preexisting social structure and to a series of material environments. But when epidemiologists control for "race," the environmental experience of racialized populations is implicitly excluded from their analysis. Such reasoning, though seemingly nonracist to the authors of the DBCP study, nonetheless had the effect of again positioning Mexicans and Mexican Americans as poorly disciplined and uniquely susceptible bodies. In this way, epidemiologic research has often reproduced a long-standing racial logic that their authors otherwise sought to avoid. Moreover, by directing attention toward the inherent qualities and individual habits of those bodies, epidemiology had the effect, however unintentional, of directing attention away from the environments in which Mexicans and Mexican

Americans typically lived and worked, such as the pesticide-laden fields of the Central Valley. It is often easier, or seems somehow more "logical," to ascribe disease to race rather than place.

In McFarland, those who rejected the conclusions of epidemiology pointed to this modernist tendency to locate disease within individual bodies, and the consequent failure to account for their own complex environmental and social histories. From the beginning, activists were skeptical of the trajectory that the McFarland research was taking. Teresa Buentello minced no words when she told a reporter, "They think we're a bunch of Mexicans, and we probably have it in our genes." Connie Rosales angrily recalled that health officials investigating the cancers had initially inquired about the use of Mexican herbs and traditional remedies instead of focusing on pesticide exposures and working conditions. Others insisted on the critical importance of the local context and astutely critiqued the laboratory paradigm on which both toxicology and epidemiology were based. As one of the mothers put it, "The pesticide DBCP showed up in our well, but they say it was below the level of danger. They know this from their rat tests. But the rats were exposed to DBCP alone. What if they gave rats the DBCP, and then exposed them to all the other chemicals we are exposed to here in McFarland? What would happen to their rats then? Nobody knows. I think they'd end up dying."[62]

The state's epidemiologists, on the other hand, were frustrated but not especially surprised by their inconclusive results. At least some of them felt that their studies of McFarland had always been driven more by the need for political leaders to demonstrate their concern for the situation than by the expectation that they would uncover meaningful information. In fact, during the 1980s, while the state's McFarland studies were under way, the attitude of the public health profession toward cancer cluster investigations had undergone a clear shift. In the 1960s and 1970s, as environmental concerns had risen to prominence, public health experts had advocated the use of epidemiology to investigate the relationship between environmental contamination and disease clusters, but by the mid-1980s the profession almost uniformly regarded such studies as a waste of effort. There were several reasons for this shift in attitude. At the federal level, the "Reagan revolution" had turned the government's attention (and dollars) away from issues of environmental contamination and public health. As government services were squeezed in the 1980s, state and local health officials were forced to make difficult decisions. In California, Governor George Deukmejian, a conservative

Republican, slashed the budgets of the state's regulatory and social service agencies, to which he was openly hostile. In that political climate, disease clusters became a low priority. Detailed environmental studies were costly, and the number of people with cancer in communities like McFarland, while statistically high, was numerically quite low. Overwhelmed by other concerns—food-borne illness, infectious disease outbreaks, and the emergence of HIV/AIDS—state and local health officials spoke of "getting more bang for their buck" and actively discouraged cluster investigations.[63]

But it was not only impoverished public health departments that shunned cluster investigations. So did university researchers. They pointed out that although a few studies had suggested links between disease clusters and environmental contaminants, none had produced definitive results. Researchers lamented that even where "true" clusters could be confirmed, it was not possible to link them conclusively to environmental conditions, given the small sample size and the insufficiency of available environmental data. Residents came and went, they behaved differently, their memories were unreliable. Likewise, the environment itself was always changing, and the relevant variables could not be controlled or even assessed. New pesticides were introduced; old ones moved into air and groundwater and were carried away. Relevant historical records, in particular records of pesticide use, were sketchy or nonexistent. With cancer, the relevant exposures likely occurred some time earlier—weeks or years or decades prior to any diagnosis. Like farmworkers, residents were unlikely to know what contaminants they had encountered over time. And whereas individuals could at least be asked about the source of their water or their food or how much they smoked, they had no idea what pesticides they might have been exposed to over the course of their lives. The decades-long failure to measure pesticide levels in the environment made it impossible to assess individuals' exposure in the quantitative manner that epidemiology required. Data, after all, are gathered for a reason. By the 1980s, when limited environmental measurements were finally available, human exposures to pesticides in the valley were so widespread that they defied complete identification much less quantification. Moreover, in cases of environmental contamination, everyone in the community or neighborhood was likely to be exposed to some degree; thus, there might be no truly unexposed persons who could serve as scientific controls.[64] The complexities of the modern environment made it profoundly difficult to create a set of research methods that could even begin to approximate the norms established in the laboratory.

Moreover, the political and liability issues that surrounded environmental contamination meant that that any such study was likely to come under intense scrutiny from industry-backed scientists and also by the courts—and many did. As more and more questions were raised about methodology, several leading epidemiologists offered their own critiques of the field. Some criticized epidemiology for its inattention to the underlying (biological) mechanisms of disease production, comparing it unfavorably to fields like microbiology. Others argued that even in comparison to animal studies of toxicity and risk, ecological studies of exposure were inadequate because they failed to control sufficiently the relevant variables. In other words, so-called environmental epidemiology was repeatedly criticized as a kind of second-rate field science, fatally hampered by the inability of investigators to control the complex social and environmental realities encountered outside the walls of the laboratory.[65] Even though the laboratory had obvious limits, particularly when the concern lay with environmental sources of disease, the laboratory paradigm remained the standard by which any model of the body and its illnesses would be measured.

Many practicing epidemiologists became openly critical of efforts to apply their methods to issues of environmental contamination, especially to cancer clusters. Several authors outlined the technical problems with such efforts in the field's leading journals, emphasizing that most suspected clusters turned out not to show elevated cancer rates on closer investigation. Richard Jackson, who had worked on the DBCP investigation for the state of California and later moved to the Centers for Disease Control (CDC), noted that among professionals cancer clusters were referred to as the "epidemiologists' fool's gold"—something you shouldn't spend your time chasing because it was scientifically valueless. As researchers refined their statistical techniques and analyses, many argued that clusters were probably chance events. In 1990 the *American Journal of Epidemiology* published a special issue dedicated to cluster studies. The journal's editor, Kenneth Rothman, opened that issue by declaring that "there is little scientific value in the study of disease clusters." That same year the CDC announced that it would no longer routinely assist in cluster investigations, citing their cost and ineffectiveness. For scientists, cancer cluster investigations offered only a research dead end. The realities of places such as McFarland did not conform to the strategies and methods of modern biomedicine, so most epidemiologists turned away from them.[66]

Thus, although the turn toward epidemiology in environmental health research initially promised a more environmental approach, in most cases environmental epidemiology has ultimately reinforced modern concepts of the body, as it did in McFarland. By failing to confirm environmental contamination as the source of disease, epidemiology reinforced the belief that individual factors were the better explanation. In McFarland, some took the epidemiological work as an exoneration of the "environment" and pesticides in a broad sense and therefore as license to ascribe the cancers to genetics and "high-fat diets." But even for those researchers who genuinely want to modify the reductionist focus of biomedicine there is, as Robert Aronowitz has observed, a seemingly inescapable tendency to gradually abandon an initially "holistic" approach in favor of a more "ontological" and reductionist one. Over time, as innovative researchers seek legitimacy for their claims, they often recast their hypotheses in narrower and more specific terms. They begin to formulate hypotheses that more closely conform to the time-tested models in their field. The paradigm and limits of laboratory science subtly shape the kinds of questions that can be asked about both disease and environments.[67]

It is hardly surprising then that many who had worked in the field of environmental epidemiology would turn, at the beginning of the twenty-first century, to the emerging field of toxicogenomics. A field inspired by the human genome project on the one hand and toxicology on the other, toxicogenomics focuses on identifying the underlying genetic factors that determine an individual's susceptibility to certain toxins. The goal is to identify discrete changes in a cell's genomic profile that are attributable to a specific toxic agent.[68] Toxicogenomics might be described as both ecological and modern in its approach. It acknowledges forthrightly the permeability of the body but takes as its focus the fact that some bodies are more permeable than others. While it focuses on issues of environmental contamination, it implicitly privileges individual factors (i.e., genetic makeup) as the critical cause of disease. No doubt the contemporary enthusiasm for toxicogenomics comes in part from the fact that it promises the kinds of individual remedies that have been so successful for modern biomedicine in the past. It also leaves behind the messiness of actual social and environmental arrangements for the constrained comforts of the modern laboratory. Once again, it abstracts the "environment" into a set of specific chemicals and turns the focus back to the interior of the body. Meanwhile the environments outside the laboratory continue to change in countless ways. Complex ecological realities continue to coexist with modernist hopes.

ECOLOGICAL BODIES AND ENVIRONMENTAL JUSTICE

In the years during and after the McFarland controversy, many laypersons and social activists began to formulate a decidedly different approach to the issue of spatialized disease. In histories of environmentalism, the 1980s are noteworthy not only as the decade in which toxic contamination came to dominate the environmental agenda but also as the period in which hundreds of grassroots groups emerged whose agendas diverged from those of mainstream liberal organizations. Although often described as a "movement," the label "environmental justice" actually encompassed diverse environmental struggles on the part of those who felt themselves disenfranchised by the political and regulatory process. These new activists did not focus on conservation or federal pollution policy. Instead what tied their struggles together was their local focus, their concern with health, and their emphasis on the structural (rather than the individual) causes of disease.

Critical to the success of environmental justice was its ability to mobilize long-standing popular resistance to reigning biomedical models of disease and environment. In its attention to the ecology of bodies and the health effects of pollution, environmental justice drew indirectly on the 1960s environmentalism of Rachel Carson and others. Like Carson, environmental justice advocates insisted on the significance of ongoing but low-level exposures to pesticides over many years. Implicitly they viewed human bodies as porous, as always susceptible to insult and injury. Where Carson had both used and critiqued toxicology (the dominant science of environment and health in the 1950s and 1960s), environmental justice advocates developed a similarly ambivalent relationship to risk factor epidemiology. Although they used epidemiologic evidence to demonstrate the spatial nature of disease, they also critiqued the field's reliance on "race" and "lifestyle" as explanatory factors. Like the McFarland mothers, environmental justice activists rejected explanations of ill health in marginalized communities as the outcome of individual choice, genetic inheritance, or simply chance, insisting instead that illness should be associated with poor environmental quality where it could not be proven otherwise.[69] Activists grounded their claims to a clean environment in civil rights discourse and the legal concept of equal protection. From their perspective, the issues of biomedical causality that dominated discussions of toxicology and epidemiology, however interesting, were always secondary to the fact that modern environments were materially different.

But while it drew from earlier understandings, environmental justice advocacy also rewrote those understandings in ways particular to the late twentieth century. In the 1960s Carson had argued powerfully that everyone was vulnerable to the modern environment—an argument that was designed to appeal to the suburban and otherwise politically complacent middle-class that was her principal audience. In contrast, environmental justice activists made a point of emphasizing that disease was, in fact, spatialized in certain communities—as well-meaning sanitationists and outright racists had argued all along. But environmental justice advocacy reinterpreted the significance of that fact: rather than the outcome of unhygienic habits and susceptible bodies, the spatialization of disease reflected particular histories of land use, the legacies of racial segregation, and the geographic effects of class. This analysis pointed to the environmental and structural, rather than the individual, causes of disease. Because businesses and government had shown proportionately less concern for the environmental quality of poor and nonwhite communities, and in some cases had actively targeted those communities for polluting activities, disease itself was one outcome of racial and class discrimination.[70]

Environmental justice advocates insisted on the need to redifferentiate modern spaces in public health discourse and to acknowledge that living in one landscape, however modern, could be more far dangerous than living in another. This was contrary to the long-standing claims of public health experts that landscape modernization was itself the key to health. It had been a primary goal of modern public health to render local environments as "space"—clean, sanitary, and therefore irrelevant.[71] Environmental justice emphasized the perpetual failure of that project.

In effect, environmental justice activists reterritorialized disease while reappropriating diseased bodies as indicators of particular landscapes. Bodies themselves became (again) a means for visualizing the unseen and ultimately intimate processes of pollution and environmental rearrangement that certain communities experienced: contaminated groundwater, pesticide-laden fog, polluted air. Like the nineteenth-century settlers and physicians who chronicled their bodies' reactions to north winds and winter fogs, community health activists now tracked diagnoses of cancer, the occurrence of miscarriages, and the rates of asthma among local children—already convinced that their ill health must bear some relation to the places that they occupied. Typically they invoked both modern and ecological models of the body. Activists and community members drew on their own experience with pollution and illness, but they framed their concerns within the languages of modern science: groundwater chemistry, laboratory toxicology, and risk factor epidemiology.[72] In these formulations, afflicted bodies were indictments of both certain environments and a capitalist society that fostered a spatialized inequality.

Ironically, as the movement for environmental justice gained national attention in the 1990s, the struggle in McFarland slipped from political and public view. The waning of overt conflict was not, however, a victory for the methods of modern science and public health. Rather it represented the exhaustion of a community that could not generate continuous and public resistance to the institutionalized view of environment and health. Divisions within the community—between Anglos and Latinos, farmworkers and the middle class, the UFW and the affected families—made it difficult for residents to remain united in their response to the cancers. Politicians who had helped highlight the McFarland situation moved on to other issues, and many of the affected families either moved away or simply wanted to move on from the tragedy of cancer. And although most agency representatives continued to insist that studies had shown McFarland's environment to be "safe," few people in the community felt reassured. Instead even the normal illnesses of childhood had become a perpetual cause for anxiety.[73]

In 1995, four years after the state of California ended its investigations, several current and former residents petitioned the U.S. EPA and the Agency for Toxic Substances and Disease Registry to reopen studies of McFarland's environment, this time framing their request in terms of environmental justice. In the meantime, questions from the press forced the California Department of Health Services to acknowledge publicly that seven new cancer cases had occurred between 1990 and 1996 and that the incidence of childhood cancer in McFarland remained unusually high. Although the EPA had agreed to undertake additional studies of McFarland's environment, the local health department, along with the town's Anglo leaders, were strongly opposed. Within the CDHS opinions were mixed: some argued that interviewing the recent cases might turn up new leads; others maintained that their methods were as unlikely to succeed the second time around as they had been the first. To do anything, this group argued, would only raise false hopes in the community while inviting more public scrutiny. In the end, the skeptics won. The CDHS released a fact sheet stating that science could "not provide an answer."[74] Nonetheless, more than a decade after the discovery of the cancer cluster, suspicions remained and studies continued.

Whatever ambiguities remained in McFarland, the developments there and elsewhere marked the decline of the sanitation paradigm in public health while exposing the weaknesses of a regulatory system built exclusively on that model. The belief in sanitation has always hinged on the overly simple belief that the environment could be made, once and for all, into a passive space. Modernist discourses of sanitation and health had generated the expectation of a benign and controllable landscape and an impervious and autonomous body—expectations that, in the end, modern public health could not deliver. As suspected carcinogens initially applied to crops turned up in air, fog, soil, water, and, eventually, human bodies, it became obvious that health and environmental experts could neither sanitize nor materially contain the new contaminants. Even modern places were not the same, and they were certainly not under anyone's "control."[75]

By the 1980s the supposed distance between bodies and environments that had helped underwrite the project of modernization had become increasingly difficult to sustain in the face of ongoing pollution and unexplained illness. As both science and experience revealed a more complicated and interconnected environment, experts could not deny outright its potential to affect health—even when their own methods failed to reveal any link between environmental conditions and disease. In fact, Proposition 65—California's "right-to-know' legislation—took as its starting point the fact that there are different ways of interpreting chronic disease and the risks posed by toxic chemicals, and therefore the public required information with which to make their own decisions. Even some experts now openly acknowledged the limits of their traditional approach to these issues. "I'd hate to think that anybody is getting the impression that we're giving the environment in McFarland a clean bill of health," one researcher ruefully remarked at the close of the state's investigation.[76] Modern understandings of disease could not encompass the experience of McFarland.

Yet modern understandings of health and environment did not disappear. It would be more accurate to say that they were recalibrated to the new realities. The enthusiasm for toxicogenomics was a case in point. While acknowledging the permeability of the body, it also followed the laboratory paradigm, reduced the environment to a set of discrete toxins, and took as its focus the differential susceptibility of certain bodies. Moreover, it was not only experts who continued to invoke modern understandings. McFarland's residents constructed their relationship to the environment in modern as well as other-than-modern ways. "Modern" in the sense that they sought an unobtainable purity, a landscape and a body free from disease-causing contaminants and pathogens; "other-than-modern" in the sense that they believed their bodies to be in constant interaction with, and deeply dependent on, the larger environment.

Although simple dichotomies fail to capture the history, it is still true that there were significantly different ways of interpreting the relationship between health and environment in McFarland. To assert a cancer "cluster" in the late twentieth century was to articulate an understanding of the relationship between bodies and environments, humans and nature, that insisted on the relevance of local ecologies. Those who pushed for the recognition and investigation of clusters assumed that local environments had an ongoing effect on health that biomedical science too often failed to reveal and that government institutions were unwilling to fully investigate. They asserted the relevance of sick bodies to understanding the larger environment and the changes it was undergoing. Whether they knew it or not, they were heirs of both Thomas Logan and Rachel Carson. Their opponents, meanwhile, marshaled the modern notion of an independent and self-contained body, insisting, among other things, that a lifetime of pesticide exposures had had no effect on their physical selves. As Michelle Murphy has argued, at any one time, there may exist "different ways of being a body in the world."[77] Certainly that was true in the Central Valley at the end of the twentieth century. But as the residents of McFarland came to realize, only certain understandings have the explicit backing of the state.

It is not surprising that those whose economic and social interests depended on the continuing use of chemicals have a significant stake in asserting the uncertainty or the irrelevance of any connections between environment and health. For those interests, the modern model works well. Yet while economic interests were part of the McFarland story, it was not agribusiness that produced epidemiologic studies or adopted a standards-based regulatory regime. Epistemology is intertwined with, but not necessarily determined by, politics—that would be too simple of a reading. Whatever claims growers and politicians might want to make about health, those claims could only be legitimated by the work of disinterested scientists and government experts who had some claim to neutrality and specialized knowledge. But however much experts might want to take into account the environmental causes of disease (and many did), they also remained professionally vested in the modern model.[78] Science and regulation are themselves products of both culture and history. Culture shapes how scientists and laypeople alike classify and think about "disease" and the "environment," and the history of knowledge production shapes both regulatory structures and contemporary scientific inquiry. It is worth remembering that the basic strategies and assumptions of modern public health were formed in the early twentieth century, when germ theory enjoyed its greatest authority. When new kinds of knowledge—about the ability of the environment to harbor pathogens, about the mobility of chemicals, about the sensitivity of bodies to repeated environmental exposures—challenged modern understandings that insisted on the separation of human beings from nature, it remained difficult if not impossible for those in public health to relinquish, perhaps understandably so, a model that had underwritten the major successes of their field as well as their own professional training. At some point, being "scientific" had come to mean trusting to a specific way of conceptualizing the relationship between bodies and their environments—even though many experts admitted that their methods were woefully inadequate. And so public health experts continued to reconcile complex environmental realities with reductionist methods. That, after all, had worked for them before. Laypeople, meanwhile, were far more willing to abandon the modern model when it failed to explain their experience.

What remains is the tension between modern and ecological ideas of health. To cast that tension as simply a conflict between rationality and irrationality, or between honesty and deceit, is to vastly oversimplify the situation and to ignore a long history of environmental health concerns. And though that tension remained submerged in McFarland in the late 1990s, it was most likely only a matter of time before it reemerged in some form. In fact, more than twenty years after the discovery of the cluster, residents still expressed concern about the environment, while health officials remained sensitive about their role in the McFarland investigations and wary of outsiders who wanted to return to the topic of the cancers.[79] Modernist ideas of bodies and environments remained in place, but their instability was now an open secret.

Conclusion

The people who have inhabited California have reshaped the landscape in countless ways. The evidence of human-induced change is everywhere; indeed, it is overwhelming. The Central Valley, like all of North America, is now a complicated mixture of human and nonhuman elements, a hybrid landscape: aquifers and aqueducts, soils and chemicals, native plants and commercial crops. But change did not occur in only one direction. As people have shaped the landscape, the landscape has shaped the bodies of its inhabitants. To the extent that health is dependent on the water that people drink, the food that they eat, the air that they breathe, and the organisms that they live alongside, human beings are not materially separable from their environments. Even as the processes of capitalism and the mediations of technology have allowed Americans to become ever more physically and intellectually alienated from their local ecologies, even the most privileged have never fully escaped or superseded them. Persistent concerns about environmental illness mark that fact.

If modernity is, as Bruno Latour suggests, a story we have told ourselves about the separation of human beings from nature, then the history of health is one more place where that story is undermined. It is not simply that what we think of as "nature" is really a complex mixture of nature and culture; what we call "human" is similarly mixed. Not only have humans mixed their labor with nature to create hybrid landscapes; nature—already a mixture of human and nonhuman elements—has intermixed with human bodies, without anyone's consent or control, and often without anyone's knowledge. The malaria plasmodium came to California in the bodies of English trappers who did not realize they carried it, found a home among California's anopheline mosquitoes, and eventually entered into the bodies of unsuspecting American immigrants, who then propagated the parasite through their agricultural practices. Farmers sprayed their trees with chemicals, which clung to branches and leaves, only to be transformed by the local climate into deadly toxins, which then entered the bodies of workers, shifting the balance of biological chemicals, making bodies tremble and sweat, and sometimes killing them. Similar if less acute changes have occurred in bodies across North America. In the mid-1980s, as the McFarland controversy was unfolding, a study conducted by the U.S. EPA found measurable levels of chemicals in all adults tested in the United States. Most people would no doubt be surprised and disturbed to learn that they have measurable amounts of chlorobenzene, benzene, ethyl benzene, toluene, and polychlorinated biphenols in their fatty tissues.[1] Although the significance of these transactions can be debated, the material connections between bodies and environments to which they attest are undeniable. Ironically, it has often been in the most industrialized landscapes—those landscapes that are typically taken as symbolic of the human alienation from nature—that these connections become most clear. As humans have industrialized the land, the land has, in turn, industrialized them. Neither the realm of nature nor the realm of the human remains pure.

Nineteenth-century medicine seems strange to contemporary readers in part because it made the intermixture of bodies and environments explicit. Admittedly that is also the source of my own attraction to it. Health and medicine remained realms of knowledge in which the modern separation between humans and nature was by no means complete. Both lay and professional understandings held bodies to be permeable. People worried openly about thick fogs, swampy ground, and the effects of mining deposits. Long before the advent of modern ecology, they understood themselves as organisms that were connected to their environment in a multitude of ways. Medical practitioners diagnosed landscapes as well as bodies, reading health from the thermometer, the rain gauge, the timing of fish runs. In that world, health was not a quality that individual bodies possessed or lacked but a state that emerged when a given body was in harmony with a particular landscape.

Although nineteenth-century medicine acknowledged the critical role of the environment, the desire to separate the environment from health and disease was already present. Ecological understandings of health always existed in tension with the desire to transform the land. Germ theory would help resolve that tension in favor of modernization. If early-twentieth-century public health professionals were overly zealous in their pursuit of germs, it was in part because the new theories served many needs. The advent of germ theory and "modern" medicine coalesced with a growing desire to downplay the environment's role in health, furthering the intellectual separation of human beings from the rest of nature. The healthy body was now a pure body, one that had not been penetrated by pathogens or parasites. The avoidance of these bodily contaminants depended on proper hygiene and sanitation, which supposedly worked to neutralize the surrounding environment: the isolation of wastes, the destruction of insects, the eradication of dirt. Once properly sanitized, the environment could then be ignored, or so some people thought.

When early-twentieth-century public health experts raised the profile of the "germ," they simultaneously described the broader environment as a passive and homogeneous space. The only actors in their story were human beings and certain microscopic pathogens. The ability to produce health now lay with the doctor on the one hand and the individual on the other—an individual who was conceptualized as an autonomous and effective actor who was now urged to rid her house of flies, wash her hands, eat right, and exercise regularly. Changes in the nonhuman environment were no longer something with which doctors or patients need concern themselves. Environmental concerns could thus be left to others—engineers, entomologists, hydrologists, agronomists. The modern construction of the body as a discrete and bounded entity encouraged the isolation of medical from environmental concerns and enabled the further specialization of professionals.

But that isolation has always been somewhat tenuous. In fact, the environment's role in disease could be professionally marginalized, but it could not be denied. Within medicine and public health, environmental concerns continued to find a home in particular subspecialties— including those of tropical medicine, disease ecology, occupational health, sanitary engineering, and modern epidemiology. Medical discourse is itself multifaceted, and some specialties have attended to the environment's influence on disease even when the most dominant specialties have not. But perhaps more important, outside of medicine the decline of Hippocratic ideas was much more gradual and incomplete, and the distinction between environmental and health concerns has often blurred.[2] Even while individuals adopted many of the tenets of a professionalized medical discourse and often deferred to experts, those understandings never fully encompassed individuals' experience of disease or place, nor have they necessarily erased older understandings. As evidence emerged in the postwar period that environmental change was affecting health, modern assumptions about the environment's irrelevance were repeatedly undermined, and the arrival of sanitary modernity was again deferred.

The notion of the body that has underwritten environmental activism since the 1960s is an uneasy combination of these preceding configurations. It juxtaposes an awareness of the body's permeability and susceptibility to environmental change with the modernist desire for bodily purity. In the early twenty-first century, the ecological body is both dependent on its environmental context and exceedingly vulnerable to contaminants. Because the ecological model of health succeeded the modern construction of the body, it seemed, in some ways, radically new. But, like their nineteenth-century predecessors, those who mobilized an ecological construction of the body saw pathology in landscapes as well as individuals; they viewed bodies as exceedingly porous and vulnerable; and they understood both health and disease as the outcome of many interacting factors. From this perspective, it is the earlier erasure of the environment's role in disease that requires explanation.

Implicit in these different constructions of the body, ecological and modern, have been differing conceptions of the environment and, at root, different ideas about agency. Ideas of bodies and environments, humans and nature, are necessarily intertwined. When nineteenth-century settlers understood their bodies as porous, they also understood the land as a set of unique ecologies, each of which had to be closely studied for its health effects. Disease was not located in a single and discrete pathogenic agent; rather, it was distributed among a host of factors, no one of which could be considered in isolation from the rest. The corollary was that human agency with respect to health was understood as far more constrained and partial. In this world, doctors were necessarily environmental scientists, and the contributions of physicians vis-à-vis nature in the process of healing were always ambiguous.

In retrospect, it may seem only logical that a society based on radical and incessant changes to the landscape would not long sustain an ecological view of the body. Nevertheless, such a view has persisted, even if it has not been the dominant understanding; and such a view has emerged in certain places and moments as the basis for advocating certain environmental actions while proscribing others. Most obviously, much of the modern environmental movement emerged from the experience of health and disease in specific postwar landscapes, like those of farmworkers in the Central Valley. While market capitalism and industrial technology have increasingly distanced Americans from their environments, their experience and understanding of health still, at certain moments, connects them to the land. No matter how invested someone may be in the processes of capitalist expansion, questions of health can force a reconsideration—as when nineteenth-century boosters warned prospective immigrants to avoid the Central Valley, or successful and otherwise modern farmers began to reconsider their use of pesticides in the wake of a family experience with cancer. Such instances suggest that, at the end of the twentieth century, human alienation from the landscape, even the highly industrialized landscape of the Central Valley, remained incomplete. All this is not to say that understandings of health and environment are independent of the social and economic contexts in which they arise; clearly they are not. But neither are they fully determined by those contexts. They must also incorporate the lived experience of illness and the often recalcitrant materiality of the nonhuman world.[3]

In a world where outright resistance to the processes of unrestrained capitalism has become incredibly difficult to sustain, concerns about illness and its relationship to an industrialized landscape provide an important means for galvanizing different kinds of people to question the trajectory of their own modernization. Political battles over health and understandings of the body in the Central Valley have, for good reason, also been battles over space. Although both postwar environmentalism and contemporary environmental justice have many social and political sources, they also find their basis in an ecological understanding of the body and an environmental understanding of health. Concerns about environmental health often cross lines of class and race, as well as the boundaries of continents and nations. Whether in the late nineteenth century or the late twentieth, debates over disease have also been arguments about how people should live in the landscape, what shape that landscape should take, and how it should be understood.[4] As Henri Lefebvre pointed out some time ago, for those that want to forge different understandings of space and nature, the reappropriation of the body and its conceptualization may be the first critical step.[5]

Concerns about environment and health remain, in the Central Valley and elsewhere. The health effects of pesticide use among both farmworkers and the public constitute an ongoing regulatory debate and, for many, a perpetual source of anxiety. The valley itself has generated a plethora of local environmental justice struggles. In 1989 the California Rural Assistance League, a group that had lobbied for farmworker rights for more than two decades, formed the Center on Race, Poverty, and the Environment to address the links between environment and health in California's poor and minority communities, and they immediately opened an office in the valley. In 1996 a coalition of cancer and environmental groups formed Californians for Pesticide Reform to lobby for bans on highly toxic pesticides and stricter regulations. In addition, air pollution from sources other than pesticides has emerged as a major health issue in the San Joaquin Valley, which now has one of the highest rates of asthma in the nation. Access to high-quality, uncontaminated drinking water continues to be an issue, a subject of both ongoing litigation and legislation. Like the harsh realities of farm labor in the region, issues of contamination and environmental health have helped recast the image of modern farming in the minds of most outsiders. By the early twenty-first century, the contradictions embedded in the landscape of rural California have become readily apparent: immense productivity and scathing poverty, perfect fruit and toxic pesticides, beautiful landscapes and the fear of cancer.

The history of health and environment as I have told it is not a linear story. It does not conform to either the progressive narrative of medical discovery or the declensionist story of relentless scientific reductionism. There have been both discoveries and reductionism, as there will continue to be. There is no single "right" way to conceptualize human and nonhuman natures, though there are better and worse ways. The environmentalist demand for absolute purity was itself an outgrowth of sanitary modernity—and equally unobtainable, as its critics are quick to point out. But if purity is unobtainable, so, too, is specific etiology. The emphasis on toxicity testing and genetic profiling offers scant hope for finally understanding how complex and ever-changing environments affect the health of their occupants and even less hope for managing those relationships. The question then is, what to do? How do we want to think about health, disease, and environment, and what kinds of practices make the most sense, both socially and biologically?[6] I am not advocating a return to medical topography or the miasma theory, yet it is worth questioning why so much of contemporary biomedicine is divorced from any study of the larger environment and why individual solutions to disease such as improving one's diet are quickly institutionalized while other, more difficult social and environmental questions are not even discussed. Ultimately we cannot escape the environments in which we live, but it should be possible to foster a science and regulatory structure that strives to make visible the density of that connection. In a cultural moment marked by seemingly unrestrained technological exuberance, recalling and rethinking our physical natures and biological dependence has never been more necessary. This book is a contribution toward that rethinking and part of an ongoing conversation about how best to understand the relationship between our own species and the larger world.

Notes

ABBREVIATIONS

BC	*Bakersfield Californian*
BL	Bancroft Library
BSE	Bureau of Sanitary Engineering
CDA	California Department of Agriculture
CDHS	California Department of Health Services
CDPH	California Department of Public Health
CSA	California State Archives
CSBH	California State Board of Health
CSJM	*California State Journal of Medicine*
CSL	California State Library
JAMA	*Journal of the American Medical Association*
KCDPH	Kern County Department of Public Health
LAT	*Los Angeles Times*
NYT	*New York Times*
SFC	*San Francisco Chronicle*
SB	*Sacramento Bee*
SSMI	Sacramento Society for Medical Improvement
WRCA	Water Resources Center Archives

INTRODUCTION

1. On the social history of the valley, see Walter Goldschmidt, *As You Sow* (New York: Harcourt, Brace, 1947); Cletus Daniel, *Bitter Harvest: A History of California Farmworkers, 1870–1941* (Berkeley: University of California Press, 1981); Devra Weber, *Dark Sweat, White Gold: California Farm Workers, Cotton, and the New Deal* (Berkeley: University of California Press, 1996); Sucheng Chan, *This Bittersweet Soil: The Chinese in California Agriculture, 1869–1910* (Berkeley: University of California Press, 1986); Steven Stoll, *The Fruits of Natural Advantage: Making the Industrial Countryside in California* (Berkeley: University of California Press, 1998); Carey McWilliams, *Factories in the Field: The Story of Migratory Farm Labor in California* (Boston: Little, Brown, 1939); Don Mitchell, *The Lie of the Land: Migrant Workers and the California Landscape* (Minneapolis: University of Minnesota Press, 1996); Linda C. Majka and Theo J. Majka, *Farm Workers, Agribusiness, and the State* (Philadelphia: Temple University Press, 1982). On environmental history, see Donald J. Pisani, *From the Family Farm to Agribusiness: The Irrigation Crusade in California and the West, 1850–1931* (Berkeley: University of California Press, 1984); Robert Kelley, *Battling the Inland Sea: American Political Culture, Public Policy, and the Sacramento Valley, 1850–1986* (Berkeley: University of California Press, 1989); Raymond Dasmann, *The Destruction of California* (New York: Collier Books, 1966); Donald Worster, *Rivers of Empire: Water, Aridity, and the Growth of the American West* (New York: Pantheon, 1985); William L. Preston, *Vanishing Landscapes: Land and Life in the Tulare Lake Basin* (Berkeley: University of California Press, 1981); David Igler, *Industrial Cowboys: Miller & Lux and the Transformation of the Far West, 1850–1920* (Berkeley: University of California Press, 2001). For an interesting exception to this literature on the valley, see Ian Tyrell, *True Gardens of the Gods: Californian-Australian Environmental Reform, 1860–1930* (Berkeley: University of California Press, 1999). For a popular account of the valley that also resists the narrative of alienation, see Stephen Johnson, Robert Dawson, and Gerald Haslam, *The Great Central Valley: California's Heartland* (Berkeley: University of California Press, 1993).

2. Worster, *Rivers of Empire*, 5.

3. Conevery Bolton Valenčius, *The Health of the Country: How American Settlers Understood Themselves and Their Land* (New York: Basic Books, 2002), 143. Valenčius's is the only full-length work to address the cultural connection between health and environment. For a collection of recent essays, see Gregg Mitman, Michele Murphy, and Christopher Sellers, eds., *Landscapes of Exposure: Knowledge and Illness in Modern Environments*, *Osiris* 19 (2004). Also see Gregg Mitman, "Hay Fever Holiday: Health, Leisure and Place in Gilded Age America," *Bulletin of the History of Medicine* 77 (2003): 600–635. On eighteenth- and nineteenth-century medical geography, see Nicolaas A. Rupke, ed., *Medical Geography in Historical Perspective*, Supplement No. 20 to *Medical History* (London: Wellcome Trust Center for the History of Medicine, 2000); Frank A. Barrett, "Daniel Drake's Medical Geography," *Social Science of Medicine* 42 (1996): 791–800.

4. For example, Kenneth Thompson, "Insalubrious California: Perception and Reality," *Annals of the Association of American Geographers* 59 (March 1969): 50–64. On "modern medicine," see Richard Harrison Shyrock, *The Development*

of Modern Medicine: An Interpretation of the Social and Scientific Factors Involved (London: Victor Gollanz, 1948).

5. For these concerns, see Michelle Murphy, "The 'Elsewhere within Here' and Environmental Illness; or, How to Build Yourself a Body in a Safe Space," *Configurations* 8 (Winter 2000): 87–120; Steve Kroll-Smith and H. Hugh Floyd, *Bodies in Protest: Environmental Illness and the Struggle over Medical Knowledge* (New York: New York University Press, 1997); Phil Brown and Edwin J. Mikkelsen, *No Safe Place: Toxic Waste, Leukemia, and Community Action* (Berkeley: University of California Press, 1990); Morton Lippmann, *Environmental Toxicants: Human Exposures and Their Health Effects* (New York: Wiley-Interscience, 2000).

6. David B. Morris, "Environment: The White Noise of Health," *Literature and Medicine* 15 (Spring 1996): 1–15. Here and in the remainder of the book, for lack of a better term, I follow Morris and others who use the words *modern* and *modernist* to refer to those beliefs and constructions about the body and its diseases that were based on germ theory and which emphasized the separation of bodies from their environmental contexts. In this view, the body is understood as self-contained, illness is strictly an attribute of the body, and the source of illness lies in discrete pathogens or parasites. I am, however, aware of debates over these terms and the argument that *modernist* and *modernism* should be restricted to those artists and art forms associated with a critique of modernization in the late nineteenth and early twentieth century; however, by restricting *modernist* in this way, we are left with no term to describe the ideological and aesthetic positions of those who embraced modernization less critically. I use *modernist* to describe the positions of those in medicine and public health who self-consciously positioned themselves and their ideas as modern, in comparison to the ideas of those who came before. For a good recent discussion of this issue, see Frederick Cooper, *Colonialism in Question: Theory, Knowledge, History* (Berkeley: University of California Press, 2005), 113–52.

7. Rachel Carson, *Silent Spring* (Boston: Houghton Mifflin, 1962); Ralph H. Lutts, "Chemical Fallout: Rachel Carson's *Silent Spring*, Radioactive Fallout and the Environmental Movement," *Environmental Review* 9 (1985): 210–25; Samuel P. Hays, *Beauty, Health, and Permanence: Environmental Politics in the United States, 1955–1985* (New York: Cambridge University Press, 1987). Although historians have recognized that health concerns have had a substantial role in the development of modern environmentalism, the standard histories have given the issue far too little attention. But see Hays and also Adam Rome, *The Bulldozer in the Countryside: Suburban Sprawl and the Rise of American Environmentalism* (New York: Cambridge University Press, 2001).

8. Timothy Mitchell, *Rule of Experts: Egypt, Techno-Politics, Modernity* (Berkeley: University of California Press, 2002); Bruno Latour, *We Have Never Been Modern*, trans. Catherine Porter (Cambridge, MA: Harvard University Press, 1993); Gregg Mitman, "In Search of Health: Landscape and Disease in American Environmental History," *Environmental History* 10 (April 2005): 184–210. Of course, I rely on the dichotomies as well. As I have suggested, the human/nature distinction is indispensable to history, even as it is limiting. But it may still be possible to the use the tools of history to tell a different kind of story.

9. Exceptions to this include some of the critical work in public health history

that focuses on specific cities: Judith Walzer Leavitt, *The Healthiest City: Milwaukee and the Politics of Health Reform* (Princeton, NJ: Princeton University Press, 1982); Stuart Galishoff, *Newark: The Nation's Unhealthiest City, 1832–1895* (New Brunswick, NJ: Rutgers University Press, 1988). More recent attempts to tell medical histories from specific places include Warwick Anderson, *The Cultivation of Whiteness: Science, Health and Racial Destiny in Australia* (New York: Basic Books, 2003); Martha L. Hildreth and Bruce T. Moran, eds., *Disease and Medical Care in the Mountain West: Essays on Region, History, and Practice* (Reno: University of Nevada Press, 1998).

10. Bruno Latour, *The Pasteurization of France,* trans. Alan Sheridan (Cambridge, MA: Harvard University Press, 1988); Andy Pickering, ed., *Science as Practice and Culture* (Chicago: University of Chicago Press, 1992); Steven Shapin and Simon Schaffer, *Leviathan and the Air-Pump: Hobbes, Boyle, and the Experimental Life* (Princeton, NJ: Princeton University Press, 1985).

11. Mitchell, *Rule of Experts,* esp. 19–53; Alan Irwin and Brian Wynne, "Conclusions," in *Misunderstanding Science? The Public Reconstruction of Science and Technology,* ed. Alan Irwin and Brian Wynne (Cambridge: Cambridge University Press, 1996), 218.

12. Joan W. Scott, "The Evidence of Experience," *Critical Inquiry* 18 (Spring 1991): 773–97; Margaret Lock, *Encounters with Aging: Mythologies of Menopause in Japan and North America* (Berkeley: University of California Press, 1993); Annemarie Mol, *The Body Multiple: Ontology in Medical Practice* (Durham, NC: Duke University Press, 2002); Barbara Duden, *The Woman beneath the Skin: A Doctor's Patients in Eighteenth-Century Germany,* trans. Thomas Dunlap (Cambridge, MA: Harvard University Press, 1991).

13. Gregg Mitman, Michelle Murphy, and Christopher Sellers, "Introduction: A Cloud over History," *Osiris* 19 (2004): 9; Valenčius, *Health of the Country,* esp. 53–84; Emily Martin, *Flexible Bodies: The Role of Immunity in American Culture from the Days of Polio to the Age of AIDS* (Boston: Beacon Press, 1994); David B. Morris, *Illness and Culture in the Postmodern Age* (Berkeley: University of California Press, 1998); Thomas Laqueur, *Making Sex: Body and Gender from the Greeks to Freud* (Cambridge, MA: Harvard University Press, 1990); Londa Schiebinger, *Nature's Body: Gender and the Making of Modern Science* (Boston: Beacon Press, 1993); Jonathan Crary and Sanford Kwinter, eds., *Incorporations* (New York: Zone, 1992). As several of these authors suggest, the ability to begin to see the modern body as a historical creation may be because our own conceptions of the body are again changing in fundamental ways. The discourse of immune systems, the interest in alternative therapies, the availability of artificial reproduction, the development of artificial intelligence, developments in modern biology, the awareness of environmental disease—all these are signs that the "modern" body is being remade and that the "zoocentric" model is giving way. See especially Dorion Sagan, "Metametazoa: Biology and Multiplicity," in Crary and Kwinter, *Incorporations,* 362–85.

14. On germ theory, see Andrew Cunningham and Perry Williams, eds., *The Laboratory Revolution in Medicine* (Cambridge: Cambridge University Press, 1992); Michael Worboys, *Spreading Germs: Disease Theories and Medical Practice in Britain, 1865–1900* (Cambridge: Cambridge University Press, 2000); Margaret Pelling, "Contagion/Germ Theory/Specificity," in *Companion Ency-*

clopedia of the History of Medicine, ed. W.F. Bynum and Roy Porter (New York: Routledge, 1993), 309–34; Nancy Tomes and John Harley Warner, eds., "Introduction to Special Issue on Rethinking the Germ Theory of Disease: Comparative Perspectives," *Journal of the History of Medicine and the Allied Sciences* 52 (January 1997): 7–16.

15. On the construction of race and the colonial project, see Matthew Frye Jacobsen, *Whiteness of a Different Color: European Immigrants and the Alchemy of Race* (Cambridge, MA: Harvard University Press, 1998); Anderson, *Cultivation of Whiteness;* Mark Harrison, *Climates and Constitutions: Health, Race, Environment, and British Imperialism in India, 1600–1850* (New York: Oxford University Press, 1999); David N. Livingstone, "Human Acclimatization: Perspectives on a Contested Field of Inquiry in Science, Medicine and Geography," *History of Science* 25 (1987): 359–94; John Comaroff and Jean Comaroff, "Medicine, Colonialism, and the Black Body," in *Ethnography and the Historical Imagination* (Boulder: Westview Press, 1993), 215–34; Dane Kennedy, "The Perils of the Midday Sun: Climatic Anxieties in the Colonial Tropics," in *Imperialism and the Natural World,* ed. John M. MacKenzie (Manchester: University of Manchester Press, 1990), 118–40. On race in nineteenth-century California, see Tomás Almaguer, *Racial Fault Lines: The Historical Origins of White Supremacy in California* (Berkeley: University of California Press, 1994); Susan Lee Johnson, *Roaring Camp: The Social World of the California Gold Rush* (New York: Norton, 2000). A relevant reading of race and whiteness in a different colonial context is Ann Stoler, "Sexual Affronts and Racial Frontiers: European Identities and the Cultural Politics of Exclusion in Colonial Southeast Asia," *Comparative Studies in Society and History* 34 (July 1992): 514–51.

16. These correlations would lie at the root of the environmental justice movement as it emerged in California and elsewhere in the 1980s. I discuss this in chapter 5, but see also Robert D. Bullard, *Dumping in Dixie: Race, Class, and Environmental Quality,* 3d ed. (Boulder: Westview Press, 2000); David E. Camacho, ed., *Environmental Injustices, Political Struggles: Race, Class, and the Environment* (Durham, NC: Duke University Press, 1998).

17. See also Anderson, *Cultivation of Whiteness;* Nayan Shah, *Contagious Divides: Epidemics and Race in San Francisco's Chinatown* (Berkeley: University of California Press, 2001).

1. BODY AND ENVIRONMENT IN AN ERA OF COLONIZATION

1. Alfred W. Crosby, "Virgin Soil Epidemics as a Factor in the Aboriginal Depopulation in America," *William and Mary Quarterly,* 3d ser., 33 (April 1976): 289–99; Alfred W. Crosby, *Ecological Imperialism: The Biological Expansion of Europe, 900–1900* (Cambridge: Cambridge University Press, 1986). On disease and culture, see Alan Bewell, *Romanticism and Colonial Disease* (Baltimore: Johns Hopkins University Press, 1999); David S. Barnes, *The Making of a Social Disease: Tuberculosis in Nineteenth-Century France* (Berkeley: University of California Press, 1995).

2. The most striking example of this is Jared Diamond, *Guns, Germs, and Steel: The Fates of Human Societies* (New York: Norton, 1997). It is perhaps not

coincidental that the emphasis on the role of disease in enabling European colonization in North America came to prominence in the 1970s, in the wake of Vietnam. At that point, the explanation of superior technology no longer seemed adequate in itself, and, moreover, Americans had lost their appetite for their history of colonial violence. See also Joyce Chaplin, *Subject Matter: Technology, the Body, and Science on the Anglo-American Frontier, 1500–1676* (Cambridge, MA: Harvard University Press, 2001).

3. Valenčius, *Health of the Country.* See also Mitman, "Hay Fever Holiday."

4. Virgil J. Vogel, *American Indian Medicine* (Norman: University of Oklahoma Press, 1970), 83; Doyce B. Nunis Jr., "Medicine in Spanish California," *Southern California Quarterly* 76 (1994): 31–58. For the possibility that diseases preceded colonization, see William L. Preston, Jon M. Erlandson, and Kevin Bartoy, "Protohistoric California: Paradise or Pandemic?" *Proceedings of the Society for California Archaeology* 9 (1996): 304–9; William L. Preston, "Serpent in Eden: Dispersal of Foreign Diseases into Pre-Mission California," *Journal of California and Great Basin Anthropology* 18 (1996): 2–37.

5. Maynard Geiger and Clement W. Meighan, *As the Padres Saw Them: California Indian Life and Customs as Reported by the Franciscan Missionaries, 1813–1815* (Santa Barbara, CA: Santa Barbara Mission Archive Library, 1976); Sherburne F. Cook, *The Conflict between the California Indian and White Civilization* (Berkeley: University of California Press, 1943); Ralph L. Beals Jr. and Joseph A. Hester, eds., *California Indians* (New York: Garland Publishing, 1974), 1:86–88; Cook's estimates in Sherburne F. Cook, "Historical Demography," in *Handbook of North American Indians,* vol. 2, *California,* ed. Robert F. Heizer (Washington, DC: Smithsonian Institution Press, 1978), 91–98; Edward D. Castillo, "The Impact of Euro-American Exploration and Settlement," in Heizer, *Handbook of North American Indians,* 2:99–127.

6. Charles N. Rudkin, ed., *The First French Expedition to California: Laperouse in 1768* (Los Angeles: G. Dawson, 1959); Nunis, "Medicine in Spanish California."

7. Sherburne F. Cook, "The Monterey Surgeons during the Spanish Period in California," *Bulletin of the History of Medicine* 5 (1937): 43–72; José Benites, "California's First Medical Survey: Report of Surgeon-General José Benites," ed. Sherburne F. Cook, *California and Western Medicine* 45 (October 1931): 352–54; Georg H. von Langsdorff, *Langsdorff's Narrative of the Rezanov Voyage to Nueva California in 1806* (Fairfield, WA: Ye Galleon Press, 1988); Nunis, "Medicine in Spanish California."

8. Sherburne F. Cook, "Smallpox in Spanish and Mexican California, 1770–1845," *Bulletin of the History of Medicine* 7 (1939): 153–91.

9. John Work, *Fur Brigade to the Bonaventura: John Work's California Expedition, 1832–1833, for the Hudson's Bay Company,* ed. Alice Bay Maloney (San Francisco: California Historical Society, 1945), 69–70, 72.

10. Work, *Fur Brigade,* 76, 71; *Kern County Weekly Courier* (Bakersfield), 29 August 1874; Stephen Powers, *Tribes of California* (1877; reprint, Berkeley: University of California Press, 1976), 380; John Work, "Letter of 24 February 1834," *Washington Historical Quarterly* 2 (1908): 163–64.

11. Although the existing anthropological and scholarly literature focuses on the arrival of malaria in the 1830s, Rollin's account, written in 1786, notes the presence of "ephemeral and intermittent fevers" in the spring and autumn, which suggests that malaria might have already been present. Certainly it could have been introduced by the Spanish from the Port of San Blas. However, the cool temperatures on the California coast would have kept malaria from becoming epidemic. Rudkin, *First French Expedition;* Sherburne F. Cook, "The Epidemic of 1830–33 in California and Oregon," *University of California Publications in American Archaeology and Ethnology* 43 (1955): 303–26; Robert Boyd, *The Coming of the Spirit of Pestilence: Introduced Infectious Diseases and Population Decline among Northwest Coast Indians, 1774–1874* (Seattle: University of Washington Press, 1999), 84–115, quote on 84; Harold Farnsworth Gray and Russell E. Fontaine, "A History of Malaria in California," in *Proceedings and Papers of the Twenty-fifth Conference of the California Mosquito Control Association* (Turlock, CA: California Mosquito Control Association, 1957). Little attention has been given to the possibility of influenza as well, though there was a concurrent epidemic of that disease in Britain at the time. See Herbert C. Taylor Jr. and Lester L. Hoaglin Jr., "The 'Intermittent Fever' Epidemic of the 1830's on the Lower Columbia River," *Ethnohistory* 9 (1962): 160–78; Cook, "Historical Demography." It is worth noting that Cook estimated Indian population declines to be still greater in the 1845–55 period, when Indians suffered directly the effects of white violence and settlement.

12. The most obvious reason that white mortality was lower was the use of quinine, although Indian methods of treatment may have exacerbated the illness in some cases. Boyd, *Coming of Pestilence,* 91–92, 99–100.

13. For a general history, see Margaret Humphreys, *Malaria: Poverty, Race, and Public Health in the United States* (Baltimore, MD: Johns Hopkins University Press, 2001). On the nineteenth-century Midwest, see Erwin H. Ackerknecht, *Malaria in the Upper Mississippi Valley, 1760–1900,* Supplement No. 4 to the *Bulletin of the History of Medicine* (1945). On the effect of the disease on eighteenth-century colonists in the Chesapeake, see Darrett B. Rutman and Anita H. Rutman, "Of Agues and Fevers: Malaria in the Early Chesapeake," in *Biological Consequences of European Expansion, 1450–1800,* ed. Kenneth F. Kiple and Stephen V. Beck (1976; reprint, Aldershot: Ashgate, 1997). For a summary of the historical incidence of malaria in the United States, see Ernest Carroll Faust, "Malaria Incidence in North America," in *Malariology: A Comprehensive Survey of All Aspects of This Group of Diseases from a Global Standpoint,* ed. Mark F. Boyd (Philadelphia: W. B. Saunders, 1949), 749–63.

14. A. L. Kroeber, *Handbook of the Indians of California,* Bureau of American Ethnology Bulletin 78 (Washington, DC: Smithsonian Institution, 1925), 361, 513, 851. For contemporary accounts, see U.S. Congress, Senate, *Statistical Report on Sickness and Mortality in the Army of the United States,* 36th Cong., 1st sess., 1860, S. Exec. Doc. No. 52, 242; "Medicine among the Mormons and the Indians of North America," *Pacific Medical and Surgical Journal* 4 (May 1861): 343–47.

15. Work, *Fur Brigade,* 73.

16. Vivian Nutton, "Humoralism," in *Companion Encyclopedia to the History of Medicine,* ed. W. F. Bynum and Roy Porter (New York: Routledge, 1993), 281–91.

17. Senate, *Statistical Report on Sickness and Mortality,* 1860, 238.

18. On environment and the body, see Duden, *Woman beneath the Skin;* Valenčius, *Health of the Country,* esp. 53–84; Anderson, *Cultivation of Whiteness,* 11–40; Trudy Eden, "Food, Assimilation, and the Malleability of the Human Body," in *A Centre of Wonders: The Body in Early America,* ed. Janet Moore Lindman and Michele Lise Tarter (Ithaca, NY: Cornell University Press, 2001), 29–42; Kennedy, "Perils of the Midday Sun," 118–40; Mitman, "Hay Fever Holiday."

19. The Spanish reportedly introduced smallpox vaccination into California in 1817. Robert J. Moes, "Manuel Quijano and Waning Spanish California," *California History* 67 (1988): 78–93; Rosemary Keupper Valle, "Prevention of Smallpox in Alta California during the Franciscan Mission Period (1769–1833)," *California Medicine* 119 (July 1973): 73–77.

20. Erwin H. Ackerknecht, "Anticontagionism between 1821 and 1867," *Bulletin of the History of Medicine* 22 (1954): 562–93; Cook, "Smallpox," 156 (quotation from letter of Comandante General Ugarte y Loyola).

21. [Thomas M. Logan et al.], "Report on the Medical Topography and Epidemics of California," *Transactions of the American Medical Association* 16 (1865): 560. On contagion, see Pelling, "Contagion/Germ Theory/Specificity," 309–34; Valenčius, *Health of the Country,* 124–27; Charles Rosenberg, "The Cause of Cholera: Aspects of Etiological Thought in Nineteenth-Century America," *Bulletin of the History of Medicine* 34 (1960): 331–54.

22. Charles E. Rosenberg, "Body and Mind in Nineteenth-Century Medicine: Some Clinical Origins of the Neurosis Concept," in *Explaining Epidemics and Other Studies in the History of Medicine* (New York: Cambridge University Press, 1992), 77. "Climatological relations" from Lorin Blodget, *Climatology of the United States, and of the Temperate Latitudes of the North American Continent* (Philadelphia: J. B. Lippincott, 1857), 456.

23. Thomas M. Logan, *Medical History of the Year 1868, in California* (Sacramento: Sacramento Society for Medical Improvement, 1868), 12; F. W. Hatch, "Report on the Epidemics of California in 1868," *Transactions of the American Medical Association* 20 (1869): 527. Also see E. Malcolm Morse, "Something about the Small-Pox Epidemic," *California Medical Gazette* 1 (January 1869): 130–31.

24. Clarence J. Glacken, *Traces on the Rhodian Shore: Nature and Culture in Western Thought from Ancient Times to the End of the Eighteenth Century* (Berkeley: University of California Press, 1967), 551–622; Rupke, *Medical Geography in Historical Perspective.*

25. Both the British and the American armies began systematizing troop mortality and climate data in the 1830s. See Philip D. Curtin, *Death by Migration: Europe's Encounter with the Tropical World in the Nineteenth Century* (Cambridge: Cambridge University Press, 1989), 1–3; James H. Cassedy, *Medicine and American Growth, 1800–1860* (Madison: University of Wisconsin Press, 1986), 45–46. On tropics, see Nancy Stepan, *Picturing Tropical Nature* (Ithaca,

NY: Cornell University Press, 2001); David Arnold, ed., *Warm Climates and Western Medicine: The Emergence of Tropical Medicine, 1500–1900* (Amsterdam: Rodopi, 1996); James Lind, *An Essay on Diseases Incidental to Europeans in Hot Climates, with the Method of Preventing Their Fatal Consequences,* 1st Amer. ed. (Philadelphia: W. Duane, 1811); James Johnson and James Ranald Martin, *The Influence of Tropical Climates on European Constitutions,* 6th ed. (New York: Samuel S. and William Wood, 1846).

26. Johnson and Martin, *Influence of Tropical Climates,* 608.

27. M. Nicolson, "Alexander von Humboldt, Humboldtian Science and the Origins of the Study of Vegetation," *History of Science* 25 (1987): 167–94; Nicolaas A. Rupke, "Humboldtian Medicine," *Medical History* 40 (July 1996): 293–310.

28. On these debates, see Livingstone, "Human Acclimatization."

29. Johnson and Martin, *Influence of Tropical Climates,* 14.

30. Stoler, "Sexual Affronts and Racial Frontiers"; Jacobsen, *Whiteness of a Different Color;* Livingstone, "Human Acclimatization." For a representative primary source, see Charles Pickering, *The Races of Man: And Their Geographical Distribution* (London: John Chapman, 1849).

31. Karen Ordahl Kupperman, "Fear of Hot Climates in the Anglo-American Colonial Experience," *William and Mary Quarterly* 41 (1984): 213–40; Chaplin, *Subject Matter.*

32. Thomas Jefferson, *Notes on the State of Virginia,* ed. William Peden (Chapel Hill: University of North Carolina Press, 1954); Gilbert Chinard, "Eighteenth-Century Theories on America as Human Habitat," *Proceedings of the American Philosophical Society* 91 (February 1947): 27–57; Kupperman, "Fear of Hot Climates," 213–40. On temperate climates and "civilization," see Glacken, *Traces on the Rhodian Shore,* 537–50.

33. Chaplin, *Subject Matter.*

34. Reginald Horsman, *Race and Manifest Destiny: The Origins of American Racial Anglo-Saxonism* (Cambridge, MA: Harvard University Press, 1981); Ackerknecht, *Malaria in the Upper Mississippi Valley,* 16, 19; Blodget, *Climatology,* 460.

35. Stepan, *Picturing Tropical Nature,* 38–39; Samuel Forry, *The Climate of the United States and Its Endemic Influences* (New York: J. & H. G. Langley, 1842); Cassedy, *Medicine and American Growth,* 45.

36. Daniel Drake, *A Systematic Treatise Historical, Etiological, and Practical on the Principal Diseases of the Interior Valley of North America as They Appear in the Caucasian, African, Indian, and Esquimaux Varieties of Its Population,* 2 vols. (New York: Burt Franklin, 1850–54); Barrett, "Daniel Drake's Medical Geography"; Otto Juettner, *Daniel Drake and His Followers: Historical and Biographical Sketches* (Cincinnati, OH: Harvey Publishing Company, 1909); James Rodger Fleming, *Meteorology in America, 1800–1870* (Baltimore, MD: Johns Hopkins University Press, 1990), 75–88, 110–115; Blodget, *Climatology.* For physicians' embrace of Blodget, see George W. Lawrence, "Report of the Committee on Climatology, Etc., of Arkansas," *Transactions of the American Medical Association* 23 (1872): 399.

37. On western booster literature, see David M. Wrobel, *Promised Lands: Promotion, Memory, and the Creation of the American West* (Lawrence: University Press of Kansas, 2002); Robert J. Orsi, "Selling the Golden State: A Study of Boosterism in Nineteenth-Century California" (Ph.D. diss., University of Wisconsin–Madison, 1973).

38. Richard Henry Dana, *Two Years before the Mast; a Personal Narrative of Life at Sea* (1842; reprint, Los Angeles: Ward Ritchie Press, 1964), 216; John Marsh, "Unpublished Letters of Dr. Marsh," *Overland Monthly* 15 (February 1890): 213–20; Lansford W. Hastings, *The Emigrant's Guide to Oregon and California* (Cincinnati: George Conclin, 1845), 44. On Marsh treating people for fever and ague, see Gray and Fontaine, "History of Malaria," 23.

39. Fourgeaud quoted in J. B. de C. M. Saunders, *Humboldtian Physicians in California* (Davis: University of California, Davis, 1971), 13. See also the account in chapter 38 of Edwin Bryant, *What I Saw in California — Being a Journal of a Tour of the Emigrant Route and South Pass of the Rocky Mountains across the Continent of North American, the Great Desert Basin, and through California, in 1846 and '47* (1848; reprint, Santa Ana, CA: Fine Arts Press, 1936).

40. Richard B. Rice, William A. Bullough, and Richard J. Orsi, *The Elusive Eden: A New History of California* (Boston: McGraw Hill, 2002), 194.

41. Letter to Julia Ann Baker, 21 December 1853, John W. H. Baker Letters, BL; Bayard Taylor, *Eldorado, or Adventures in the Path of Empire* (New York: Alfred A. Knopf, 1949), 165; George F. Kent, "Life in California in 1849, as Described in the 'Journal' of George F. Kent," ed. John Walton Caughey, *California Historical Society Quarterly* 20 (March 1941): 26–46, 36; Thomas Kerr, "An Irishman in the Gold Rush: The Journal of Thomas Kerr," ed. Charles L. Camp, *California Historical Society Quarterly* 8 (June 1929): 180; William M'Collum, *California as I Saw It* (1850; reprint, Los Gatos, CA: Talisman Press, 1960), 133. Gunnell cited in Mitchel Roth, "Cholera, Community, and Public Health in Gold Rush Sacramento and San Francisco," *Pacific Historical Review* 66 (1997): 550. Also see J. R. Black, "On the Ultimate Causes of Malarial Disease," *New York Journal of Medicine*, n.s., 14 (March 1854): 200.

42. Kent, "Life in California," 35; Alonzo Delano, *Alonzo Delano's California Correspondence: Being Letters Hitherto Uncollected from the Ottawa (Illinois) Free Trader and the New Orleans True Delta, 1849–1852*, ed. Irving McKee (Sacramento, CA: Sacramento Book Collectors Club, 1952), 54; Bryant, *What I Saw in California*; J. D. B. Stillman, *The Gold Rush Letters of J. D. B. Stillman* (Palo Alto, CA: Lewis Osborne, 1967), 51; Thomas M. Logan, "Letters from California," *New York Journal of Medicine*, n.s. 13 (March 1851), 278–83, quote on 279; Thomas M. Logan, "Letters from California," *New York Journal of Medicine*, n.s., 13 (May 1851): 421–26, quote on 426.

43. *Fourth Biennial Report of the CSBH* (Sacramento, 1877), 38; Henry Gibbons, "Report on Practical Medicine," *Transactions of the Medical Society of the State of California, 1875–76* [6th]: 36.

44. Death rates in the early 1850s among troops were about 18 or 19 per 1,000. By comparison, Philip Curtin estimated the mortality of European troops stationed in the Caribbean in the same period at 47 per 1,000. However, the

California statistics may not be comparable simply because men were unlikely to remain long in the region; and if they were ill, they were likely to be transferred home. U.S. Congress, Senate, *Statistical Report on the Sickness and Mortality in the Army of the United States,* 34th Cong., 1st sess., 1856, S. Exec. Doc. No. 96, 449, 452; Curtin, *Death by Migration,* 29.

45. Sacramento statistics appeared in the *California State Medical Journal* 1 (October 1856) and are reproduced in George W. Groh, *Gold Fever: Being a True Account, Both Horrifying and Hilarious, of the Art of Healing (So-Called) during the California Gold Rush* (New York: William Morrow, 1966), 173–74.

46. *Second Biennial Report,* 54–68, quote on 54; "The Insane in California," *Pacific Medical and Surgical Journal* 11 (August 1868): 138; Henry Gibbons, "Insanity and Disease in California," *Pacific Medical and Surgical Journal* 14 (April 1872): 496–503.

47. Thomas M. Logan, "Report on the Medical Topography and Epidemics of California," *Transactions of the American Medical Association* 12 (1859): 91.

48. J. P. Leonard, "Health in California," *Boston Medical and Surgical Journal* 41 (1850): 323–24; J. P. Leonard, "Letter from California," *Boston Medical and Surgical Journal* 41 (1850): 394–99.

49. By the 1870s British colonial policy assumed that lengthy stays in the tropics were to be avoided at all costs. Mark Harrison, "'The Tender Frame of Man': Disease, Climate and Racial Difference in India and the West Indies, 1760–1869," *Bulletin of the History of Medicine* 70 (1996): 68–93.

50. In the 1840s American racial discourse began to emphasize the concept of an "Anglo-Saxon" subdivision of the white race. Horsman, *Race and Manifest Destiny,* 158–228.

51. [Logan et al.], "Report on the Medical Topography" (1865), 549–50; U.S. Congress, *Report of the Joint Special Committee to Investigate Chinese Immigration,* 44th Cong., 2d sess., 1877, S.R. 689, 78. Most Chinese immigrants to California came from South China where malaria had long been endemic; thus many had acquired some immunity to the disease. For contemporary comments on immunity, see Philip King Brown, "The Malarial Fevers of the Sacramento and San Joaquin Valleys," *Transactions of the Medical Society of the State of California, 1899,* 29: 275. On concerns over race in the West, see Wrobel, *Promised Lands,* 157–80.

52. Nott cited in Livingston, "Human Acclimatization," 369. Robert Knox, *The Races of Men: A Fragment* (1850; reprint, Miami, FL: Mnemosyne, 1969), 44 (original emphasis); Louis Agassiz, "Sketch of the Natural Provinces of the Animal World and Their Relation to the Different Types of Man," in *Types of Mankind,* ed. J. C. Nott and George R. Gliddon (Philadelphia: Lippincott, Gambo, 1854), lviii–lxxvi; Horsman, *Race and Manifest Destiny;* George M. Fredrickson, *The Black Image in the White Mind: The Debate on Afro-American Character and Destiny, 1817–1914* (New York: Harper and Row, 1971); Nancy Stepan, "Biological Degeneration: Races and Proper Places," in *Degeneration: The Dark Side of Progress,* ed. J. Edward Chamberlain and Sander L. Gilman (New York: Columbia University Press, 1985), 97–120, esp. 100–101.

53. Hastings quoted in Horsman, *Race and Manifest Destiny,* 211. On the population of European descent in California during the Spanish and Mexican

periods, see Daniel J. Garr, "A Rare and Desolate Land: Population and Race in Hispanic California," *Western Historical Quarterly* 6 (April 1975): 133–48. James Ranald Martin, *The Influence of Tropical Climates on European Constitutions*, 7th ed. (London: John Churchill, 1856), 138.

54. M'Collum, *California as I Saw It*, 88. On argonaut attitudes toward whites in Latin America, see also Brian Roberts, *American Alchemy: The California Gold Rush and Middle-Class Culture* (Chapel Hill: University of North Carolina Press, 2000), 122–24. On European fears of degeneration in hot climates, see Stepan, "Biological Degeneration"; Daniel Pick, *Faces of Degeneration: A European Disorder, c. 1848–1918* (Cambridge: Cambridge University Press, 1989). On degeneration and the western frontier, see Louis S. Warren, "Buffalo Bill Meets Dracula: William F. Cody, Bram Stoker, and the Frontiers of Racial Decay," *American Historical Review* 107 (October 2002): 1124–57. Quote in letter from John Baker to Julia Ann Baker, 20 September 1853, John W. H. Baker Letters, BL.

55. Blodget, *Climatology*, 273–74. As one observer wrote of Fort Yuma in the southern California desert, it had not "received its full share of justice as the hottest military post on the continent of North America, and, perhaps, in the world." Senate, *Report on Sickness and Mortality* (1856), 438.

56. "Pacific Coast Diseases," *California Medical Gazette* 1 (January 1857): 323; *Second Biennial Report*, 18–19.

57. F. W. Hatch, "On the Climate of the Valley of the Sacramento, California," *New York Journal of Medicine*, n.s., 15 (July 1855): 33–34; John S. Griffin, *A Doctor Comes to California: The Diary of John S. Griffin, Assistant Surgeon with Kearny's Dragoons, 1846–1847*, ed. George Walcott Ames Jr. (San Francisco: California Historical Society, 1943), 43; Peter C. Remondino, *Longevity and Climate* (San Francisco: Woodward & Co., [1890]), 34–35.

58. Dr. Henry Gibbons made a similar argument about the Irish in this period. Trying to explain the high rates of consumption among Irish immigrants to California, Gibbons claimed they were subject to illness when removed from a harsh life to one of "ease and indulgence," such as the one they had supposedly found on the West Coast. *Third Biennial Report*, 241. More generally, Horsman, *Race and Manifest Destiny*, 189–207.

59. James Blake, "On the Climate and Diseases of California," *American Journal of the Medical Sciences* 24 (1852): 63; Senate, *Report on Sickness and Mortality* (1856), 442; J. S. Hittel, *Resources of California*, 3d ed. (San Francisco: A. Roman and Co., 1867), 368; Horatio R. Storer, "Female Hygiene," appendix to *First Biennial Report*. Retort to Cole quoted in Guy P. Jones, "Thomas M. Logan, M.D., Organizer of California State Board of Health," *California's Health* 2 (15 March 1945): 132. Alan Bewell cites a similar concern over the vulnerability of white females among the nineteenth-century British in *Romanticism and Colonial Disease*, 280–82. The assumption that white Americans would outbreed others in the West was a common assumption among American expansionists in the period. See Horsman, *Race and Manifest Destiny*, 243.

60. J. H. Stallard, *Female Health and Hygiene on the Pacific Coast* (San Francisco: Bonnard & Daly, 1876), 5; *First Biennial Report*, 6; Charles Nordhoff, *Nordhoff's West Coast California, Oregon, and Hawaii* (London: KPI, 1987),

198–99. On positive effects of climate on fertility, Remondino, *Longevity and Climate,* 20.

61. Charles Loring Brace, *The New West: Or, California in 1867–1868* (New York: G. P. Putnam & Son, 1869), 367–71; Remondino, *Longevity and Climate,* 21. Late-nineteenth-century boosters also frequently published pictures of healthy white babies to substantiate their claims. See, Wrobel, *Promised Lands,* 174.

62. This point has been elegantly developed in Valenčius, *Health of the Country.* On doctor-naturalists, see Nunis, "Medicine in Spanish California."

63. For example, Gibbons, "Practical Medicine," 29; "Climate and Disease," *Pacific Medical and Surgical Journal* 14 (February 1872): 421 (for quote).

64. Richard Harrison Shyrock (*Medical Licensing in America, 1650–1965* [Baltimore, MD: Johns Hopkins University Press, 1967, 31–32]) cites a study in eastern Tennessee that found that only 17 percent of physicians had taken a regular degree. On Logan's background, see J. M. Toner, "Life and Professional Labors of Thomas Muldrup Logan, M.D., of California," *Transactions of the Medical Society of the State of California, 1875–76* [6th]: 136–43; Jones, "Thomas M. Logan"; Henry Harris, *California's Medical Story* (San Francisco: J. W. Stacey, 1932). For Logan's views on the profession, see Thomas M. Logan, *Valedictory Address in Behalf of the Faculty of the Medical Department of the University of California* (San Francisco, 1874). Quote from [Logan et al.], "Report on Medical Topography" (1865), 543.

65. Quote is from "Climate and Disease," 421. Hatch, "Climate of the Valley of the Sacramento"; F. W. Hatch, "Report on Climatology and Diseases of California," *Transactions of the American Medical Association* 23 (1872): 335–67. See also the description of the meteorological interests of Henry Gibbons's brother, William Gibbons, in Harris, *California's Medical Story,* 328.

66. Logan quoted in J. Roy Jones, *Memories, Men and Medicine: A History of Medicine in Sacramento, California* (Sacramento: Sacramento Society for Medical Improvement, 1950), 53–54; *First Biennial Report,* 16–20. Thomas Logan collected the most extensive meteorological data for the city of Sacramento, which were used by contemporary engineers and eventually incorporated into the Smithsonian Institution's meteorological records. See *Annual Report of the California Surveyor-General, 1861–62* (Sacramento, 1862) and Thomas M. Logan, "Contributions to the Physics, Hygiene and Thermology of the Sacramento River," *Pacific Medical and Surgical Journal* 7 (1864): 145–51.

67. Thomas M. Logan, "Contributions to the Medical History of California," *California Medical Gazette* 1 (October 1856): 192–93. In fact, the formation of the California State Board of Health coincided with concerns over the pace of white immigration and the state's relatively slow growth in (white) population. Orsi, "Selling the Golden State," 11–59.

2. PLACING HEALTH AND DISEASE

1. *Sixth Biennial Report of the California State Board of Health* (Sacramento, 1880), 31.

2. SSMI, "Minutes," 15 August 1882, 19 August 1884, CSL.

3. Pelling, "Contagion/Germ Theory/Specificity"; SSMI, *Minutes,* 19 October 1881, 19 October 1880; J. Campbell Shorb, "The Miasmatic Diseases of California," pt. 1, *California Medical Gazette* 1 (July 1868–January 1869): 53.

4. M. M. Chipman, "Micro-Organisms and Their Relations to Human and Animal Life," *Transactions of the Medical Society of the State of California, 1889* [19th]: 163–83; B. M. Gill, "The Medical Topography of Northern California," *Transactions of the Medical Society of the State of California, 1895* [15th]: 174.

5. *Second Biennial Report of the California State Board of Health* (Sacramento, 1873), 52–54.

6. Stillman, *Gold Rush Letters,* 73.

7. "Diseases Incident to the Season."

8. Logan, "Report on Medical Topography" (1859), 113.

9. Logan, *Medical History of the Year 1868,* 10.

10. Hatch, "Report on the Epidemics of California in 1868," 513–14; Logan, *Medical History of the Year 1868.*

11. Gregg Mitman makes a similar observation with respect to hay fever sufferers in "Hay Fever Holiday."

12. *Third Biennial Report of the California State Board of Health* (Sacramento, 1875), 14.

13. Thomas Logan, *Circular* (Sacramento, 1856).

14. *Fourth Biennial Report of the California State Board of Health* (Sacramento, 1877), 26–27, 58.

15. John Harley Warner, "Medical Sectarianism, Therapeutic Conflict, and the Shaping of Orthodox Professional Identity in Antebellum American Medicine," in *Medical Fringe and Medical Orthodoxy, 1750–1850,* ed. W. F. Bynum and Roy Porter (London: Croom Helm, 1987), 234–60; John Harley Warner, *The Therapeutic Perspective: Medical Practice, Knowledge, and Identity in America, 1820–1885* (Cambridge, MA: Harvard University Press, 1986); John H. Warner, "'The Nature-Trusting Heresy': American Physicians and the Concept of the Healing Power of Nature in the 1850s and 1860s," *Perspectives in American History* 11 (1977–78): 295–96 (for quote).

16. SSMI, "Minutes," 21 July 1868; Logan, "Report on Medical Topography" (1859), 120; A. B. Nixon, "Random Thoughts on 'Conservative Medicine,'" *California Medical Gazette* 2 (June 1870): 207–8; A. B. Nixon, "Nature in Disease, or Conservative Medicine," *Pacific Medical and Surgical Journal* 15 (April 1873): 528–35; "The Place of Nature in Therapeutics," *Pacific Medical and Surgical Journal* 15 (May 1873): 573–92.

17. Kenneth Thompson, "Climatotherapy in California," *California Historical Quarterly* 50 (1971): 111–130; Kenneth Thompson, "Wilderness and Health in the Nineteenth Century," *Journal of Historical Geography* 2 (1976): 145–61; "The Hot Air Bath in Therapeutics: Sun-Stroke," *Pacific Medical and Surgical Journal* 11 (September 1868): 174–76; "Concerning Sanitary Science," *Pacific Rural Press,* 4 August 1883; SSMI, "Minutes," 21 December 1872; E. Gould Buffum, *Six Months in the Gold Mines: From a Journal of Three Years' Residence in Upper and Lower California, 1847–8–9,* ed. John W. Caughey ([Los Angeles]: Ward Ritchie Press, 1959), 114 (for quote on hydropathy).

18. M. M. Chipman, "Development of Botany in California," *Transactions of the Medical Society of the State of California, 1891* [21st], 284–90.

19. *First Biennial Report of the California State Board of Health* (Sacramento, 1871), 2; "Medical Topography in California," *California Medical Gazette* 2 (October 1869): 33.

20. Mol, *Body Multiple;* Latour, *Pasteurization of France.*

21. *Third Biennial Report,* 14.

22. *Fourth Biennial Report,* 60.

23. J. P. Widney, "Report of Committee on Medical Topography, Meteorology, Endemics, and Epidemics," *Transactions of the Medical Society of the State of California, 1889* [19th]: 213–14 (for quote). On the interest of early California physicians in medical geography and topography, see Saunders, *Humboldtian Physicians in California.* For landscape as the basis of therapeutics, see *Third Biennial Report,* 29.

24. Hatch, "Report on Climatology and Diseases of California," 362–67; *Second Biennial Report,* 109–38; Susan E. Cayleff, *Wash and Be Healed: The Water-Cure Movement and Women's Health* (Philadelphia: Temple University Press, 1987).

25. On climatotherapy generally, see René Dubos and Jean Dubos, *The White Plague: Tuberculosis, Man, and Society* (Boston: Little, Brown, 1952), 173–74; Thompson, "Wilderness and Health"; Thompson, "Climatotherapy in California." On consumption, see Samuel George Morton, *Illustrations of Pulmonary Consumption, Its Anatomical Characters, Symptoms, and Treatment* (Philadelphia: Key and Biddle, 1834); John E. Baur, *The Health Seekers of Southern California, 1870–1900* (San Marino, CA: The Huntington Library, 1959). For primary accounts of California, see James Blake, "On the Climate of California in Its Relations to the Treatment of Pulmonary Consumption," *Pacific Medical and Surgical Journal* 3 (1860): 263–64; F. W. Hatch, "Some of the Health Resorts of the Coast Range of Mountains," *Transactions of the Medical Society of the State of California, 1881–1882* [12th]: 225–43; M. H. Biggs, "Medical Topography of Santa Barbara," *Transactions of the California Medical Society, 1870–1871* [1st]: 133–37; H. S. Orme, "Topography, Climate, and Diseases of Los Angeles County, Valley, and City," *Transactions of the Medical Society of the State of California, 1874–1875* [5th]: 84–89; F. W. Hatch, "The Seaside Health Resorts of California," in *Seventh Biennial Report of the California State Board of Health* (Sacramento, 1882), 85–93.

26. J. D. B. Stillman, "Observations on the Medical Topography and Diseases (Especially Diarrhoea) of the Sacramento Valley, California, during the Years 1849–50," *New York Journal of Medicine,* n.s., 13 (July 1851), offprint, BL. Also Johnson, *Roaring Camp,* 127–30; Dr. J. Praslow, *The State of California: A Medico-Geographical Account* (1857; reprint, trans. Frederick C. Cordes, San Francisco: John J. Newbegin, 1939), 58–59; James L. Tyson, *Diary of a Physician in California, Being the Results of Actual Experience Including Notes of the Journey by Land and Water, and Observations on the Climate, Soil, Resources of the Country, Etc.* (1850; reprint, Oakland, CA: Biobooks, 1955), 66 (for quote); Kerr, "Irishman in the Gold Rush," 180–82. Also, several undated newspaper quotes from mining towns appear in Thomas M. Logan,

"Malarial Fevers and Consumption in California," in *Third Biennial Report,* 117–18.

27. John W. Audubon, *Audubon's Western Journal: 1849–1850* (1906; reprint, Tucson: University of Arizona Press, 1984), 234, 237; Stillman, *Gold Rush Letters,* 296.

28. George H. Derby, "The Topographical Reports of George H. Derby," pt. 1, ed. Francis P. Farquhar, *California Historical Society Quarterly* 11 (June 1932): 110; Senate, *Sickness and Mortality,* 1856, 449; Audubon, *Audubon's Western Journal,* 234.

29. Victor J. Fourgeaud, "Two Fourgeaud Letters," ed. John Francis McDermott, *California Historical Society Quarterly* 20 (June 1941): 119; Senate, *Sickness and Mortality,* 1856, 445; W. P. Gibbons, "Notes on Topography, and on the Distribution of Plants in California," in *Tenth Biennial Report of the California State Board of Health* (Sacramento, 1888), 184; F. W. Hatch, *Sixth Anniversary Address before the Sacramento Society for Medical Improvement* (San Francisco: Joseph Winterburn & Company, 1874), 8; Logan, "Report on Medical Topography" (1859), 83, 86. Also, Derby, "Topographical Reports," pt. 1, 111–12.

30. Johnson and Martin, *Influence of Tropical Climates,* 573 (original emphasis). On the discourse of tropicality more generally, see David Arnold, *The Problem of Nature: Environment, Culture and European Expansion* (Oxford: Blackwell, 1996); Bewell, *Romanticism and Colonial Disease;* Stepan, *Picturing Tropical Nature;* Anderson, *Cultivation of Whiteness,* 73–94.

31. Hatch, "Report on the Epidemics," 527.

32. Kerr, "Irishman in the Gold Rush," 178; Derby, "Topographical Reports," pt. 1, 111–12. On labor in tropical climates, see Johnson and Martin, *Influence of Tropical Climates,* 591. On white vulnerability, Warwick Anderson, "The Trespass Speaks: White Masculinity and Colonial Breakdown," *American Historical Review* 102 (December 1997): 1343–70.

33. [Logan et al.], "Report on Medical Topography" (1865), 539.

34. Hatch, "Climate of the Valley of the Sacramento," 18.

35. J. H. C. Bonte, "The Northerly Winds of California," *Transactions of the California State Agricultural Society, 1881* (1881) (Sacramento): 211.

36. Orme, "Topography, Climate, and Diseases of Los Angeles," 86 (for quote); Senate, *Sickness and Mortality,* 1856, 445; *Second Biennial Report,* 158. Logan, "Letters from California" (March 1851), 282. On temperature and malaria, "What Is Malaria?" *Pacific Medical and Surgical Journal* 14 (November 1871): 283. See also Michael Worboys, "Germs, Malaria, and the Invention of Mansonian Tropical Medicine: From 'Disease in the Tropics' to 'Tropical Diseases,'" in *Warm Climates and Western Medicine: The Emergence of Tropical Medicine, 1500–1900,* ed. David Arnold (Amsterdam: Rodopi, 1996), 181–207.

37. *Settlers' Experience in Kern County, California as Related by Themselves with Advice to Newcomers* (Bakersfield, CA, 1894), 10; *Ninth Biennial Report of the California State Board of Health* (Sacramento, 1886), 71.

38. [Logan et al.], "Report on Medical Topography" (1865), 549–50.

39. *Fifth Biennial Report of the California State Board of Health* (Sacramento, 1879), 16–17; Shorb, "Miasmatic Diseases of California," 5; "Diseases Incident to the Season," *Pacific Rural Press,* 30 September 1871; SSMI, "Min-

utes," 21 November 1899; "Moisture as the Cause of Periodic Fever," *Pacific Rural Press*, 7 March 1874; Powers, *Tribes of California*, 417–18; Kroeber, *Handbook of the Indians of California*, 354; A. H. Gayton, "Culture-Environment Integration: External References in Yokuts Life," *Southwestern Journal of Anthropology* 2 (Autumn 1946): 252–68, 257. Given that these accounts of Indian practices are from the mid-nineteenth century, it is unclear whether seasonal migration away from the rivers began only after the assumed introduction of malaria in the 1830s.

40. P. B. M. Miller, "Epidemic Relapsing Fever among the Chinese at Oroville," *Pacific Medical and Surgical Journal* 16 (January 1875): 370–75; *Third Biennial Report*, 117, 119–21.

41. Dale C. Smith, "The Rise and Fall of Typhomalarial Fever," *Journal of the History of Medicine and Allied Sciences* 37 (1982): 182–220, 287–321. Although *vivax* was most likely the predominant form, it is likely that *falciparum* and *malariae* were also present in the late nineteenth century. Harold Farnsworth Gray, "The Confusing Epidemiology of Malaria in California," *American Journal of Tropical Medicine and Hygiene* 5 (May 1956): 411–18.

42. "Diseases Incident to the Season," *Pacific Rural Press*, 30 September 1871; "Ague or Intermittent Fever," *Pacific Rural Press*, 28 December 1872; Shorb, "Miasmatic Diseases," 55.

43. Taylor's account reprinted in Guy P. Jones, "Early Public Health in California," *California's Health* 2 (15 April 1945): 146–49, 146 (for quote).

44. [Logan et al.], "Report on Medical Topography" (1865), 539.

45. Bonte, "Northerly Winds"; *Eighth Biennial Report*, 206; George H. Derby, "The Topographical Reports of George H. Derby," pt. 2, ed. Francis P. Farquhar, *California Historical Society Quarterly* 11 (September 1932): 248; [Logan et al.], "Report on Medical Topography" (1865), 536 (for discussion of winds); H. W. Harkness, "Northers, or North-West Winds of California," *Pacific Medical and Surgical Journal* 11 (May 1869): 545–50; *Fourth Biennial Report*, 19, 57; "The North Wind," *Pacific Rural Press*, 19 June 1880. On antebellum attitudes toward the atmosphere and disease, see Valenčius, *Health of the Country*, 109–32.

46. Johnson and Martin, *Influence of Tropical Climates*, 38–39; *Second Biennial Report*, 108–9; *Third Biennial Report*, 158; *Tenth Biennial Report of the California State Board of Health* (Sacramento, 1888), 224; Logan, "Malarial Fevers and Consumption," 121.

47. Edward Belcher, *Narrative of a Voyage Round the World, Performed in Her Majesty's Ship Sulphur during the Years 1836–1842* (London: Henry Colburn, 1843), 2:123–24.

48. For a summary of swampland estimates, see Phyllis Fox, "Rebuttal to David R. Dawdy Exhibit 3 in Regard to Freshwater Inflow to San Francisco Bay under Natural Conditions," State Water Contractors Exhibit No. 276, November 1987, California Water Resources Control Board, Bay-Delta Hearings Transcripts, 9–20, MS 87/3, WRCA. Quote from U.S. War Department, *Reports of Explorations and Surveys, to Ascertain the Most Practicable and Economical Route for a Railroad from the Mississippi River to the Pacific Ocean. Made*

under the Direction of the Secretary of War, in 1853–[6]. (Washington, DC: A. O. P. Nicholson, Printer, 1855–61), 5:2, 10.

49. War Department, *Reports of Explorations and Surveys,* 5:2, 191–92; Fox, "Rebuttal" (especially appendix A for eyewitness accounts of tule acreage); *Annual Report of the California Surveyor-General for 1856* (Sacramento, 1856), 240.

50. Washington Ayer, "Topography and Meteorology," *Transactions of the State of California Medical Society, 1880–1881* [11th]: 43; Tyson, *Diary of a Physician,* 66; Titus Fey Cronise, *The Natural Wealth of California* (San Francisco: H. H. Bancroft & Co., 1868), 384–85; Audubon, *Audubon's Western Journal,* 234.

51. [Logan et al.], "Report on Medical Topography" (1865), 509 (original emphasis).

52. Malaria, literally "bad air" in Italian, was used interchangeably with *miasma* by some writers, whereas others used *malaria* to designate the disease and *miasma* to refer to the cause. There was considerable confusion on this issue in the medical literature of the mid-nineteenth century. On California diseases attributable to miasma, see *Eighth Biennial Report,* 90.

53. Logan, "Malarial Fevers," 115.

54. J. P. Widney, "Irrigation and Drainage," in *Seventh Biennial Report of the California State Board of Health* (Sacramento, 1882), 104–6.

55. *Kern County Weekly Courier* (Bakersfield, CA), 6 June 1874; Shorb, "Miasmatic Diseases," 54.

56. H. S. Orme, "Irrigation and Forestry Considered in Connection with Malarial Diseases," in *Tenth Biennial Report,* 224; Orme, "Irrigation—Its Influence on Health, Etc.," in *Eighth Biennial Report,* 59 (Rowell quote); Miller, "Epidemic Relapsing Fever"; "Origin of Disease," *Pacific Rural Press,* 29 June 1872; *The Family Health Annual* (Oakland, CA: Pacific Press, 1878), 21.

57. "Diseases Incident to the Season."

58. Will Green cited in Orme, "Irrigation," 52; *Settlers' Experience in Kern County,* 48; M. M. Chipman, "Importance of Forest Preservation and Timber Cultivation," *Transactions of the Medical Society of the State of California, 1882–1883* [13th]: 276.

59. Frank Adams, *The Historical Background of California Agriculture* (Berkeley: California Agricultural Experiment Station, 1946); Paul W. Rhode, "Learning, Capital Accumulation, and the Transformation of California Agriculture," *Journal of Economic History* 55 (December 1995): 773–800; Walton Bean, *California: An Interpretive History* (New York: McGraw-Hill, 1978), 226–38. On agrarian ideology in California, see Tyrell, *True Gardens of the Gods.*

60. Stillman, "Observations on Medical Topography," 297: Valenčius, *Health of the Country,* 217–28.

61. Calvin Brown, "On the Utilization of Water, and Its Sources, in California," in *Annual Report of the California Surveyor-General, 1861–62* (Sacramento, 1862), 32 (emphasis added).

62. *Kern County Weekly Courier* (Bakersfield, CA), 6 June 1874; *Third Biennial Report,* 130–31.

63. Rhode, "Learning, Capital Accumulation"; Stoll, *Fruits of Natural*

Advantage; David Vaught, *Cultivating California: Growers, Specialty Crops, and Labor, 1875–1920* (Baltimore, MD: Johns Hopkins University Press, 1999); Tyrell, *True Gardens of the Gods.*

64. On invalids, Baur, *Health-Seekers,* 110–19. On women, "Practical Gardening and Rural Aesthetics," *California Farmer,* 19 April 1855; Sheila Rothman, *Living in the Shadow of Death: Tuberculosis and the Social Experience of Illness in American History* (New York: Basic Books, 1994), 167–68. On consumption of fruit, R. W. C. Farnsworth, ed., *A Southern California Paradise* (Pasadena, CA: R. W. C. Farnsworth, 1883), 100; "The Use of Fruits, and How to Use," *California Farmer,* 27 July 1855; "Fruit Instead of Medicine," *Pacific Rural Press,* 10 June 1871; "Use of Fruit," *Pacific Rural Press,* 13 July 1872; Elwood Cooper, *A Treatise on Olive Culture* (San Francisco: Cubery & Company, 1882); [J. L. Howland], *The Olive in California* (Pomona, CA: Pomona Nursery, 1892); Felix Gillet, *Fragariculture; or the Culture of the Strawberry* (San Francisco: Spaulding & Barton, Steam Book and Job Printers, 1876).

65. Shorb, "Miasmatic Diseases," 31; W. P. Gibbons, "Report on Indigenous Botany," 136–37; Chipman, "Importance of Forest Preservation," 268; L. C. Lane, "Notes of Travel in the Interior," *San Francisco Medical Press* 3 (1862): 223–24; "Plants as Doctors," *California Horticulturist and Floral Magazine* 3 (1872): 353; Thompson, "Wilderness and Health."

66. Kenneth Thompson, "The Australian Fever Tree in California: Eucalypts and Malaria Prophylaxis," *Annals of the Association of American Geographers* 60 (June 1970): 230–44; Elwood Cooper, *Forest Culture and Eucalyptus Trees* (San Francisco: Cubery & Co., 1876), 44; W. P. Gibbons, "Report on Indigenous Botany," *Transactions of the Medical Society of the State of California, 1878–1879* [9th]: 140; Chipman, "Importance of Forest Preservation," 278; *Report of the Commissioner of Agriculture for the Year 1874* (Washington, DC, 1875), 273.

67. *Second Biennial Report,* 38–39; *Third Biennial Report,* 29.

68. Bonte, "The Northerly Winds," 213; J. P. Widney, "Climatic Changes Which Man Is Working in Southern California," *Southern California Practitioner* 1 (October 1886): 389–93; SSMI, "Minutes," 21 February 1899.

69. Cronise, *Natural Wealth of California,* 350–51; "Influence of Agriculture upon the General Interests of the People," *California Farmer,* 4 January 1855; Will S. Green, *The History of Colusa County, California and the General History of the State, with Supplement* (Sacramento: Sacramento Lithograph Co., 1950), viii. (Originally this appeared in *Hutchings Illustrated California Magazine* 1 [April 1857]).

70. Benjamin C. Truman, *Semi-Tropical California: Its Climate, Healthfulness, Productiveness, and Scenery* (San Francisco: A. L. Bancroft, 1874), 38.

71. William Hammond Hall Diary, 3 September 1878, box 1137/1, William Hammond Hall Papers, CSL. Hall himself became quite interested in issues of local health and sanitation. See the materials in box 1/4, MS 913 (William Hammond Hall Papers), CHS.

72. Logan, "Letters from California," 423; W. P. Gibbons, "On Forest Culture as a Prophylactic to Miasmatic Diseases," *Pacific Medical and Surgical Journal* 18 (August 1875): 116; SSMI, "Minutes," 19 October 1886.

73. Logan, "Report on Medical Topography" (1859), 94–95.

74. Benjamin Rush, "An Inquiry into the Causes of the Increase of Bilious and Intermitting Fevers in Pennsylvania," *Medical Inquiries and Observations,* new ed. (Philadelphia: T. Dobson, 1797), 2:268 (original emphasis).

75. U.S. Geological Survey, *Irrigation Near Bakersfield, California,* by Carl Ewald Grunsky, Water-Supply and Irrigation Papers No. 17 (Washington, DC: U.S. GPO, 1898); U.S. Geological Survey, *Irrigation Near Fresno, California,* by Carl Ewald Grunsky, Water-Supply and Irrigation Papers No. 18 (Washington, DC: U.S. GPO, 1898).

76. "Irrigation—Effect on Health," *Pacific Medical and Surgical Journal* 14 (July 1871): 139; *Second Biennial Report,* 54; Ronald Loren Nye, "Visions of Salt: Salinity and Drainage in the San Joaquin Valley, California, 1870–1970" (Ph.D. diss., University of California, Santa Barbara, 1986).

77. A. G. Warfield, "Field Notes," February–March 1879, box 20, book 30, 11, State Engineering Department/William Hammond Hall Papers, Acc. 91–07–04, CSA; *The Family Health Annual* (Oakland, CA: Pacific Press, 1878), 19; U.S. Congress, Senate, *Report of the Special Committee of the U.S. Senate on the Irrigation and Reclamation of Arid Lands,* pt. 2, *The Great Basin Region and California,* 51st Cong., 1st sess., 1890, S.R. 928, 226–36.

78. U.S. Geological Survey, *Irrigation Near Bakersfield,* 93; Warfield, "Field Notes," 12. Statement of George Manuel, in Senate, *Special Committee on Irrigation and Reclamation,*" 229; "Irrigation—Effect on Health," 139.

79. "Statement of William Hammond Hall," in Senate, *Special Committee on Irrigation and Reclamation,* 217. There seems to have been a general consensus that the climate and health of Bakersfield had improved as a result of irrigation. For instance, in 1880 James Schuyler (an engineer and colleague of Hall's) wrote that "the change for the better in the climate of [Kern County] . . . , since the general introduction of irrigation, has been as marked as the improvement in the soil." Schuyler's report appears in California Engineer's Office, *Report of the State Engineer to the Legislature, Session of 1880–1881* (Sacramento, 1881), 2:4, 109–10.

80. Gibbons, "Indigenous Botany," 140; Logan, "Report on Medical Topography" (1859), 94–95; Gibbons, "On Forest Culture," 151; Shorb, "Miasmatic Diseases," 5. Also, Widney, "Irrigation and Drainage," in *Seventh Biennial Report,* 104; [Logan et al.], "Report on Medical Topography" (1865), 547.

81. Chipman, "Forest Preservation," 270 (for quote); M. M. Chipman, "Government Forest Reservations," *Transactions of the Medical Society of the State of California, 1893* [23d]: 264–80. Gibbons, "Indigenous Botany," 135.

82. Kenneth Thompson, "Irrigation as a Menace to Health in California," *Geographical Review* 59 (April 1969): 195–214; Orme, "Irrigation," 55. Local people's perceptions of the ill effects of irrigation are also documented in A. G. Warfield, "Field Notes" (1879).

83. Logan, "Malarial Fevers and Consumption," 117, 118.

84. M. M. Chipman, "Report on the Committee on Medical Topography, Endemics, Etc.," *Transactions of the Medical Society of the State of California, 1880–81* [11th]: 128–51 (quote on 151); SSMI, "Minutes," 15 August 1882; Chipman, "Forest Preservation," 262. On the controversy over mining debris, see Robert Kelley, *Gold vs. Grain: California's Hydraulic Mining Controversy* (Glendale, CA: Arthur H. Clark, 1959).

85. Those areas of medicine in which ecological understandings would persist most strongly—in the treatment of hay fever and tuberculosis and, later, in the new specialty of tropical medicine—would, not incidentally, be closely associated with unaltered, "natural" landscapes. See Mitman, "Hay Fever Holiday"; Warwick Anderson, "Natural Histories of Infectious Disease: Ecological Vision in Twentieth-Century Biomedical Science," *Osiris* 19 (2004): 39–61.

3. PRODUCING A SANITARY LANDSCAPE

1. Guy P. Jones, "Early Public Health in California," *California's Health* 2 (15 June 1945): 177–79; *Eighteenth Biennial Report of the CSBH* (Sacramento, 1904), 6–9; Guenter B. Risse, "'A Long Pull, a Strong Pull, and All Together': San Francisco and Bubonic Plague, 1907–1908," *Bulletin of the History of Medicine* 66 (Summer 1992): 260–82.

2. N. K. Foster Diary, 1904–9, Guy E. Jones Papers, BL; *Nineteenth Biennial Report of the CSBH* (Sacramento, 1906), 9–10, 111–12; *Twentieth Biennial Report of the CSBH* (Sacramento, 1908), 22–25; *Twenty-second Biennial Report of the CSBH* (Sacramento, 1913), 1–3; *Twenty-fifth Biennial Report of the CSBH* (Sacramento, 1918), 12–13, 132–22; *Thirtieth Biennial Report of the CDPH* (Sacramento, 1929), 7–9; "Appropriation Needs," *CSBH Monthly Bulletin* 8 (December 1912): 96–101; William F. Snow, "Some of California's New Health Laws," *CSBH Monthly Bulletin* 9 (October 1913): 53–59.

3. On progressivism generally, see Robert H. Wiebe, *The Search for Order* (New York: Hill and Wang, 1967); Sidney M. Milkis and Jerome M. Mileur, eds., *Progressivism and the New Democracy* (Amherst: University of Massachusetts Press, 1999). On American public health in this period, see Elizabeth Fee and Dorothy Porter, "Public Health, Preventive Medicine and Professionalization: England and America in the Nineteenth Century," in *Medicine in Society: Historical Essays,* ed. Andrew Wear (Cambridge: Cambridge University Press, 1992), 249–75; Nancy Tomes, *The Gospel of Germs: Men, Women, and the Microbe in American Life* (Cambridge, MA: Harvard University Press, 1998); Leavitt, *The Healthiest City;* John Duffy, *The Sanitarians: A History of American Public Health* (Urbana: University of Illinois Press, 1990); Barbara Gutmann Rosenkrantz, *Public Health and the State: Changing Views in Massachusetts, 1842–1936* (Cambridge, MA: Harvard University Press, 1972). For a transnational view, see Dorothy Porter, *The History of Public Health and the Modern State* (Amsterdam: Rodopi, 1994).

4. On germ theory generally, see Worboys, *Spreading Germs;* Cunningham and Williams, *Laboratory Revolution in Medicine;* Tomes and Warner, "Introduction to Special Issue on Rethinking the Germ Theory of Disease." On "modern" medicine, see Shyrock, *Development of Modern Medicine.*

5. Tropical environments would, however, continue to be pathologized. See Arnold, *Problem of Nature,* 171–75; Stepan, *Picturing Tropical Nature;* Anderson, *Cultivation of Whiteness,* 73–94.

6. Although the environment's role in etiology was minimized, it was still acknowledged to have a role in the healing of certain diseases, notably tuberculosis for which there were no alternative therapies. Frank B. Rogers, "The Rise

and Decline of the Altitude Therapy of Tuberculosis," *Bulletin of the History of Medicine* 43 (1969): 1–16; Rothman, *Living in the Shadow of Death;* Francis Marion Pottenger, *The Fight against Tuberculosis: An Autobiography* (New York: Henry Schuman, 1952).

7. "Apostles of the germ" is from Tomes, *Gospel of Germs,* 23.

8. On modernist narratives of disease, see Charles L. Briggs and Clara Mantini-Briggs, *Stories in the Time of Cholera: Racial Profiling during a Medical Nightmare* (Berkeley: University of California Press, 2003).

9. Rhode, "Learning, Capital Accumulation"; Alan M. Paterson, *Land, Water, and Power: A History of the Turlock Irrigation District, 1887–1987* (Glendale, CA: Arthur H. Clark, 1987), 109–38; Joseph A. McGowan, *History of the Sacramento Valley,* 3 vols. (New York: Lewis Historical Publishing Company, 1961), 181–237; Preston, *Vanishing Landscapes,* 170.

10. McGowan, *History of the Sacramento Valley;* Rhode, "Learning, Capital Accumulation"; Wallace Smith, *Garden of the Sun: A History of the San Joaquin Valley, 1772–1939,* 2d ed., ed. William B. Secrest Jr. (Fresno, CA: Linden Publishing, 2004); Preston, *Vanishing Landscapes;* Stephen Johnson, Robert Dawson, and Gerald Haslam, *The Great Central Valley: California's Heartland* (Berkeley: University of California Press, 1993).

11. Lawrence J. Jelinek, *Harvest Empire: A History of California Agriculture* (San Francisco: Boyd & Fraser, 1979), 61–77; Paterson, *Land, Water, and Power;* Preston, *Vanishing Landscapes.*

12. http://fisher.lib.virginia.edu/collections/stats/histcensus (accessed 20 September 2005).

13. Arthur T. Johnson, *California: An Englishman's Impressions of the Golden State* (London: Stanley Paul & Co., [1913]), 265–66.

14. "Unsavory reputation" from *Twenty-fourth Biennial Report of the CSBH* (Sacramento, 1916), 13. Typhoid in major Central Valley towns in the first decade of the century averaged 31.5 per hundred thousand, compared to only 15.2 in San Francisco. *CSBH Monthly Bulletin* 5 (July 1909): 19–20; *CSBH Monthly Bulletin* 10 (April 1915): 259.

15. Overall, in 1915 California's malaria infection rate was only 2 percent, compared to a 4.8 percent rate for the nation as a whole. William B. Herms, "Successful Methods of Attack on Malaria in California," *CSJM* 13 (May 1915): 187; William B. Herms, *Medical Entomology, with Special Reference to the Health and Well-Being of Man and Animals,* 3d ed. (New York: Macmillan, 1939), 2–3; William B. Herms, "Malaria Control," *CSBH Monthly Bulletin* 16 (November 1920): 76; California Bureau of Vector Control, *Mosquito Abatement in California* (1951), 5; Harold Farnsworth Gray and Russell E. Fontaine, "A History of Malaria in California," *Proceedings and Papers of the California Mosquito Control Association* 25 (1957): 35; Harold F. Gray, "Malaria Control in California," *American Journal of Public Health* 2 (1912): 452–55. Physician comment in *CSBH Monthly Bulletin* 10 (June 1915): 315. On quinine treatment, see *CSBH Monthly Bulletin* 11 (February 1916): 370.

16. Harry E. Butler, "A History of the First Malaria Mosquito Control Campaign" (1945, typescript), CSL; William B. Herms, *Malaria: Cause and Control* (New York: Macmillan, 1913), 89–103; Harold Farnsworth Gray, "The Cost of

Malaria: A Study of Economic Loss Sustained by the Anderson-Cottonwood Irrigation District, Shasta County, California," *JAMA* 72 (24 May 1919): 1533–35.

17. In fact, nineteenth-century practitioners had often diagnosed a disease that they labeled "typho-malarial fever." Smith, "Rise and Fall of Typhomalarial Fever." William Sedgewick of the Massachusetts State Board of Health is credited with establishing the role of water supplies in the transmission of typhoid. Rosenkrantz, *Public Health and the State,* 104–5. On mosquito vector, William B. Herms, *Medical Entomology, with Special Reference to the Health and Well-Being of Man and Animals,* 3d ed. (New York: Macmillan, 1939), 2–3.

18. As Michael Worboys has put it, "*contamination* became more important than [environmental] *configuration.*" Worboys, *Spreading Germs,* 288 (original emphasis); James H. Cassedy, *Charles V. Chapin and the Public Health Movement* (Cambridge, MA: Harvard University Press, 1962), 126–42; Charles V. Chapin, *The Sources and Modes of Infection* (New York: John Wiley & Sons, 1910); Hibbert Winslow Hill, *The New Public Health* (New York: Macmillan, 1916), 8; Hibbert Winslow Hill, "What Is the New Public Health?" *CSBH Weekly Bulletin* 2 (3 November 1923): 149–50. For endorsement by California officials, Walter M. Dickie, "Public Health Today," *CSBH Weekly Bulletin* 3 (11 October 1924): 137–39.

19. Worboys, *Spreading Germs,* 285; Owsei Temkin, "Health and Disease," in *The Double Face of Janus and Other Essays in the History of Medicine* (Baltimore: Johns Hopkins University Press, 1977), 435–37.

20. Henri Lefebvre, *The Production of Space,* trans. Donald Nicholson (Oxford: Blackwell, 1991), 287; L. S. McClung, and K. F. Meyer, "The Beginnings of Bacteriology in California," *Bacteriological Reviews* 38 (September 1974): 251–71. Christopher Sellers has also noted that more complex models of disease persisted in the field of occupational health, a subject that I take up in the next chapter. See Christopher Sellers, "The Public Health Service's Office of Industrial Hygiene and the Transformation of Industrial Medicine," *Bulletin of the History of Medicine* 65 (Spring 1991): 42–73.

21. Herms, *Malaria: Cause and Control.*

22. *CSBH Monthly Bulletin* 2 (January 1907): 70; 5 (March 1910): 219.

23. *CSBH Monthly Bulletin* 5 (July 1909): 17–18; 5 (March 1910): 218–19.

24. Harvey Monroe Hall, "Walnut Pollen as a Cause of Hay Fever," *Science* 47 (24 May 1918): 516–17; Gregg Mitman, "Natural History and the Clinic: The Regional Ecology of Allergy in America," *Studies in the History and Philosophy of Biology and Biomedical Science* 34 (2003): 491–510. I am indebted to Gregg Mitman for calling my attention to this issue.

25. Martin V. Melosi, *The Sanitary City: Urban Infrastructure in America from Colonial Times to the Present* (Baltimore, MD: Johns Hopkins University Press, 2000), 45–47, 103–16; Tomes, *Gospel of Germs;* Suellen Hoy, *Chasing Dirt: The American Pursuit of Cleanliness* (New York: Oxford University Press, 1995); Leavitt, *Healthiest City.*

26. Frank Morton Todd, *Eradicating Plague from San Francisco: Report of the Citizens' Health Committee and an Account of Its Work* (San Francisco:

C. A. Murdock & Co., 1909), 155; Susan Craddock, *City of Plagues: Disease, Poverty, and Deviance in San Francisco* (Minneapolis: University of Minnesota Press, 2000), 124–60; Shah, *Contagious Divides,* 120–57; Risse, "'A Long Pull, a Strong Pull, and All Together.'"

27. *Twenty-fourth Biennial Report,* 91; *Twenty-seventh Biennial Report of the CSBH* (Sacramento, 1923), 65.

28. The growing interest in rural sanitation coalesced with a broader effort to improve all aspects of farm life, which was given its most prominent statement in 1909 by President Theodore Roosevelt's Commission on Country Life. What would subsequently be called "the country life movement" espoused the moral and physical virtues of rural life and argued that the condition of the countryside was critical to the nation as a whole. That movement was largely articulated by eastern elites and, like so much Progressive reform, was underwritten by anxieties over the way in which urbanization and immigration were reshaping the nation. See U.S. Congress, Senate, *Report of the Country Life Commission,* 60th Cong., 2d sess., 1909, S. Doc. 705; William L. Bowers, *The Country Life Movement in America, 1900–1920* (Port Washington, NY: Kennikat Press, 1974). On rural hygiene, William B. Herms, "Rural Hygiene and Sanitation," *CSBH Monthly Bulletin* 15 (February 1920): 247–54; William B. Herms, "Health on the Farm," *University of California Journal of Agriculture* 2 (December 1914): 136–37; Henry N. Ogden, *Rural Hygiene* (New York: Macmillan, 1913); Allen W. Freeman, "The Farm: The Next Point of Attack in Sanitary Progress," *JAMA* 55 (27 August 1910): 736–38; Tomes, *Gospel of Germs,* 195–204; John Ettling, *The Germ of Laziness: Rockefeller Philanthropy and Public Health in the New South* (Cambridge, MA: Harvard University Press, 1981).

29. On urban sanitation in this period, see Martin V. Melosi, ed., *Pollution and Reform in American Cities, 1870–1930* (Austin: University of Texas Press, 1980); Melosi, *Sanitary City;* Hoy, *Chasing Dirt;* Joel A. Tarr, *The Search for the Ultimate Sink: Urban Pollution in Historical Perspective* (Akron: University of Ohio Press, 1996); Leavitt, *Healthiest City;* Galishoff, *Newark: The Nation's Unhealthiest City.*

30. For instance, in 1908 federal statistics for California showed cities to have a death rate of 20.3 per thousand and rural areas a rate of 16.8 per thousand. *CSBH Monthly Bulletin* 6 (December 1910): 13–14.

31. *CSBH Monthly Bulletin* 10 (April 1915): 263.

32. Ralph Chester Williams, *The United States Public Health Service, 1798–1950* (Washington, DC: U.S. Public Health Service, 1951), 369–70; *CSBH Monthly Bulletin* 15 (February 1920): 247 (for quote); Freeman, "The Farm"; various letters in "Malaria Control" file, box 3/1, f3204, CDPH records, CSA.

33. *CSBH Monthly Bulletin* 6 (November 1910): 276; Gray, "Malaria Control in California."

34. Daniel, *Bitter Harvest,* 88–96; Mitchell, *Lie of the Land,* 36–57; Anne Marie Woo-Sam, "Domesticating the Immigrant: California's Commission of Immigration and Housing and the Domestic Immigration Policy Movement, 1910–1945" (Ph.D. diss., University of California, Berkeley, 1999), 248–301.

35. Alisa Klaus, "Depopulation and Race Suicide: Maternalism and Prona-

talist Ideologies in France and the United States," in *Mothers of a New World: Maternalist Politics and the Origins of Welfare States,* ed. Seth Koven and Sonya Michel (New York: Routledge, 1993), 188–212. For concerns about "race suicide" in California, see *Nineteenth Biennial Report of the California State Board of Health* (Sacramento, 1906), 59.

36. Chester H. Rowell, "Chinese and Japanese Immigrants—A Comparison," *Annals of the American Academy of Political and Social Science* 34 (September 1909): 230. As V. S. McClatchy put it before the U.S. Senate Committee on Immigration in 1924, "California regards herself as a frontier state. She has been making for 20 years the fight of the nation against incoming of alien races whose peaceful penetration must in time with absolute certainty drive the white race to the wall." Quoted in Roger Daniels, *The Politics of Prejudice: The Anti-Japanese Movement in California and the Struggle for Japanese Exclusion* (Berkeley: University of California Press, 1977), 99. Elwood Mead convinced the California Legislature to approve his personal pet project—two planned agricultural settlements in the Central Valley that were limited to whites and favored married couples. These settlements were intended both to attract white colonists to California and to demonstrate the feasibility of keeping whites on the land under the right conditions. See Elwood Mead, "Social Needs of Farm Life," 1919, carton 12/127, Elwood Mead Papers, BL (quote on 5–6); James R. Kluger, *Turning on Water with a Shovel: The Career of Elwood Mead* (Albuquerque: University of New Mexico Press, 1992), 85–101.

37. W. H. S. Jones, *Malaria, a Neglected Factor in the History of Greece and Rome* (Cambridge: Macmillan and Bowes, 1907); Ronald Ross, "Malaria in Greece," *Annual Report of the Smithsonian Institution for 1908* (Washington, DC: U.S. GPO, 1909), 697–710; Humphreys, *Malaria,* 83 (for U.S. Public Health Service quote); CSBH, *Malaria in California: A Detailed Survey of Nine Widely Separated California Communities with Recommendations and Estimates,* by Louva G. Lenert and Edward T. Ross (Sacramento, 1923), 65; "Rural Sanitation: Effective Cooperation," *CSBH Monthly Bulletin* 12 (August 1916): 74–81; Stanley B. Freeborn, "Malaria Control: A Report of Demonstration Studies at Anderson, California," *CSBH Monthly Bulletin* 15 (March 1920): 287.

38. Quoted in William F. Snow, "Malaria, the Minotaur of California," *CSBH Monthly Bulletin* 5 (August 1909): 109; Herms, "Malaria Control," 76 (for Butler quote); Walter W. Weir, "The Drainage Situation in the Rice Growing Areas of the Sacramento Valley" (1921, typescript), WRCA; "The Malaria Problem," *Transactions of the Commonwealth Club of California* 11 (March 1916): 15.

39. Freeborn, "Malaria Control," 279. The racialized nature of antimalarial work was obvious from its inception. In his volume on mosquito brigades, Ronald Ross had stipulated that malaria control work should be undertaken only in those areas that had a significant white population, under the assumption that nonwhite settlements lacked enough "civilization" to make the effort worthwhile (Ronald Ross, *Mosquito Brigades and How to Organize Them* [London: George Philip & Son, 1902], 44–45). On racialized discourse and public health, see also Shah, *Contagious Divides;* Warwick Anderson, "Excremental Colonial-

ism: Public Health and the Poetics of Pollution," *Critical Inquiry* 21 (Spring 1995): 640–69; Alexandra Minna Stern, "Buildings, Boundaries, and Blood: Medicalization and Nation-building on the U.S.-Mexico Border, 1910–1930," *Hispanic American Historical Review* 79 (February 1999): 41–82.

40. Logan quoted in Jones, *Memories, Men, and Medicine,* 54 (emphasis added); William Herms, "Flies—Their Habits and Control," *CSBH Monthly Bulletin* 11 (March 1916): 463. Also, Anderson, "Excremental Colonialism."

41. Gyan Prakash, *Another Reason: Science and the Imagination of India* (Princeton: Princeton University Press, 1999), 132; Reynaldo Illeto, "Cholera and Origins of American Sanitary Order in the Philippines," in *Discrepant Histories: Translocal Essays on Filipino Cultures,* ed. Vincente Rafael (Philadelphia: Temple University Press, 1995), 51–82; Harriet Deacon, "Racial Segregation and Medical Discourse in Nineteenth-Century Cape Town," *Journal of Southern African Studies* 22 (June 1996): 287–308; Warwick Anderson, "The Third-World Body," in *Medicine in the Twentieth Century,* ed. Roger Cooter and John Pickstone (Amsterdam: Harwood Academic Publishers, 2000), 235–45.

42. Two notable exceptions are Alan I. Marcus, "Physicians Open a Can of Worms: American Nationality and Hookworm in the United States, 1893–1909," *American Studies* 30 (Fall 1989): 103–22; Warwick Anderson, "Going through the Motions: American Public Health and Colonial 'Mimicry,'" *American Literary History* (2002): 686–719.

43. Deborah Fitzgerald, "Accounting for Change: Farmers and the Modernizing State," in *The Countryside and the Age of the Modern State: Political Histories of Rural America,* ed. Catherine Stock and Robert D. Johnson (Ithaca, NY: Cornell University Press, 2001), 202–3 (for quote); David B. Danbom, *The Resisted Revolution: Urban America and the Industrialization of Agriculture, 1900–1930* (Ames: University of Iowa Press, 1979); Herms, "Health on the Farm."

44. *Twenty-seventh Biennial Report,* 16, 65; "Sanitary Inspection Reports, no. 419–472," box 2/2, CDPH Records, f3204, CSA; Letter from CSBH to Carson C. Cook, 27 January 1917, box 1/24, CDPH Records, f3204, CSA.

45. William W. Cort, "Dangers to California from Oriental and Tropical Parasitic Diseases," *CSBH Monthly Bulletin* 14 (July 1918): 11–12.

46. Ann Stoler has argued that "cultural racism" emerged in the early twentieth century in European and American colonies as legal categories of race were abrogated. My point here complements hers, by showing how new medical concepts helped to bring about new formulations of "race," not only in the colonies, but also in the United States (Stoler, "Sexual Affronts and Racial Frontiers"). For other examinations of the relationship between race and public health, see Briggs and Martini-Briggs, *Stories in a Time of Cholera;* Shah, *Contagious Divides;* Anderson, "Excremental Colonialism"; Stern, "Buildings, Boundaries and Blood."

47. "The Sanitary Engineer and Public Health," *CSBH Monthly Bulletin* 6 (March 1911): 539–41; Hill, "What Is the New Public Health?"; Melosi, *Sanitary City,* 103–16 (115 for quote by engineer George Whipple); Fee and Porter, "Public Health, Preventive Medicine and Professionalization"; Paul Starr, *The Social Transformation of American Medicine: The Rise of the Sovereign Profes-*

sion and the Making of a Vast Industry (New York: Basic Books, 1982), 180–97; Worboys, *Spreading Germs*, 235–36.

48. Mol, *Body Multiple*, 87–117 (quote on 96).

49. *CSBH Monthly Bulletin* 4 (July 1908): 25–27; *CSBH Monthly Bulletin* 2 (January 1907): 70; Christopher Hamlin, *A Science of Impurity: Water Analysis in Nineteenth-Century Britain* (Berkeley: University of California Press, 1990). Patrick Gurian and Joel Tarr note that many engineers opposed the adoption of bacteriological standards for water quality because they felt that general standards could not be designed to fit local conditions (Patrick Gurian and Joel A. Tarr, "The First Federal Drinking Water Standards and Their Evolution: A History from 1914 to 1974," in *Improving Regulation: Cases in Environment, Health, and Safety,* ed. Paul S. Fischbeck and R. Scott Farrow [Washington, DC: Resources for the Future, 2001], 43–69). The State Board of Health actually employed the sanitary engineer Charles Gilman Hyde as a consultant beginning in 1907. Hyde was a professor at the University of California, Berkeley, and a leading engineer of his generation. He subsequently trained nearly all of the engineers that worked for the board in the first half of the century; the State Bureau of Sanitary Engineering was initially housed in the civil engineering building at the University of California, next to Hyde's office (C. G. Gillespie, "Origins and Early Years of the Bureau of Sanitary Engineering," interview by Malca Chall, 27 January 1970).

50. William B. Herms, "How to Control the Common House Fly," *CSBH Monthly Bulletin* 5 (May 1910): 274; "Sanitary Inspections," *CSBH Monthly Bulletin* 10 (June 1915): 318–41; "Sanitary Inspection Reports, 1917–1918," box 2, CDPH Records, f3204, CSA.

51. "Sanitary Inspections."

52. "The Outbreak of Typhoid Fever at Hanford," *CSBH Monthly Bulletin* 9 (May 1914): 269–72; "A Typhoid Outbreak and a Clean-up Campaign," *CSBH Monthly Bulletin* 12 (August 1916): 82–83.

53. *CSBH Monthly Bulletin* 9 (May 1914): 268–272; *Twenty-fifth Biennial Report*, 33.

54. On the temporalization of space and the spatialization of time as critical aspects of modernity, see Timothy Mitchell, "The Stage of Modernity," in *Questions of Modernity*, ed. Timothy Mitchell (Minneapolis: University of Minnesota Press, 2000), 1–34, esp. 14–17; "Sanitary Inspections," 318 (for quote).

55. W. F. Bynum, "'Reasons for Contentment': Malaria in India, 1900–1920," *Parassitologia* 40 (1998): 19–27.

56. *CSBH Monthly Bulletin* 1 (August 1905): 17 (for quote). On debates between doctors and engineers and entomologists, see the statements of William Herms, Karl Meyer, and C. E. Grunsky in *Transactions of the Commonwealth Club of California* 11 (March 1916): 1–40. Early medical entomologists such as L. O. Howard, chief of the U.S. Department of Agriculture's Bureau of Entomology, struggled to keep insect-borne diseases within the purview of entomology rather than medicine, arguing for the critical role of entomologists (like himself) who understood insect structure and behavior. On these professional struggles, see Hae-Gyung Geong, "Exerting Control: Biology and Bureaucracy in the Development of American Entomology, 1870–1930" (Ph.D. diss., University of Wisconsin–Madison, 1999), 247–99.

57. On the Panama campaign, see Charles Francis Adams, *The Panama Canal Zone: An Epochal Event in Sanitation* (Boston: Massachusetts Historical Society, 1911); Joseph Bucklin Bishop, *The Panama Gateway* (New York: Scribner, 1913); W. C. Gorgas, "The Conquest of the Tropics for the White Race," *JAMA* 52 (19 June 1909): 1967–69; Joseph A. LePrince and A. J. Orenstein, *Mosquito Control in Panama: The Eradication of Malaria and Yellow Fever in Cuba and Panama* (New York: G. P. Putnam's Sons, 1916). On colonies as laboratories, see Frederick Cooper and Ann Stoler, eds., *Tensions of Empire: Colonial Cultures in a Bourgeois World* (Berkeley: University of California Press, 1997); Mitchell, "The Stage of Modernity." For reception in California, Snow, "Malaria, the Minotaur"; "The Mosquito Pest Can Be Stamped Out," *CSBH Monthly Bulletin* 14 (February 1919): 273–74; *CSBH Weekly Bulletin* 1 (17 June 1922): 71.

58. K. F. Meyer, "Source of Malaria in California," *Transactions of the Commonwealth Club of California* 11 (March 1916): 25.

59. Gray and Fontaine, "History of Malaria"; William B. Herms and Harold Farnsworth Gray, *Mosquito Control: Practical Methods for Abatement of Disease Vectors and Pests* (New York: Commonwealth Fund, 1940); Earl W. Mortenson, "A Historical Review of Mosquito Prevention in California, Part I (1904–1946)," *Proceedings of the California Mosquito and Vector Control Association* 53 (1985): 41–46. On Herms, see "William Brodbeck Herms, Entomology and Parasitology: Berkeley, 1876–1949," at http://dynaweb.oac.cdlib.org:8088/dynaweb/uchis/public/inmemoriam/inmemoriam1949.

60. LePrince and Orenstein, *Mosquito Control in Panama,* 63; Mortenson, "Historical Review of Mosquito Prevention"; "An Intensive Anti-mosquito Campaign," *CSBH Monthly Bulletin* 7 (January 1912): 177.

61. Pisani, *From the Family Farm to Agribusiness,* 379–80; W. S. Guildford, "Irrigation Practice in the Sacramento Valley" (Sacramento Valley Irrigation Company, [1914], typescript), 10, WRCA; Harold Farnsworth Gray, "Historical Highlights of 'Permanent' Mosquito Control in California," *Proceedings of the California Mosquito Control Association* 20 (1952): 61–62.

62. Harold Farnsworth Gray et al., "Discussion on 'Prevention of Mosquito Breeding,' by Spencer Miller," *Transactions of the American Society of Civil Engineers* 76 (1913): 767; Herms and Gray, *Mosquito Control,* 117; Clark C. Spence, "The Golden Age of Dredging: The Development of an Industry and Its Environmental Impact," *Western Historical Quarterly* 11 (October 1980): 401–414; CSBH, *Malaria in California: A Detailed Survey of Nine Widely Separated California Communities with Recommendations and Estimates* (Sacramento, 1923).

63. Freeborn, "Malaria Control"; Herms and Gray, *Mosquito Control;* Richard F. Peters, "Development of Mosquito Control in California through the Window of the Department of Health Services," *Proceedings of the California Vector Control Association* 52 (1984): 35–38; Harold Farnsworth Gray, "Which Way Now?" *Proceedings of the California Mosquito Control Association* 18 (1950): 3–4.

64. CSBH, *Malaria and Mosquito Control* (Sacramento, 1923), 16; CSBH, *Malaria in California.*

65. Herms, *Mosquito Control,* 2.

66. Various letters on gambusia in box 3/1, CDPH Records, f3204; Henry R. Rupp, "Adverse Assessments of *Gambusia Affinis:* An Alternate View for Mosquito Control Practitioners," *Journal of the American Mosquito Control Association* (June 1996): 155–66; Johnson A. Neff, "Impressions of Mosquito Control vs. Wildlife," *Proceedings and Papers of the Annual Conference of Mosquito Abatement Officials in California* 10 (1939): 13–18. In response to Neff's paper, one entomologist suggested that since mosquitoes might be vectors of animal and bird diseases, the concern of wildlife officials was misplaced (statement of Dr. M. A. Stewart, in *Proceedings and Papers,* 18–19). Also, Dick Peters Oral History, interviewed by Ernie Lusk, Earl Mortenson, and Bob Schoeppner, 14 September 1994 and 22 August 1995 (this oral history was recorded and transcribed by the Historical Archives Committee of the California Mosquito and Vector Control Association and was kindly made available to me by Dr. Noor Tietze). On the interest of ecologists in "land health" in this period, see Mitman, "In Search of Health."

67. Edward Halford Ross, *The Reduction of Domestic Mosquitoes: Instructions for the Use of Municipalities, Town Councils, Health Officers, Sanitary Inspectors, and Residents in Warm Climates* (Philadelphia: P. Blakiston's Son & Co., 1911), 1–13; Spencer Miller, "Discussion on 'Prevention of Mosquito Breeding,' by Spencer Miller," *Transactions of the American Society of Civil Engineers* 76 (1913): 780; *CSBH Monthly Bulletin* 2 (March 1907): 97; L. O. Howard, *The House Fly—Disease Carrier: An Account of Its Dangerous Activities and of the Means of Destroying It,* 2d ed. (New York: Frederick A. Stokes Co., 1911), xvi–xvii; W. B. Herms, *The House Fly in Its Relation to Public Health,* California Agricultural Experiment Station Bulletin 215 (Sacramento, 1911), 533; L. O. Howard, *The House Fly and How to Suppress It* (Washington, DC: U.S. Department of Agriculture, 1924); William Dwight Pierce, *Sanitary Entomology: The Entomology of Disease, Hygiene and Sanitation* (Boston: Richard G. Badger, 1921), 20–22. For a more skeptical view of the role of insects from the same period, see George H. F. Nuttall, "On the Role of Insects, Arachnids and Myriapods as Carriers in the Spread of Bacterial and Parasitic Diseases of Man and Animals: A Critical and Historical Study," *Johns Hopkins Hospital Reports* 8 (1900): 1–154. Also Naomi Rogers, "Germs with Legs: Flies, Disease, and the New Public Health," *Bulletin of the History of Medicine* 63 (1989): 599–617.

68. Herms and Gray, *Mosquito Control,* 9; "Plague in Squirrels," *CSBH Monthly Bulletin* 4 (July 1908): 15–16; *CSBH Monthly Bulletin* 5 (November 1909): 88–91; 11 (January 1916): 329; 11 (June 1916): 618.

69. William B. Herms, *Medical and Veterinary Entomology* (New York: Macmillan, 1923), 8–9. This was also evident in the resistance of California public health officials to recognizing the possibility that plague was endemic to the state and that their strategy of vector (i.e., squirrel) eradication was misplaced, which I discuss below.

70. Herms, *The House Fly,* 532–33.

71. Senate, *Report of the Country Life Commission,* 45–46.

72. Herms, "Rural Hygiene and Sanitation," 247; *CSBH Monthly Bulletin* 15 (February 1920): 247; Charles Gilman Hyde, "A Report upon the Sewerage

and Sewage Disposal of the City of Gridley, Butte County, California," November 1913, file 29.3, BSE Records, MS 80/3, WRCA. Jean-Pierre Goubert points out that French peasants in the early twentieth century still believed that dirt provided protection against disease (Jean-Pierre Goubert, *The Conquest of Water: The Advent of Health in the Industrial Age* [Princeton, NJ: Princeton University Press, 1989], 216–18).

73. Herms, *Malaria: Cause and Control,* 9; Duffy, *The Sanitarians,* 206.

74. Herms, *Mosquito Control,* 92; Herms, *The House Fly,* 533; Herms, "Rural Hygiene and Sanitation," 253.

75. "Sanitation in Country Districts," *CSBH Monthly Bulletin* 6 (November 1910): 347. On state's approach, see also Woo-Sam, "Domesticating the Immigrant," 248–331.

76. "The California Sanitation Exhibit," *CSBH Monthly Bulletin* 4 (March 1909): 107–11; H. O. Jenkins, "A Traveling Sanitation Exhibit Directed by the State Board of Health of California in 1909" (Massachusetts Institute of Technology, Boston, 1910, mimeograph). The sanitation car was subsequently copied by several other state health departments, most notably Louisiana's. It may also have been the model for a similar exhibit in the colonial Philippines. Duffy, *The Sanitarians,* 226; Anderson, "Excremental Colonialism."

77. Herms, *The House Fly,* 533; Herms, *Medical Entomology,* 291–92; L. O. Howard, *A History of Applied Entomology (Somewhat Anecdotal)* (Washington, DC: Smithsonian Institution, 1930), 474–75; Rogers, "Germs with Legs."

78. Herms, *Malaria: Cause and Control,* 72–78; Gray and Fontaine, "History of Malaria," 34; Tomes, *Gospel of Germs,* 135–54; Hoy, *Chasing Dirt;* Marilyn Irvin Holt, *Linoleum, Better Babies and the Modern Farm Woman, 1890–1930* (Albuquerque: University of New Mexico Press, 1995).

79. Comment by Thomas H. Means on Spencer Miller, "Preventing Mosquito Breeding," *Transactions of the American Society of Civil Engineers* 76 (1913): 779; Bowers, *Country Life Movement,* 124.

80. CSBH, *Malaria in California,* 52; Stanley F. Bailey, D. C. Baerg, and H. A. Cristensen, "Seasonal Distribution and Behavior of California Anopheline Mosquitoes," *Proceedings of the California Mosquito Control Association* 40 (1972): 92–101; Thomas McKeown, *The Modern Rise of Population* (New York: Academic Press, 1976); Worboys, *Spreading Germs.* On technical and scientific power emerging out of complex material contexts, see Mitchell, *Rule of Experts,* 19–53.

81. "Plague in Squirrels," 15; William F. Snow, "When Commerce and Health Unite," *CSBH Monthly Bulletin* 6 (February 1911): 520–21; *Twenty-third Biennial Report of the CSBH* (Sacramento, 1914), 10; *Twenty-second Biennial Report,* 12; Dan C. Cavanaugh, "K. F. Meyer's Work on Plague. Biographical Notes," *Journal of Infectious Diseases* 129, Supplement (May 1974): 411–13; K. F. Meyer, "The Prevention of Plague in Light of Newer Knowledge," *Annals of the New York Academy of Sciences* 48 (1947): 429–66; "Four University of California Bioscientists: Karl Friedrich Meyer (1884–1974)," http://bancroft.berkeley.edu/Exhibits/Biotech/meyer.html (accessed 15 September 2005).

82. Cort, "Dangers to California," 6.

83. Worboys, "Germs, Malaria, and the Invention of Mansonian Tropical Medicine."

84. William F. Snow, "Some Other Tropical Diseases Occasionally Seen in California," *CSBH Monthly Bulletin* 5 (December 1909): 117–19; *CSBH Monthly Bulletin* 13 (May 1918): inside cover.

85. Herbert Gunn, "Uncinariasis in California, Based on Observations of Sixty-two Cases," *CSJM* 3 (July 1905): 213; Creighton Wellman, "Comments on Tropical Medicine," *CSJM* 8 (January 1910): 23–24.

86. Growers, however, insistently denied that rice was to blame and pointed instead to the region's long history of malaria. Over the next several decades, the state of California would continue to fund research into the ecology of rice fields in the hope that the malaria problem could be contained. Stanley B. Freeborn, "The Rice Fields as a Factor in the Control of Malaria," *Journal of Economic Entomology* 10 (June 1917): 354–59; Stanley B. Freeborn, "Rice, Mosquitoes and Malaria," *CSBH Monthly Bulletin* 12 (November 1916): 247–52; Basil G. Markos, "Distribution and Control of Mosquitoes in Rice Fields," *Journal of the National Malaria Society* 10 (1951): 233–47; Weir, "Drainage Situation"; Stanley B. Freeborn, "The Malaria Problem in the Rice Fields," *CSJM* 15 (October 1917): 413.

87. Walter M. Dickie, "Migration and the Spread of Disease," *CSBH Weekly Bulletin* 3 (31 January 1925): 201–2; Henry du R. Phelan, "Trypanosome and Its Relation to Certain Diseases," *CSJM* 3 (November 1905): 351–52; Alan M. Kraut, *Silent Travelers: Germs, Genes, and the "Immigrant Menace"* (New York: Basic Books, 1994).

88. *CSBH Weekly Bulletin* 7 (23 June 1928): 77; George E. Ebright, "Plans for Malaria Control under the New Mosquito Abatement Act," *CSBH Monthly Bulletin* 11 (December 1916): 251; Butler, "History of the First Mosquito Control Campaign." Others, however, laid the blame for malaria more generally on the presence of African Americans in the United States ("Statement by Ray Lyman Wilbur," *Transactions of the Commonwealth Club of California* 11 (March 1916): 12.

89. Cort, "Dangers to California," 6–10; *CSBH Monthly Bulletin* 14 (April 1919): 347 (for houseboat); Guy P. Jones, "Typhus Fever in California," *CSBH Monthly Bulletin* 12 (October 1916): 180–85; Shah, *Contagious Divides*, 179–203; Stern, "Buildings, Boundaries, and Blood."

90. Anderson, "Third-World Body"; Anderson, "Excremental Colonialism"; Lefebvre, *Production of Space,* 343; Cindi Katz, "Major/Minor: Theory, Nature, and Politics," *Annals of the Association of American Geographers* 85 (1995): 164–68; Cindi Katz and Andrew Kirby, "In the Nature of Things: The Environment and Everyday Life," *Transactions of the Institute of British Geographers* 16 (1991): 259–71.

91. California Mosquito Control Association, *Proceedings and Papers of the Annual Conference* 11 (1941): 3–14; California Assembly, Public Health Committee, *A Research Report on Encephalitis in California* (Sacramento, 1953); Mortenson, "Historical Review of Mosquito Prevention"; Peters, "Development of Mosquito Control in California"; Harlin L. Wynns, "The Danger to Civilian Populations on the Pacific Coast from Mosquito-Transmitted Infections in

Returning Military Personnel," *Proceedings of the California Mosquito and Vector Control Association* 13 (1944): 22–29. On history of disease ecology, Anderson, "Natural Histories of Infectious Disease"; Helen Tilley, "Ecologies of Complexity: Tropical Environments, African Trypanosomiasis, and the Science of Disease Control Strategies in British Colonial Africa, 1900–1940," *Osiris* 19 (2004): 21–38.

92. Helmut Kloos, "Valley Fever *(Coccidioidomycosis):* Changing Concepts of a 'California Disease,'" *Southern California Quarterly* 55 (Spring 1973): 59–88; C. W. Emmons, "A Reservoir of Coccidioidomycosis in Wild Rodents," *Journal of Bacteriology* 45 (1943): 306.

93. K. F. Meyer, "Why Epidemics," *CDPH Weekly Bulletin* 16 (5 June 1937): 73–75; (12 June 1937): 77–79; (19 June 1937): 81–83; (3 July 1937): 89–91. An entomologist at the University of California complained in 1941 that the study of encephalitis in California had "been too one sided. Either the attention of the worker has been devoted entirely to the disease itself . . . ," he wrote, "or else it has been fixed on transmission experiments with the 'insect vector.' These two lines of endeavor should be brought together and studied in the field epidemiologically" (Thomas H. C. Aitken, "The Relationship of the Distribution of Cases of Equine Encephalomyelitis [Human and Equine] and Mosquitoes in California," *Proceedings of the California Mosquito Control Association* 11 [1941]: 8–15). On Mexican laborers, see Henry P. Anderson, *The Bracero Program in California, with Particular Reference to Health Status, Attitudes, and Practices* (Berkeley: University of California, School of Public Health, 1961).

4. MODERN LANDSCAPES AND ECOLOGICAL BODIES

1. Griffith E. Quinby and Allen B. Lemmon, "Parathion Residues as a Cause of Poisoning in Crop Workers," *JAMA* 166 (15 February 1958): 740–46.

2. Herbert K. Abrams, "Public Health Aspects of Agricultural Chemicals," *California's Health* 6 (15 January 1949): 97–102.

3. Walter M. Dickie, "Health of the Migrant," *CSBH Weekly Bulletin* 17 (18 June 1938): 81–83; Stanford F. Farnsworth, "Malaria and the Migratory Problem in California," *Proceedings and Papers of the Annual Conference of Mosquito Abatement Officials in California* 9 (1938): 39–46; "Migration and Communicable Diseases," *CDPH Weekly Bulletin* 17 (4 June 1938): 73–74; U.S. Public Health Service, *A Study of Medical Problems Associated with Transients* by C. F. Blankenship and F. Safer, Public Health Bulletin No. 258 (Washington, DC: U.S. GPO, 1940); Malcolm H. Merrill, "Health Conditions and Services in California for Domestic Seasonal Agricultural Workers and Their Families," parts 1 and 2, *California's Health* 18 (1 February 1961): 113–17 and 18 (15 February 1961): 121–24.

4. On these shifts, see also Christopher Sellers, "Body, Place, and the State: The Making of an 'Environmentalist' Imaginary in the Post–World War II U.S.," *Radical History Review* 74 (1999): 31–64; Maril Hazlett, "Voices from the *Spring: Silent Spring* and the Ecological Turn in American Health," in *Seeing Nature through Gender,* ed. Virginia J. Scharff (Lawrence: University Press of

Kansas, 2003), 103–28; Nancy Langston, "Gender Transformed: Endocrine Disruptors in the Environment," in *Seeing Nature through Gender,* 129–68.

5. Stoll, *Fruits of Natural Advantage;* Miriam J. Wells, *Strawberry Fields: Politics, Class, and Work in California Agriculture* (Ithaca, NY: Cornell University Press, 1996), 19–37; Rhode, "Learning, Capital Accumulation," 778; Ernesto Galarza, *Merchants of Labor: The Mexican Bracero Story* (Santa Barbara, CA: McNalley and Loftin, 1964), 109–10; Preston, *Vanishing Landscapes,* 211–37; Ernesto Galarza, *Farm Workers and Agri-business in California, 1947–1960* (Notre Dame, IN: University of Notre Dame Press, 1977); Warren E. Johnston and Alex F. McCalla, *Whither California Agriculture: Up, Down, or Out? Some Thoughts about the Future,* Giannini Foundation Special Report No. 04–1 (Davis, CA: Giannini Foundation, University of California, 2004); Paterson, *Land, Water, and Power,* 355.

6. Mark L. Stemen, "Genetic Dreams: An Environmental History of the California Cotton Industry, 1902–1953" (Ph.D. diss., University of Iowa, 1999), 208; McGowan, *History of the Sacramento Valley,* 358; James J. Parsons, "A Geographer Looks at the San Joaquin Valley," *Geographical Review* 76 (1986): 371–89.

7. Jelinek, *Harvest Empire,* 61–94; Johnston and McCalla, *Whither California Agriculture?;* "California Farm Income Leads Other States," *SFC,* 25 February 1953; "Agriculture Reported as California's No. 1 Industry," *SFC,* 17 February 1962; "State Enjoys Year of Record Crops," *SFC,* 23 September 1962; "California Set Crop Income Record," *SFC,* 22 December 1962.

8. Wells, *Strawberry Fields,* 24; Galarza, *Farm Workers and Agri-business.*

9. Anderson, *Bracero Program,* 218; *California's Health* 18 (1 February 1961): 115; Wells, *Strawberry Fields;* Galarza, *Farm Workers and Agri-business.*

10. These examples are taken from the reports in "Agriculture—Bureau of Chemistry—General Correspondence," f3742, CDA Records, CSA.

11. According to Joseph McGowan (*History of the Sacramento Valley,* 250), the end of World War II yielded thousands of trained flyers and surplus aircraft, which were quickly employed in agricultural spraying. By 1960 there were two thousand companies in the business of aerial spraying, with seven thousand planes, servicing roughly 4.5 million acres of farmland in the western United States. On pesticide usage, Stoll, *Fruits of Natural Advantage,* 98; Martin Brown, "An Orange Is an Orange," *Environment* 17 (July–August 1975): 6–11; Memo from Allen B. Lemmon to Charles V. Dick, 22 March 1955, file 134, f3742, CDA Records. For 20 percent figure, Robert Z. Rollins, "Federal and State Regulation of Pesticides," *American Journal of Public Health* 53 (1963): 1427–31.

12. Edmund P. Russell, *War and Nature: Fighting Humans and Insects with Chemicals from World War I to "Silent Spring"* (New York: Cambridge University Press, 2001); Thomas R. Dunlap, *DDT: Scientists, Citizens, and Public Policy* (Princeton, NJ: Princeton University Press, 1981); Margaret Humphreys, "Kicking a Dying Dog: DDT and the Demise of Malaria in the American South, 1942–1950," *Isis* 87 (March 1996): 1–17.

13. Lakshman Karalliedde, Stanley Feldman, John Henry, and Timothy Marrs, eds., introduction to *Organophosphates and Health* (London: Imperial College Press, 2001), xxi–xxii. The introduction of various chemicals into California agriculture can be tracked in part through the CDPH's annual reports,

Reports of Occupational Disease in California Attributed to Pesticides and Agricultural Chemicals (1950–73).

14. California Assembly, Interim Committee on General Research, Subcommittee on Pesticides, *Government and Pesticides in California* (Sacramento, 1965), 25–26; CDHS, "Pesticide Use Reports and Restricted Materials," [198?], CDHS-McFarland files.

15. California Assembly, *Research Report on Encephalitis in California.*

16. James J. Kalstrom to Allen B. Lemmon, "Annual Report—Agricultural Pest Control—1955," 6 January 1956, file 125, f3742, CDA Records; "Death of Fish in Canals and Ditches Adjoining Rice Fields in Sacramento River Blamed on Use of Insecticides," *SFC*, 6 June 1956; "Report by Fish and Game Department Listing Fish and Wildlife Killed by Pesticides and Herbicides," *SFC*, 6 August 1969; California Department of Fish and Game, *Pesticides: Their Use and Toxicity in Relation to Wildlife*, by Robert L. Rudd and Richard E. Genelly, Game Bulletin No. 7 (Sacramento, 1956).

17. Concerns about crop and livestock damage and the destruction of bee colonies had first raised opposition to the metallic sprays in use before World War II. In particular, the use of airplanes to apply calcium arsenate and the resulting "drift" had brought pesticide issues to the attention of the state legislature in the early 1940s, leading to the passage of a 1949 law that established a permit process for so-called economic poisons. California Senate, Joint Legislative Committee on Agriculture and Livestock Problems, *Special Report on Enforcement of State Laws Relating to Agricultural Pest Control Operators and Their Use of Injurious Materials* (Sacramento, 1953).

18. Ethelbert Johnson to R. Z. Rollins, "Annual Report—Pest Control—1957," 6 January 1958, 3, file 125, f3742, CDA Records; Letter from Mrs. Stella L. Stender to State Board of Health, 27 August 1957, file 144, CDA Records; Memorandum from Allen B. Lemmon to Chas. V. Dick, 9 June 1955 and 11 October 1955, file 134, CDA Records; Robert Z. Rollins to Dr. Christine Einert, 21 March 1957, file 144, CDA Records. Also, Letter from Mr. and Mrs. Donald L. Thatcher to Malcolm Merrill, 17 February 1960, and Letter from Bradford M. Crittenden, California Highway Patrol to Malcolm H. Merrill, 9 March 1960, box 11/35, CDPH Records, CSA. In Tucson, Arizona, rural residents complained of health effects following the spraying of parathion on agricultural crops (California Senate, Fact Finding Committee on Agriculture, *Hearings*, 1964 [19 June, San Francisco], 13–14).

19. Irma West, "Occupational Disease of Farm Workers," *Archives of Environmental Health* 9 (July 1964): 92–98; CDPH, *Reports of Occupational Disease Attributed to Pesticides and Agricultural Chemicals, California, 1950* (Berkeley, 195[1]).

20. Thomas H. Milby, Fred Ottoboni, and Howard W. Mitchell, "Parathion Residue Poisoning among Orchard Workers," *JAMA* 189 (3 August 1964): 351–56.

21. Ephraim Kahn, "Pesticide Related Illness in California Farm Workers," *Journal of Occupational Medicine* 18 (October 1976): 693–96.

22. Howard W. Chambers, "Organophosphorus Compounds: An Overview,"

Organophosphates: Chemistry, Fate, and Effects, ed. J. E. Chambers and P. E. Levi (San Diego, CA: Academic Press, 1992), 3–17; Marion Moses, "Pesticide-related Health Problems and Farmworkers," *AAOHN Journal* 37 (March 1989): 115–30; Devra Lee Davis, Aaron Blair, and David G. Hoel, "Agricultural Exposures and Cancer Trends in Developed Countries," *Environmental Health Perspectives* 100 (April 1992): 39–44; Aaron Blair and Shelia Hoar Zahm, "Cancer among Farmers," *Occupational Medicine: State of the Art Reviews* 6 (1991): 335–54; Scott Shane, "Chemicals Sickened Gulf War Veterans, Latest Study Finds," *NYT,* 14 October 2004. The first evidence that OP pesticides could cause long-term neuropsychological problems was reported in 1961 by Australian researchers (Karalliedde et al., *Organophosphates and Health,* xxiii).

23. Anderson, *Bracero Program,* 213, 286, 288.

24. CDPH, "Community Studies on Pesticides," 15 December 1970 (CDPH, Bureau of Adult Health, Berkeley, photocopy), 14; Anderson, *Bracero Program,* 215, 217. For a similar observation in a different context, see Murphy, "The 'Elsewhere within Here' and Environmental Illness."

25. Anderson, *Bracero Program,* 215.

26. Irma West, "Pesticides and Other Agricultural Chemicals as a Public Health Problem with Special Reference to Occupational Disease in California," Report to the United States Senate Committee on Government Operations, Subcommittee on Reorganization and International Organizations (18 July 1963, typescript), CSL; Mary K. Farinholt, *The New Masked Man in Agriculture: Pesticides and the Health of Agricultural Users* (Cleveland, OH: National Consumers Committee for Research and Education, 196[2]).

27. Quinby and Lemmon, "Parathion Residues," 740. For workers' accounts, see CDPH, "Community Studies on Pesticides," 15 December 1970; statement of Lee Mizrahi in U.S. House, Committee on Education and Labor, *Occupational Safety and Health Act of 1969: Hearings on H.R. 843, H.R. 3809, H.R. 4294, H.R. 13373,* 91st Cong., 1st sess., 1969 [hereafter *OSHA Hearings*], 1449–50, 1453 (for quote); statement of Thomas Milby in U.S. House, *OSHA Hearings,* 1387. On popular portrayals of disease in the 1940s and 1950s, see Martin, *Flexible Bodies,* 21–44.

28. On immigrants labeled as disease carriers, see Anderson, *Bracero Program,* 223–24; Shah, *Contagious Divides;* Kraut, *Silent Travelers.* On racial stereotypes associated with Mexicans and Mexican Americans, including the discourse of hygiene, see David Montejano, *Anglos and Mexicans in the Making of Texas, 1836–1986* (Austin: University of Texas Press, 1987), 220–34; Stern, "Buildings, Boundaries, and Blood." On the susceptibility of Mexicans in California to disease, see C. R. Kroeger, "Significance of Tuberculosis in the Itinerant Mexican Laborer," (1953), box 11/16, f3160, CDPH Records. Quote is from statement of Ralph Teall in California Senate, Fact Finding Committee on Agriculture, *Hearings,* 1964 (16 June, El Centro), 12. Comment on need for supervision is from West, "Occupational Disease of Farm Workers," 98.

29. Anderson, *Bracero Program;* CDPH, *Health Conditions and Services in California for Domestic Seasonal Agricultural Workers and Their Families* (Sacramento, CA, 1960). Even as late as 1966 in hearings before the U.S. Senate

on migratory labor, the threat of pesticides was ignored and the focus remained confined to traditional labor issues: the need for collective bargaining, a minimum wage, adequate housing, and sanitary facilities. See U.S. Senate, Subcommittee on Migratory Labor of the Committee on Labor and Public Welfare, *Amending Migratory Labor Laws*, 89th Cong., 1st and 2d sess., 1965–66.

30. Christopher Sellers, *Hazards of the Job: From Industrial Disease to Environmental Health Science* (Chapel Hill: University of North Carolina Press, 1997); Jacqueline Karnell Corn, *Response to Occupational Health Hazards: A Historical Perspective* (New York: Van Nostrand Reinhold, 1992). For a brief history of pesticide regulation in California, see http://www.calepa.ca.gov/About/History01/dpr .htm (accessed 27 May 2003); Rollins, "Federal and State Regulation of Pesticides"; Thomas Milby, M.D., interview by author, tape recording, 1 October 2002.

31. Sellers, *Hazards of the Job*, esp. 141–86; Jeffrey M. Paull, "The Origin and Basis of Threshold Limit Values," *American Journal of Industrial Medicine* 5 (1984): 227–38.

32. Karin Knorr Cetina, *Epistemic Cultures: How the Sciences Make Knowledge* (Cambridge, MA: Harvard University Press, 1999), 26–45; Robert E. Kohler, *Landscapes and Labscapes: Exploring the Lab-Field Border in Biology* (Chicago: University of Chicago Press, 2002), 1–22; Bruno Latour and Steven Woolgar, *Laboratory Life: The Social Construction of Scientific Facts* (Beverly Hills, CA: Sage, 1979); Pelling, "Contagion/Germ Theory/Specificity"; Steven Shapin, "The House of Experiment in Seventeenth-Century England," *Isis* 79 (1988): 373–404. On how practices create their objects, Latour, *Pasteurization of France*; Mol, *The Body Multiple*.

33. Sellers, *Hazards of the Job*, 141–86; Koehler, *Landscapes and Labscapes*; Shapin, "House of Experiment." Even Irma West, the CDPH's leading expert on pesticides, reportedly shared this dislike of field research (Milby interview).

34. Robert C. Spear, David L. Jenkins, and Thomas H. Milby, "Pesticide Residues and Field Workers," *Environmental Science and Technology* 9 (April 1975): 308–13; William J. Popendorf and John T. Leffingwell, "Regulating OP Pesticide Residues for Farmworker Protection," *Residue Reviews* 82 (1982): 125–202.

35. Paull, "Origin of Threshold Limit Values"; Popendorf and Leffingwell, "Regulating OP Pesticide Residues," 126 (for quote).

36. U.S. Geological Survey, *Pesticides in the Atmosphere: Distribution, Trends, and Governing Factors*, by Michael S. Majewski and Paul D. Capel, Open-File Report 94–506 (Sacramento, 1995); U.S. Geological Survey, *Occurrence of Nitrate and Pesticides in Ground Water Beneath Three Agricultural Land-Use Settings in the Eastern San Joaquin Valley, California, 1993–1995*, by Karen R. Burow, Jennifer L. Shelton, and Neil M. Dubrovsky, Water-Resources Investigations Report 97–4284 (Sacramento, 1998); California, Joint Legislative Committee on Agriculture and Livestock Problems, *Special Report on Agricultural Use of Aircraft, Regulation of Pest Control Operators, and Use and Application of Hazardous Materials* (May 1949).

37. F. A. Gunther, "Insecticide Residues in California Citrus Fruits and Products," *Residue Reviews* 28 (1968): 1–120; Milby, "Parathion Residue Poisoning"; Robert C. Spear, "Report of the Status of Research into the Pesticide

Residue Intoxication Problem in the Central Valley of California," in *Pesticide Residue Hazards to Farm Workers: Proceedings of a Workshop Held February 9–10, 1976* (Salt Lake City: U.S. Department of Health, Education and Welfare, National Institute for Occupational Safety and Health, 1976), 43–62; California Assembly, Office of Research, *California's Pesticide Regulatory Program and Farmworker Safety Issues Related to the Use of Organophosphate Pesticides: A Background Report* ([Sacramento?], 1977).

38. Popendorf and Leffingwell, "Regulating OP Pesticide Residues"; Homer R. Wolfe and John F. Armstrong, "Exposure of Workers to Pesticides," *Archives of Environmental Health* 14 (April 1967): 622–33.

39. This research is summarized in Popendorf and Leffingwell, "Regulating OP Pesticide Residues." Also, Spear, Jenkins, and Milby, "Pesticide Residues and Field Workers."

40. Popendorf and Leffingwell, "Regulating OP Pesticide Residues," 191.

41. According to Miriam J. Wells, in the 1990s an estimated 78 percent of all farm work in California was performed by hired laborers (Wells, *Strawberry Fields,* 24). For history and working conditions in California agriculture, see Stoll, *Fruits of Natural Advantage,* 124–54; McWilliams, *Factories in the Field;* Daniel, *Bitter Harvest;* Galarza, *Farm Workers and Agri-business;* Anderson, *Bracero Program.*

42. Dr. Irma West lamented the "wasted opportunity for research." See West, "Pesticides and Other Agricultural Chemicals," 10–11.

43. Paull, "Threshold Limit Values"; Robert Proctor, *Cancer Wars: How Politics Shapes What We Know and Don't Know about Cancer* (New York: Basic Books, 1995), 153–73; W. Popendorf, "Exploring Citrus Harvesters' Exposure to Pesticide Contaminated Foliar Dust," *American Industrial Hygiene Association Journal* 41 (September 1980): 652–59; Spear, Jenkins, and Milby, "Pesticide Residues and Field Workers." For comment on cumulative effects of parathion exposure, see CDPH, *Occupational Disease in California Attributed to Pesticides and Agricultural Chemicals, 1959* (Berkeley, 1961), 7.

44. Paradoxically, pesticides both bolstered and undermined the notion of a single causative agent. Through their ability to control insect vectors and control diseases like malaria, pesticides allowed experts to narrowly locate these diseases within insect bodies. However, pesticide-induced illness subsequently underscored the multiple causes of disease.

45. Cholinesterase is a chemical that occurs naturally in the body but whose concentrations are affected by organophosphates. Early work on organophosphates and cholinesterase was conducted in the late 1940s (D. Grob, W. L. Garlick, and A. M. Harvey, "The Toxic Effects in Man of the Anti-Cholinesterase Insecticide Parathion [p-Nitrophenyl Diethylthionophosphate]," *Johns Hopkins Hospital Bulletin* 81 [1950]: 106–29). On debates over significance of cholinesterase, see *Pesticide Residue Hazards,* 73–76, 89.

46. Popendorf and Leffingwell, "Regulating OP Pesticide Residues," 126.

47. For discussions of modern conceptions of the body, see Emily Martin, "The Body at Work: Boundaries and Collectivities in the Late Twentieth Century," in *The Social and Political Body,* ed. Theodore R. Schatzki and Wolfgang Natter (London: Guilford Press, 1996), 145–59; Martin, *Flexible Bodies;* and

Donna J. Haraway, "The Politics of Postmodern Bodies: Constitutions of Self in Immune System Discourse," in *Simians, Cyborgs, and Women* (New York: Routledge, 1991), 203–30.

48. Statement of Thomas Milby in House, *OSHA Hearings*, 1389–90.

49. On the ability to manage and quantify respiratory exposures, see Popendorf and Leffingwell, "Regulating OP Pesticide Residues," 152; Keith T. Maddy, "Current Considerations on the Relative Importance of Conducting Additional Studies on Hazards of Field Worker Exposure to Pesticide Residues," in *Pesticide Residue Hazards*, 134–35. On working conditions, including direct spraying of workers, see CDPH, "Community Studies on Pesticides," 15 December 1970, pt. 1, esp. 20–22. Later research made possible by new imaging technologies has revealed that pesticides routinely penetrate chemical protective clothing. R. A. Fenske et al., "Fluorescent Tracer Evaluation of Chemical Protective Clothing during Pesticide Applications in Central Florida Citrus Groves," *Journal of Agricultural Safety and Health* 8 (2002): 319–31.

50. In the face of this recognition, some argued that workers' health should be managed through routine testing of cholinesterase. Those with significant cholinesterase depression would be kept away from organophosphates, while others would be allowed to continue their exposure. Implicit here, as at least one investigator pointed out, was the notion that bodies could be "titrated" to their environments. See comment of Dr. Robert Spear in *Pesticide Residue Hazards*, 37.

51. William J. Popendorf and Robert C. Spear, "Preliminary Survey of Factors Affecting the Exposure of Harvesters to Pesticide Residues," *American Industrial Hygiene Association Journal* 35 (June 1974): 374–80; Popendorf, "Exploring Citrus Harvesters' Exposure."

52. Haraway, "Politics of Postmodern Bodies"; Martin, *Flexible Bodies*, 23–44.

53. CDPH, *Air Pollution: Effects Reported by California Residents from the California Health Survey* (Berkeley, CA, [1958?]), 29; Scott Hamilton Dewey, *Don't Breathe the Air: Air Pollution and U.S. Environmental Politics, 1945–1970* (College Station: Texas A&M University Press, 2000), 94.

54. Lutts, "Chemical Fallout"; Memorandum from John A. Maga to Malcolm H. Merrill, "Radioactive Rainout and Fallout," 1 April 1958, file 12/27, R:020, CDPH Records.

55. James Whorton, *Before Silent Spring: Pesticides and Public Health in Pre-DDT America* (Princeton, NJ: Princeton University Press, 1974); John Wargo, *Our Children's Toxic Legacy: How Science and Law Fail to Protect Us from Pesticides*, 2d ed. (New Haven, CT: Yale University Press, 1998), 70–78; Christopher J. Bosso, *Pesticides and Politics: The Life Cycle of a Public Issue* (Pittsburgh: University of Pittsburgh Press, 1987), 61–78; California Governor's Special Committee on Public Policy Regarding Agricultural Chemicals, *Report on Agricultural Chemicals and Recommendations for Public Policy* (Sacramento, 1960); Raymond Coppock, "Pesticide Residue Problem Gets Close Attention in State," *SB*, 5 June 1960. On cranberries, Bosso, *Pesticides and Politics*, 94–100.

56. Frank M. Stead, "Health Problems as Affected by Irrigation Agricul-

ture," 30 April 1957, reprinted in Frank Stead, "Earl Warren and the State Department of Public Health," interview by Gabrielle Morris (Berkeley, CA: Bancroft Library, Regional Oral History Office, 1973); U.S. House, Committee on Appropriations, Subcommittee on Departments of Labor and Health, Education and Welfare, and Related Agencies Appropriations, *Report on Environmental Health Problems,* 86th Cong., 2d sess., 1960; U.S. Public Health Service, *Report of the Committee on Environmental Health Problems to the Surgeon General* (Washington, DC: U.S. GPO, 1962).

57. Malcolm H. Merrill, "The Sanitarian in Our Changing Environment," *California's Health* 19 (15 May 1962): 161–64; Malcolm H. Merrill, "Recent Trends in Public Health," *California Medicine* 72 (January 1950): 22–25; Frank M. Stead, "Historical Concept of Environmental Sanitation," in *Managing Man's Environment in the San Francisco Bay Area* (Berkeley: University of California, Institute of Governmental Studies, 1963), 3–17. These developments, in particular the concern with air pollution, would lead the California Department of Public Health to establish a section on epidemiology, to track the incidence of noninfectious disease in specific populations in the mid-1950s, and also to expand the Bureau of Sanitary Engineering and give it new responsibilities in air pollution and radiologic health. In 1963, in the wake of the Gross Report, the Division of Environmental Sanitation would be reorganized again and renamed the Division of Environmental Health. "Air Pollution Medical Research, 1954–1960," files 67–69, R :020, CDPH Records; CSBH, "Minutes," 13 December 1963, f3204, CDPH Records.

58. Irma West, "Biological Effects of Pesticides in the Environment," in *Organic Pesticides in the Environment,* ed. Aaron A. Rosen and H. F. Kraybill (Washington, DC: American Chemical Society, 1966), 38.

59. Robert Coudy to Malcolm H. Merrill, 10 April 1958, file 12/27, f3487, CDPH Records; Letter from Malcolm H. Merrill to Mrs. Laura Tallian, 30 March 1962, box 7/39, R: 018, CDPH Records; Letter from William H. Clark to Mr. Robert Coudy, 22 April 1958, box 12/27, R: 020, CDPH Records; Letter from Malcolm H. Merrill to Christine V. Agur, 7 September 1961, box 12/13, f3204, CDPH Records; Theron G. Randolph, *Environmental Medicine: Beginnings and Bibliographies of Clinical Ecology* (Fort Collins, CO: Clinical Ecology Publications, 1987); Francis M. Pottenger and Bernard Krohn, "Poisoning from DDT and Other Chlorinated Hydrocarbon Pesticides: Pathogenesis, Diagnosis, and Treatment," *Journal of Applied Nutrition* 14 (1961): 126–39; U.S. House, *OSHA Hearings,* 1356.

60. Rachel Carson, *Silent Spring* (Boston: Houghton Mifflin, 1962); Linda Lear, *Rachel Carson: Witness for Nature* (New York: Henry Holt, 1997), 339–40, 365; Rachel Carson as cited in Dunlap, *DDT,* 100.

61. On the rhetoric of *Silent Spring,* see Craig Waddell, ed., *And No Birds Sing: Rhetorical Analyses of Rachel Carson's "Silent Spring"* (Carbondale: Southern Illinois University Press, 2000). For an example of popular understandings of ecological health, statement of Betty Morales in California Senate, Fact Finding Commission on Agriculture, *Hearings,* 1963 (23 October, Sacramento), 381–83.

62. Carson, *Silent Spring,* 47–49. Although Carson was principally inter-

ested in the health effects of pesticides, she was influenced by the work of several microbiologists who had been crafting a more holistic and ecological account of infectious disease, particularly René Dubos. René Dubos, *Mirage of Health: Utopias, Progress, and Biological Change* (New York: Harper & Brothers, 1959); Jill Cooper and David Mechanic, "Introduction to the Transaction Edition," in René Dubos, *So Human an Animal: How We Are Shaped by Surroundings and Events* (1968; reprint, New Brunswick, NJ: Transaction, 1998), ix–xviii; Anderson, "Natural Histories of Infectious Disease."

63. On the sources of Carson's sense of physical connection to the world and her concern that chemicals crossed the placenta, see Hazlett, "The Story of *Silent Spring* and the Ecological Turn." The discovery of the placenta's permeability is credited to N. McAlister Gregg who found that maternal infection with rubella was manifest in certain kinds of birth defects. Gregg also hypothesized that other toxic substances were likely to cross the placenta as well. Then, in 1961, an article appeared on the connection between maternal ingestion of thalidomide and fetal deformity. Ann Dally, "Thalidomide: Was the Tragedy Preventable?" *Lancet* 351 (1998): 1197–99.

64. Rosenberg, "Pathologies of Progress," 729. Also, Mary Douglas and Aaron Wildavsky, *Risk as Culture: An Essay on the Selection of Technical and Environmental Dangers* (Berkeley: University of California Press, 1982); Lawrence Buell, "Toxic Discourse," *Critical Inquiry* 24 (1988): 639–65.

65. On this dichotomy as a fundamental assumption of modernity, see Latour, *We Have Never Been Modern;* Mitchell, *Rule of Experts.*

66. Sellers, "Body, Place, and the State"; Lear, *Rachel Carson,* 312–38.

67. On spreading knowing widely, see Mol, *Body Multiple.*

68. On this issue with respect to Hueper, see Christopher Sellers, "Discovering Environmental Cancer: Wilhelm Hueper, Post–World War II Epidemiology, and the Vanishing Clinician's Eye," *American Journal of Public Health* 87 (November 1997): 1824–35. On Hargraves, see Hazlett, "The Story of *Silent Spring*," 103–4.

69. Hazlett, "Voices from the Spring"; Lear, *Rachel Carson,* 426, 446, 572 n.103. Also, testimony of Teall of the California Medical Association in California Senate, Fact Finding Committee on Agriculture, *Hearings, 1964* (19 June, San Francisco), 5–15.

70. Lear, *Rachel Carson,* 411–14, 435; Hazlitt, "The Story of *Silent Spring*," 219–30; 258–66.

71. Only in 1964 did the first article on the issue appear in a major medical journal. Milby, Ottoboni, and Mitchell, "Parathion Residue Poisoning among Orchard Workers"; Milby interview.

72. California Assembly, Office of Research, *California's Pesticide Regulatory Program and Farmworker Safety Issues Related to the Use of Organophosphate Pesticides: A Background Report* (Sacramento, 1977), 21. In 1969 the union succeeded in negotiating a health and safety clause in its contract with Perelli-Minetti that specifically addressed pesticides, forbidding the use of DDT, aldrin, dieldrin, and endrin and establishing a joint committee of union members and farm managers to establish safety guidelines for the use of other toxic compounds in the fields. "UFWOC Signs Historic Pesticide Safety Clause," *El Mal-*

criado, 1 October 1969, 3; Pulido, *Environmentalism and Economic Justice,* 115; Brown, *United Farmworkers,* 165.

73. CDPH, "Community Studies on Pesticides," 15 December 1969–1 November 1970, 32.

74. According to Farinholt (*New Masked Man in Agriculture,* 13), a medical guide on pesticides prepared by the CDC warned of this difficulty in diagnosis. On "hysteria" among workers, see Allen B. Lemmon to Chas. V. Dick, "Report for Week Ended September 3, 1955," 7 September 1955, 1, file 134, f3742, CDA Records, CSA.

75. In 1969 the CDPH tried to estimate incidence by interviewing 1,120 workers in Tulare County, asking specifically about symptoms during the previous twelve months that had caused them to seek medical treatment. They found that farmworkers experienced symptoms potentially associated with pesticide poisoning fifteen times more often than nonfarmworkers in the same area but that fewer than 6 percent of those affected had their illnesses reported via workmen's compensation. In 1973, Richard Howitt, a researcher at the University of California, Davis, interviewed workers on the Monterey Peninsula, recording symptoms during the previous thirty days that had caused workers to lose at least one-half day of work. That study found that cases definitely related to pesticide poisoning occurred three hundred times more frequently than those officially reported through workmen's compensation. Ephraim Kahn, "Pesticide Related Illness in California Farm Workers," *Journal of Occupational Medicine* 18 (October 1976): 693–96; U.S. House, *OSHA Hearings,* 1358–59, 1457.

76. CDPH, *Community Studies on Pesticides,* 15 December 1969–1 November 1970, 26 (emphasis added); CDPH, *Community Studies on Pesticides,* 15 December 1970, 16, 28.

77. On the UFW, see Brown, "United Farmworkers"; Galarza, *Farm Workers and Agri-business;* Sam Kushner, *Long Road to Delano* (New York: International Publishers, 1975). On DDT and DDE testing, U.S. House, *OSHA Hearings,* 1448.

78. "Pesticides and the Public," *El Malcriado,* 1–15 November 1970, 2 ; "Chavez Blasts FDA for Condoning Poisoned Food,"*El Malcriado,* 15–30 October 1969, 3; "The Threat of Chemical Poisons," *El Malcriado,* 1 January 1969, 5; U.S. Senate, Subcommittee on Migratory Labor, *Hearings on Migrant and Seasonal Farmworker Powerlessness,* pt. 6B, 91st Cong., 1st sess., 1969, 3196 (for quote), 3275–91. For other work on the UFW and pesticides, see Robert Gordon, "Poisons in the Fields: The United Farm Workers, Pesticides and Environmental Politics," *Pacific Historical Review* 68 (February 1999): 51–77; Pulido, *Environmentalism and Economic Justice;* Marion Moses, "Farmworkers and Pesticides," in *Confronting Environmental Racism: Voices from the Grassroots,* ed. Robert D. Bullard (Boston: South End Press, 1993), 161–78. For an example of the UFW's invocation of Carson, see "Data on DDT and Parathion," *El Malcriado,* 1–15 July 1969, 7, 11.

79. U.S. House, *OSHA Hearings,* 1360; U.S. Senate, *Hearings on Farmworker Powerlessness,* 3808–09.

80. Several white men interviewed by the CDPH in 1969–70 expressed classic gendered attitudes toward the risk of pesticides. CDPH, "Community Stud-

ies on Pesticides," 15 December 1969–1 November 1970, 13–33. See also the statement of Representative Scherle in U.S. House, *OSHA Hearings,* 1344–45; "Mothers Alarmed at DDT Danger," *El Malcriado,* 1–15 August 1969, 3. More recent sociological work has pointed to the fact that the views of hazards are more closely correlated with race and gender than with variables such as income and education. Linda Kalof, Thomas Dietz, Gregory Guagnano, and Paul C. Stern, "Race, Gender and Environmentalism: The Atypical Beliefs of White Men," *Race, Gender and Class* 9 (2002): 112–130.

81. CDPH, "Community Studies on Pesticides," 15 December 1969–1 November 1970, 23. See also the examples in Hazlett, "Voices from the *Spring*"; "Consumer Group Backs Boycott," *El Malcriado* 15 August–15 September 1969, 15; Brown, "United Farm Workers," 194; Ulrich Beck, *Risk Society: Toward a New Modernity,* trans. Mark Ritter (London, 1992), 74. Both the Consumer Federation of America and the National Consumers League endorsed the boycott.

82. The longest interval applied to parathion was subsequently extended for the heaviest applications. The history of these regulations at the federal level is quite complex. In 1972 the EPA convened a special task force on Occupational Exposure to Pesticides to suggest reentry standards, although the committee failed to recommend reentry intervals due to the opposition of agricultural interests. In frustration, some farmworker and environmental groups petitioned OSHA to issue emergency standards, which it did in 1973. Those standards were subsequently challenged by the American Farm Bureau and were overturned on appeal (*Florida Peach Growers v. U.S. Department of Labor* [1974]), which awarded jurisdiction on the issue to the EPA. That was significant, because the EPA's statute required it to take benefits as well as risks into account, whereas the OSHA statute called only for the protection of worker safety. The EPA initially adopted reentry regulations in 1974. These were updated in 1991; however, they remained relatively weak. Wargo, *Our Children's Toxic Legacy,* 256–57, 374 (n. 31, 32); Victoria Elenes, "Farmworker Pesticide Exposures: Interplay of Science and Politics in the History of Regulation (1947–1988)" (Master's thesis, University of Wisconsin–Madison, 1991); Orville E. Paynter, "Worker Reentry Safety: Viewpoint and Program of the Environmental Protection Agency," *Residue Reviews* 62 (1976): 13–20; Robert F. Wasserstrom and Richard Wiles, *Field Duty: U.S. Farmworkers and Pesticide Safety* (Washington, DC: World Resources Institute, 1985), 13–19.

83. David Arnold (*Problem of Nature,* esp. 11) has also remarked on this general shift in Western attitudes.

5. CONTESTING THE SPACE OF DISEASE

1. Jennifer Warren, "Mysterious Cancer Clusters Leave Anxiety in Three Towns," *LAT,* 12 July 1992.

2. Celene Krauss, "Challenging Power: Toxic Waste Protests and the Politicization of White, Working-Class Women," in *Community Activism and Feminist Politics: Organizing across Race, Class, and Gender,* ed. Nancy A. Naples (New York: Routledge, 1998), 129–50. The literary critic Lawrence Buell has tracked the emergence of what he terms "toxic discourse" in this decade. See

Buell, "Toxic Discourse." Also, Cynthia Deitering, "The Postnatural Novel," in *The Ecocriticism Reader: Landmarks in Literary Ecology,* ed. Cheryll Glotfelty and Harold Fromm (Athens: University of Georgia Press, 1996), 196–203.

3. California Department of Fish and Game, *Pesticides: Their Use and Toxicity;* "Investigators Report Another Fish-Dieoff in Sacramento River," *SFC,* 21 June 1963; "State Fish Warden Reports Death of 10,000 Fish," *SFC,* 1 August 1963; "Fish Poisoned in Yolo County Canals," *SFC,* 2 August 1963; "More Fish Killed Near Yuba City," *SFC,* 13 August 1963; "Another Fish Kill Reported in Yolo County," 31 August 1963; Edward D. Stetson, "Planning a Master Drain for the San Joaquin Valley of California," presentation to the 1963 Annual Meeting of the American Society of Agricultural Engineers, typescript, 17, WRCA; U.S. Federal Water Pollution Control Administration, *Effects of the San Joaquin Master Drain on Water Quality of the San Francisco Bay and Delta* (San Francisco, 1967); Nye, "Visions of Salt," 250–63, 283–84. Elsewhere in the country, testing in the 1950s had revealed that DDT was contaminating some drinking water supplies, and some experts suggested that this might pose long-term health risks (W. C. Hueper, "Cancer Hazards from Natural and Artificial Water Pollutants," in *Conference on Physiological Aspects of Water Quality,* ed. Harry A. Faber and Lena J. Bryson [Washington, DC: U.S. Public Health Service, 1960], 181–94).

4. Joseph A. Cotruvo and Chieh Wu, "Controlling Organics: Why Now?" *Journal of the American Water Works Association* 70 (November 1978): 590–94; Daniel A. Okun, "Drinking Water and Public Health Protection," in *Drinking Water Regulation and Health,* ed. Frederick W. Pontius (New York: John Wiley & Sons, 2003), 3–24. Gurian and Tarr, "First Federal Drinking Water Standards."

5. Gurian and Tarr, "First Federal Drinking Water Standards."

6. More accurately, the Safe Drinking Water Act insisted on the establishment of two sets of standards. One set, the "recommended maximum contaminant levels" (RMCLs, later renamed "maximum contaminant level goals," or MCLGs) were to be based solely on the protection of health and were not enforceable. The second set of goals, the MCLs, were enforceable but could take into account technical feasibility when necessary. William E. Cox, "Evolution of the Safe Drinking Water Act: A Search for Effective Quality Assurance Strategies and Workable Concepts of Federalism," *William and Mary Law and Policy Review* 21 (Winter 1997): 69–165.

7. An MCL was adopted for total trihalomethanes; in addition, a treatment standard was adopted for large public water supplies likely to be polluted by soluble organic compounds (SOCs) of "industrial origin." Opposition led to abandonment of the treatment standard in 1981. Congress took a much more forceful approach in 1986 amendments, mandating the adoption of MCLs for many organics, including several pesticides. Cox, "Evolution of the Safe Drinking Water Act."

8. Wargo, *Our Children's Toxic Legacy,* 147–52; R. G. Butler, G. T. Orlob, and P. H. McGauhey, "Underground Movement of Bacterial and Chemical Pollutants," *Journal of the American Water Works Association* 46 (1954): 97–113; New York State Department of Health, *Bibliography of Organic Pesticide Publi-*

cations having Relevance to Public Health and Water Pollution Problems by Patrick R. Dugan, Robert M. Pfister, and Margaret L. Sprague ([Albany?], 1963). On the lack of research, see W. F. Barthel, R. T. Murphy, W. G. Mitchell, and Calvin Corley, "The Fate of Heptachlor in the Soil Following Granular Application to the Surface," *Journal of Agriculture and Food Chemistry* 8 (November–December 1960): 445–47; William E. Stanley and Rolf Eliassen, *Status of Knowledge of Ground Water Contaminants* (Cambridge, MA: Department of Civil and Sanitary Engineering, MIT, 1960). For assumption that pesticides would not reach groundwater, see Mahfouz H. Zaki, Dennis Moran, and David Harris, "Pesticides in Groundwater: The Aldicarb Story in Suffolk County, NY," *American Journal of Public Health* 72 (December 1982): 1391–94. The California study mentioned is California Department of Water Resources, *The Fate of Pesticides Applied to Irrigated Agricultural Land,* Bulletin No. 174-1 (Sacramento, 1968). Craig Colten documents what experts did know about industrial wastes and groundwater contamination in "A Historical Perspective on Industrial Wastes and Groundwater Contamination," *Geographical Review* 81 (April 1991): 215–28; however, most of his examples focus on the underground disposal of wastes.

9. In response to industry pressure, the federal government continued to allow the use of DBCP on pineapples in Hawaii; a complete ban on the use of DBCP in agriculture did not come until 1981. Jean Seligmann and Mark Whitaker, "Industrial Sterility," *Newsweek,* 29 August 1977, 69; M. D. Whorton, R. M. Krauss, S. Marshall, and T. H. Milby, "Infertility in Male Pesticide Workers," *Lancet* 2 (1977): 1259–61; H. Babich and D. L. Davis, "Dibromochloropropane (DBCP): A Review," *Science of the Total Environment* 17 (1981): 207–21; *Federal Register* 42 (9 September 1977): 45536–49; Sharon Frey, "DBCP: A Lesson in Groundwater Management," *UCLA Journal of Environmental Law and Policy* 5 (1985): 81–99. On recommended uses, see A. L. Taylor, "Progress in Chemical Control of Nematodes," in *Plant Pathology: Problems and Progress, 1908–1958,* comp. Charles S. Holton (Madison: University of Wisconsin Press, 1959), 427–34.

10. California Department of Food and Agriculture, "Pesticide Movement to Ground Water: Survey of Ground Water Basins for DBCP, EDB, Simazine and Carbofuron," draft report (Sacramento, 1983).

11. On DBCP, S. A. Peoples, K. T. Maddy, W. Cusick, T. Jackson, C. Cooper, and A. S. Frederickson, "A Study of Samples of Well Water Collected from Selected Areas in California to Determine the Presence of DBCP and Certain Other Pesticide Residues," *Bulletin of Environmental Contamination and Toxicology* 24 (1980): 611–18; David B. Cohen, "Ground Water Contamination by Toxic Substances: A California Assessment," in *Evaluation of Pesticides in Ground Water,* ed. Willa Y. Garner, Richard C. Honeycutt, and Herbert N. Nigg (Washington, DC: American Chemical Society, 1986), 499–529; Helmut Kloos, "1,2-Dibromo-3-Chloropropane (DBCP) and Ethylene Dibromide (EDB) in Well Water in the Fresno/Clovis Metropolitan Area, California," *Archives of Environmental Health* 51 (July–August 1996): 291–99. On pesticides and groundwater generally, Zaki, Moran, and Harris, "Pesticides in Groundwater"; S. Z. Cohen, S. M. Creeger, and C. G. Enfield, "Potential Pesticide Contamination of

Groundwater from Agricultural Uses," in *Treatment and Disposal of Pesticide Wastes*, ed. Raymond F. Krueger and James N. Seiber (Washington, DC: American Chemical Society, 1984), 297–326.

12. Council on Environmental Quality, *Environmental Quality: The Eleventh Annual Report, 1980* (Washington, DC: U.S. GPO, 1980), 81–100; U.S. House, Committee on Energy and Commerce, Subcommittee on Health and the Environment and Committee on Interstate and Foreign Commerce, Subcommittee on Transportation and Commerce, *Hazardous Waste and Drinking Water: Joint Hearings*, 96th Cong., 2d sess., 1980; Harold C. Barnett, *Toxic Debts and the Superfund Dilemma* (Chapel Hill: University of North Carolina Press, 1994), 23–24; Michael R. Edelstein, *Contaminated Communities: Coping with Residential Toxic Exposure*, 2d ed. (Boulder: Westview Press, 2004).

13. In addition to the soil fumigants, the chemicals found most frequently in groundwater were aldicarb, atrazine, simazine, pentachlorophenol, maneb, ziram, and thiram. Organophosphates such as parathion—while still widely used—were only rarely found in groundwater, presumably because they usually broke down quickly in water. Kloos, "1,2-Dibromo-3-Chloropropane (DBCP)"; Peoples et al., "Samples of Well Water"; California Water Resources Control Board, *Groundwater Contamination by Pesticides: A California Assessment*, by Yoram J. Litwin, Norman N. Hantzache, and Nancy A. George (Sacramento, 1983); Cohen, "Ground Water Contamination," 503 (for quote); Paul G. Barnett, *Survey of Research on the Impacts of Pesticides on Agricultural Workers and the Rural Environment* (Davis: California Institute for Rural Studies, 1989), 21; Robert B. Gunnison, "Survey of State Wells Finds 57 Pesticides," *SFC*, 17 April 1985; Sharon Begley, Gerald C. Lubenow, and Mark Miller, "Silent Spring Revisited?" *Newsweek*, 14 July 1986; S. Z. Cohen, C. Eiden, and M. N. Lorber, "Monitoring Ground Water for Pesticides," in *Evaluation of Pesticides in Ground Water*, 170–96. The California Assembly passed the Organic Chemical Contamination Act (AB 1803), which mandated that drinking water be tested periodically for forty listed pesticides (Frey, "DBCP," 91). The transnational dimensions of the problem were exacerbated by environmental concerns in the United States because American chemical companies typically exported banned pesticides, such as DBCP, to countries in the developing world (U.S. Senate, Committee on Agriculture, Nutrition, and Forestry, *Circle of Poison: Impact of U.S. Pesticides on Third World Workers*, 102d Cong., 1st sess., 1991).

14. U.S. Geological Survey, *Nitrate and Pesticides in Ground Water in the Eastern San Joaquin Valley, California: Occurrence and Trends*, by Karen R. Burow, Sylvia V. Stork, and Neil M. Dubrovsky, Water-Resources Investigations Report No. 98–4040 (Sacramento, 1998).

15. U.S. Geological Survey, *Regional Assessment of Nonpoint-Source Pesticide Residues in Ground Water, San Joaquin Valley, California*, by Joseph L. Domagalski and Neil M. Dubrovsky, Water-Resources Investigations Report No. 91–4027 (Sacramento, 1991), 5–8; U.S. Geological Survey, *Environmental Setting of the San Joaquin-Tulare Basins, California*, by Jo Ann M. Gronberg et al., Water-Resources Investigations Report No. 97–4205 (Sacramento, 1998), 23–42. The earlier report was U.S. EPA, *Effects of Agricultural Pesticides in the Agricultural Environment, Irrigated Croplands, San Joaquin Valley* (Washington, DC: U.S. GPO, 1972).

16. U.S. Geological Survey, *Ground Water in the Central Valley, California—A Summary Report*, by G. L. Bertoldi, R. H. Johnston, and K. D. Evenson, USGS Professional Paper 1401A (Sacramento, 1991); U.S. Geological Survey, *Environmental Setting*, 23–29. On cross contamination, see California Department of Water Resources, *Water Well Standards: San Joaquin County*, Bulletin No. 74–5 (Sacramento, 1965). Although typically the chemicals present in irrigation water are derived from leaching, sometimes pesticides and herbicides are put directly into irrigation water. This technique, called "chemigation," was introduced in 1980 (Litwin, *Groundwater Contamination by Pesticides*, 109).

17. Frey, "DBCP."

18. U.S. Geological Survey, *Pesticides in the Atmosphere: Distribution, Trends, and Governing Factors*, by Michael S. Majewski and Paul D. Capel, Open-File Report NO. 94–506 (Sacramento, 1995); Lynn R. Goldman et al., "Acute Symptoms in Persons Residing Near a Field Treated with the Soil Fumigants Methyl Bromide and Chloropicrin," *Western Journal of Medicine* 147 (1987): 95–98. "Coincidental exposures" include drift and other "inadvertent" applications (Barnett, *Impact of Pesticides*, 24).

19. Zev Ross and Jonathan Kaplan, *Poisoning the Air: Airborne Pesticides in California* (San Francisco: California Public Interest Research Group and Californians for Pesticide Reform, 1998), 6; D. E. Glotfelty, J. N. Seiber, and L. A. Liljedahl, "Pesticides in Fog," *Nature* 325 (12 February 1987): 602–5; Dwight E. Glotfelty, Michael S. Majewski, and James N. Seiber, "Distribution of Several Organophosphorus Insecticides and Their Oxygen Analogues in a Foggy Atmosphere," *Environmental Science and Technology* 24 (1990): 353–57; J. M. Zabik and J. N. Seiber, "Atmospheric Transport of Organophosphate Pesticides from California's Central Valley to the Sierra Nevada Mountains," *Journal of Environmental Quality* 22 (March 1993): 80–90. For more recent concerns about pesticides and air pollution in the valley, see Sean Gray, Zev Ross, and Bill Walker, *Every Breath You Take: Airborne Pesticides in the San Joaquin Valley* (Washington, DC: Environmental Working Group, 2001), www.ewg.org.

20. In addition, twenty-six pesticides had been made subject to periodic monitoring. Ross and Kaplan, *Poisoning the Air*, 14.

21. Research conducted in the 1980s found that as little as 10 to 15 percent of applied pesticides reached their target, while the remainder moved into air, soil, and water. Marion Moses et al., "Environmental Equity and Pesticide Exposure," *Toxicology and Industrial Health* 9 (1993): 913–59.

22. "McFarland—Star of the San Joaquin," Special Historical Edition of *The McFarland Press*, 17 November 1967, vertical file, Beale Memorial Library, Bakersfield, CA.

23. Johnston and McCalla, *Whither California Agriculture*, chap. 3; Fred Krissman, "California Agribusiness and Mexican Farm Workers (1942–1992): A Bi-National Agricultural System of Production/Reproduction" (Ph.D. diss., University of California, Santa Barbara, 1996), 204. Most of the pesticide usage on grapes (by weight) is accounted for by sulfur, which poses a low hazard. However, grapes also account for significant use of high hazard insecticides, such as methomyl, carbofuran, and formerly dinoseb. William S. Pease et al., *Pesticide Use in California: Strategies for Reducing Environmental Health Impacts* (Berkeley: University of California, California Policy Seminar, 1996).

24. Krissman, "California Agribusiness," 172–90; Fred Krissman, "Cycles of Deepening Poverty in Rural California: The San Joaquin Valley Towns of McFarland and Farmersville," in *The Dynamics of Hired Farm Labour: Constraints and Community Responses,* ed. Jill L. Findeis et al. (Oxon: CABI Publishing, 2002), 183–96.

25. In addition, later surveys found that the community's fetal and infant mortality rates had also increased in the early 1980s. The actual number of cancer cases was always an issue of dispute with community members, who knew of additional cases that were excluded from the tally for various reasons. Ronald B. Taylor, "Cancer Cluster Probe Focuses on Dozen Pesticides," *LAT,* 31 December 1987; "Housing Demand in Cluster High," *BC,* 3 January 1988; Ronald B. Taylor, "Officials Assailed for Slowness in Probing Cancer Clusters," *LAT,* 17 October 1987.

26. "Players in McFarland Case," *BC,* 15 October 1987; California Senate, Toxics and Public Safety Management Committee, *Childhood Cancer Incidences — McFarland* [Hearings], 23 July 1985, McFarland, CA.

27. Kern County Health Department, *Epidemiologic Study of Cancer in Children in McFarland, California, 1985–1986; Phase I; Statistical Considerations, Current Environment,* by Leon M. Hebertson et al. (Bakersfield, CA, 198[6]). High concentrations of nitrates can cause methemoglobinemia, or "blue-baby syndrome," in infants. Although nitrates themselves are not carcinogenic, they can act as procarcinogens. For instance, when populations are also exposed to l-proline along with nitrate, the potential for nitrosamine formation is much higher, and recent studies have shown an association between levels of N-nitrosoproline and certain cancers. Overall the evidence linking nitrates in drinking water with cancer has been equivocal (Kenneth P. Cantor, Carl M. Shy, and Clair Chilvers, "Water Pollution," in *Cancer Epidemiology and Prevention,* ed. David Schottenfeld Jr. and Joseph F. Fraumeni [New York: Oxford University Press, 1996], 428–29).

28. Michelle Murphy, "The 'Elsewhere within Here' and Environmental Illness"; Joan C. Martin, "Drugs of Abuse during Pregnancy: Effects upon Offspring Structure and Function," *Signs: Journal of Women and Culture in Society* 2 (Winter 1976): 357–68; Lawrence D. Longo, "Environmental Pollution and Pregnancy: Risks and Uncertainties for the Fetus and Infant," *American Journal of Obstetrics and Gynecology* 137 (15 May 1980): 162–73; Dally, "Thalidomide"; Nancy Weaver and Maria Camposeco, "4th Valley Town Added to Childhood Cancer List," *SB,* 24 September 1989.

29. "Housing Demand in Cluster High."

30. Fran Smith, "Cancer Stalks Town's Children," *San Jose Mercury News,* 30 June 1985.

31. Melosi, *Sanitary City,* 55–56.

32. Neil Pearce, "Traditional Epidemiology, Modern Epidemiology, and Public Health," *American Journal of Public Health* 86 (May 1996): 678–83; Mervyn Susser, "Choosing a Future for Epidemiology: I. Eras and Paradigms," *American Journal of Public Health* 86 (May 1996): 668–73; Nancy Krieger, "Epidemiology and the Web of Causation: Has Anyone Seen the Spider?" *Social Science and Medicine* 39 (1994): 887–903; Mervyn Susser, "Epidemiology Today: 'A Thought-Tormented World,'" *International Journal of Epidemiology* 18 (Sep-

tember 1989): 481–88; Mervyn Susser, "Epidemiology in the United States after World War II: The Evolution of Technique," *Epidemiologic Reviews* 7 (1985): 147–77; Steve Wing, "Limits of Epidemiology," *Medicine and Global Survival* 1 (1994): 74–86. Susser ("Epidemiology after World War II") points out that the focus on multiple factors reflected the rise of both probabilistic thinking and computer technology after World War II. Computers made it increasingly feasible to store and manipulate large data sets and facilitated the analyses of large numbers of variables.

33. John R. Goldsmith and Lester Breslow, "Epidemiological Aspects of Air Pollution," *Journal of the Air Pollution Control Association* 9 (November 1959): 129–32; Raymond Neutra, interview by author, 7 March 2005, Oakland, CA.

34. John Higginson and Calum S. Muir, "The Role of Epidemiology in Elucidating the Importance of Environmental Factors in Human Cancer," *Cancer Detection and Prevention* 1 (1976): 79–105; National Cancer Institute, *Atlas of Cancer Mortality for U.S. Counties: 1950–1969*, by Thomas J. Mason et al. (Bethesda, MD: U.S. Department of Health, Education, and Welfare, 1975); W. J. Blot, "Cancer Mortality in U.S. Counties with Petroleum Industries," *Science* 198 (7 October 1977): 51–53.

35. "Environmental Cancer on the Rise," *Science News,* 5 July 1980; E. M. Whelan, "What Is Environmental Cancer and How Can You Defend against It?" *Vogue,* December 1978; "Cancer and Your Environment," *Harper's Bazaar,* April 1976; "What Causes Cancer?" *Newsweek,* 26 January 1976; Samuel Epstein, *Politics of Cancer* (New York: Anchor, 1979); Proctor, *Cancer Wars,* 57–64.

36. Higginson and Muir, "Role of Epidemiology," 92; E. Boyland, "A Chemist's View of Cancer Prevention," *Proceedings of the Royal Society of Medicine* 60 (1967): 93–99; Clark W. Heath Jr., "Environmental Pollutants and the Epidemiology of Cancer," *Environmental Health Perspectives* 27 (1978): 7–10. On Love Canal, Adeline Levine, *Love Canal: Science, Politics, and People* (Lexington, MA: Lexington Books, 1982). On Woburn, Jonathan Harr, *A Civil Action* (New York: Random House, 1995); Brown and Mikkelsen, *No Safe Place.*

37. California Department of Health Services, *Literature Review on the Toxicological Aspects of DBCP and an Epidemiological Comparison of Patterns of DBCP Drinking Water Contamination with Mortality Rates from Selected Cancers in Fresno County, California 1970–1979,* by Richard J. Jackson et al. (Berkeley, 1982); "Birth Defects Linked to Contamination from Silicon Valley Company," *Seattle Times,* 17 January 1985. Later work would cast doubt on these initial conclusions, and the state would distance itself from any implication that there was a link between contamination and cancer or birth defects in these communities, which activists in the Fairchild case attributed to industry pressure. M. Donald Whorton et al., "Problems Associated with Collecting Drinking Water Quality Data for Community Studies: A Case Example, Fresno County, California," *American Journal of Public Health* 78 (1988): 47–51; Margaret Wrensch et al., "Hydrogeologic Assessment of Exposure to Solvent-Contaminated Drinking Water: Pregnancy Outcomes in Relation to Exposure," *Archives of Environmental Health* 45 (1990): 210–16; David Naguib Pellow and Lisa

Sun-Hee Park, *The Silicon Valley of Dreams: Environmental Injustice, Immigrant Workers, and the High-Tech Global Economy* (New York: New York University Press, 2002), 73–74.

38. Kern County Health Department, *Epidemiologic Study.*

39. *The Wrath of Grapes* (Keene, CA: United Farm Workers, 1986), video-recording; Michael W. Graf, "Regulating Pesticide Pollution in California under the 1986 Safe Drinking Water and Toxic Exposure Act (Proposition 65)," *Ecology Law Quarterly* 28 (September–October 2001): 663–753. Shortly afterward, Rosales and other families derided the UFW for what they saw as the union's political opportunism. Fred Setterberg and Lonny Shavelson, *Toxic Nation: The Fight to Save Our Communities from Chemical Contamination* (New York: John Wiley & Sons, 1993), 76; Ron Talbot, interview by author, 9 December 2003, Bakersfield, CA; Russell Clemings, "Cancer Probe Sought," *SB,* 6 September 1987.

40. Eliot Diringer, "State Hearing on Cancer 'Epidemic' in Central Valley," *SFC,* 17 October 1987 (Rosales quote); Setterberg and Shavelson, *Toxic Nation,* 206–11; Lloyd G. Carter, "Are Pesticides Killing the Children? Cancer Rate High in California Town," *Seattle Times,* 11 July 1988; Weisskopf, "Pesticides and Death"; Ron Harris, "Jackson to Put Campaign Focus on Cancer Cluster Town," *LAT,* 28 May 1988. Deukmejian would later veto a bill sponsored by Torres and Harvey that would have appropriated $330,000 for studies of McFarland (Memo from Clifford L. Allenby, Secretary of Health and Welfare, to David M. Caffrey, 13 January 1989, CDHS-McFarland files).

41. Weaver and Camposeco, "4th Valley Town" (Moses quote); Carter, "Are Farm Pesticides Killing the Children?" (Rosales quote); Ron Harris, "Jackson to Put Campaign focus on Cancer Cluster Town," *LAT,* 28 May 1988; Warren, "Mysterious Cancer Cluster." See also Scott, "Child Cancer Cluster Poses Puzzle."

42. Lionel Martinez, "Farm Laborers Struggle to Guard Against What's Causing Cancer," *BC,* 24 September 1988.

43. The political disenfranchisement of Latinos was a product of the fact that many were noncitizens and also a system of at-large elections that more easily allowed the white minority to maintain local political power. Krissman, "Cycles of Deepening Poverty," 183–96; Warren, "Mysterious Cancer Clusters." But as Michelle Murphy has pointed out, in the case of toxic exposures, not even whiteness is a guarantee of power ("Uncertain Exposures and the Privilege of Imperception: Activist Scientists and Race at the U.S. Environmental Protection Agency," *Osiris* 19 [2004]: 266–82).

44. One example is Paul Buxman, a farmer near Dinuba, who abandoned chemical pesticides after his son was diagnosed with leukemia. *In Our Children's Food* (Boston: WGBH, 1993), videorecording; Russell Clemings, "Couple Says It Paid High Price for Raising Alarm," *SB,* 15 February 1988; Julie Guthman, *Agrarian Dreams: The Paradox of Organic Farming in California* (Berkeley: University of California Press, 2004), 20.

45. On Deukmejian's overall disinterest, see Hal Rubin, "Strange Delays in the State's Cancer Probe," *SB,* 25 June 1989. On CDHS and CDFA, email from L. Goldman to R. Kreutzer et al. re: "Earlimart study/interest by CDFA, EPA," 15 September 1989; Memo to Lynn R. Goldman from Larry Nelson, Depart-

ment of Pesticide Regulation, Cal-EPA, re: "Comments on Phase III McFarland Report," 28 October 1991, CDHS-McFarland files.

46. Weaver and Camposeco, "4th Valley Town"; California Department of Health Services, "Epidemiologic Study of Adverse Health Effects in Children in McFarland, California—Draft Phase II Report" (Berkeley, 1988); Reynolds et al., "Four County Study of Childhood Cancer."

47. One of the clusters, Rosamond, lay outside of the valley proper; however, Rosamond was discovered to be the location of considerable toxic waste disposal. Setterberg and Shavelson, *Toxic Nation*, 215; Eliot Diringer, "5 Children of Farm Workers/New Cancer Cluster in Farm Town," *SFC*, 14 September 1989. The quote is from Daphne Washington, a KCDPH employee who had written a 1980 report on groundwater contamination in the county; it appears in Weisskopf, "Pesticides and Death."

48. Setterberg and Shavelson, *Toxic Nation*, 246–60 (description of meeting). Overall cancer rates among children in the southern San Joaquin Valley were not significantly different from rates in San Francisco and Los Angeles, and within the valley, those localities with the highest rates were the cities of Fresno and Bakersfield. However, cancer rates were slightly lower in rural areas that were not used for farming (Reynolds et al., "Four County Study" [quote on 696]). One of the lead epidemiologists on McFarland would later publish an article showing that, from a statistical standpoint, a certain number of "chance" cancer clusters were to be expected in a state the size of California. The implication was that McFarland was nothing more than a random aggregate of cases and that the cancers had no relationship to the local environment (Raymond Neutra, Shanna Swan, and Thomas Mack, "Clusters Galore: Insights about Environmental Clusters from Probability Theory," *Science of the Total Environment* 127 [1992]: 187–200).

49. Glen Martin, "Cluster: Random or Environmental?" *SFC*, 4 October 1998; Krissman, "California Agribusiness and Mexican Farmworkers," 212 n. 29. CDHS also continued to foster this discourse in subtle ways. Part of the response to the cluster had been the establishment of a child-health screening clinic, which community members had wanted as a means to have children screened for cancer and other chronic diseases. However, CDHS consistently emphasized in its newsletters that while no new cancers were detected through the screening program, "other problems" such as tooth decay, poor nutrition, and incomplete immunizations were prevalent. CDHS, "Update on Cancer among Children in McFarland," May 1996, KCDPH files.

50. Penny Newman, "Cancer Clusters among Children: The Implications of McFarland," *Journal of Pesticide Reform* 9 (Fall 1989): 10–13; Janny Scott, "Child Cancer Cluster Poses Puzzle," *NYT*, 21 September 1988; "Town's Cancer Deaths Prompt Call for Action," *SB*, 29 November 1987; "Kern Cancer Study Attacked," *SB*, 3 April 1987; Carter, "Are Farm Pesticides Killing the Children?"

51. Weaver and Camposeco, "4th Valley Town"; Warren, "Mysterious Cancer Cluster"; Mark Arax, "Cancer Mystery Still Plagues Farm Town," *LAT*, 14 August 1997; Setterberg and Shavelson, *Toxic Nation*, 258–60.

52. Lois Gibbs, "Social Policy and Social Movements," *Annals of the American Academy of Political and Social Science* 97 (November 2002): 98–109, 103.

On scientific and hermeneutic knowledge, see Bronislaw Szerszynski, Scott Lash, and Brian Wynne, "Introduction: Ecology, Realism and the Social Sciences," in *Risk, Environment and Modernity: Towards a New Ecology*, ed. Scott Lash, Bronislaw Szerszynski, and Brian Wynne (London: Sage, 1996), 1–26. On lay expertise and experience as an alternative form of knowledge, see Kroll-Smith and Floyd, *Bodies in Protest.* On the tension between experiential and abstract knowledge applied in a historical context, see Linda Nash, "The Changing Experience of Nature: Historical Encounters with a Northwest River," *Journal of American History* 86 (March 2000): 1600–29.

53. Working Group on Draft Report, McFarland Childhood Cancer Study, "Minutes," 16 March 1987, CDHS–McFarland files; "Petitions for Emergency Action in McFarland, California: Requests for Preliminary Assessment/Site Investigation, Hazard Ranking, and Emergency Removal and Abatement Actions," 1 March 1995, KCDPH files; Julie Durick, "Officials Search for Clues to McFarland Cancer Cases," *BC*, 28 June 1985. A 1989 study by CDHS revealed that people living or working near cotton fields had an increased incidence of respiratory problems. Tom Maurer, "Defoliation Stirs Up County Air," *BC*, 30 September 1993; Sally Connell, "Young Victims Recall Struggle with Cancers," *BC*, 24 September 1988 (Caudillo quote); Warren, "Mysterious Cancer Clusters" (bug quote); CDPH, *Community Studies on Pesticides*, 15 December 1969–1 November 1970, photocopy, 32 (comment about cancers).

54. Charles W. Schmidt, "Childhood Cancer: A Growing Problem," *Environmental Health Perspectives* 106 (January 1998): A18–A23; Michael Greenberg and Daniel Wartenberg, "Communicating to an Alarmed Community about Cancer Clusters: A Fifty State Survey," *Journal of Community Health* 16 (April 1991): 71–82 (1,500 requests).

55. Edelstein, *Contaminated Communities*, 161–92; Deborah Lupton, *Risk* (London: Routledge, 1999); James Flynn, Paul Slovic, and C. K. Mertz, "Gender, Race, and Perception of Environmental Health Risks," *Risk Analysis* 14 (December 1994): 1101–8.

56. Allan Mazur, *A Hazardous Inquiry: The Rashomon Effect at Love Canal* (Cambridge, MA: Harvard University Press, 1998); S. Schwartz, P. White, and R. Hughes, "Environmental Threats, Communities, and Hysteria," *Journal of Public Health Policy* 6 (March 1985): 58–77; Tim Aldrich and Thomas Sinks, "Things to Know and Do about Cancer Clusters," *Cancer Investigations* 20 (2002): 810–16; David Robinson, "Cancer Clusters: Findings vs. Feelings," *Medscape General Medicine* 4 (6 November 2002) [electronic resource].

57. Often the problem is blamed on the media, which experts argue exaggerate problems and misstate facts. However, studies of media representations of disease clusters have failed to find the kinds of bias that critics have assumed. My own extensive reading of the media in the McFarland case suggests that journalists typically presented the experts' position as authoritative while also reporting the frustrations of community members. For other analyses of the media in similar situations, see Michael Greenberg and Daniel Wartenberg, "Newspaper Coverage of Cancer Clusters," *Health Education Quarterly* 18 (Fall 1991): 363–74; Mazur, *Hazardous Inquiry;* Dorothy Nelkin, *Selling Science: How the Press Covers Science and Technology*, rev. ed. (New York: W. H. Freeman, 1995).

58. I am oversimplifying here to some extent. In fact, epidemiology recognizes the problem of "interaction" among risk factors. Yet interaction is typically subordinated to independent disease-exposure relationships. Hence the overwhelming emphasis on controlling confounding "bias" through study design and appropriate statistical analysis. K. J. Rothman, "Causes," *American Journal of Epidemiology* 104 (1976): 587–92. Moreover, the more recent emphasis on "meta-analysis" (the combination of results from a series of studies conducted among different groups) implicitly assumes that there is an underlying universal exposure-disease relationship that is not context dependent. On all of this, see Wing, "Limits of Epidemiology"; Naomar Almeida-Filho, "The Epistemological Crisis of Contemporary Epidemiology: Paradigms in Perspective," *Sante Culture Health* 8 (1991): 145–66.

59. On individualism in epidemiology, Sylvia Noble Tesh, *Hidden Arguments: Political Ideology and Disease Prevention Policy* (New Brunswick, NJ: Rutgers University Press, 1988); Robert A. Aronowitz, *Making Sense of Illness: Science, Society, and Disease* (New York: Cambridge University Press, 1998); Wing, "Limits of Epidemiology."

60. CDHS, *Toxicological Aspects of DBCP;* R. Cooper and R. David, "The Biological Concept of Race and Its Application to Public Health and Epidemiology," *Journal of Health Politics, Policy, and Law* 11 (1986): 97–116.

61. Bullard, *Dumping in Dixie;* Edelstein, *Contaminated Communities,* 233–41; U.S. Government Accounting Office, *Siting of Hazardous Waste Landfills and Their Correlation with Racial and Economic Status of Surrounding Communities* (Washington, DC, 1983), www.gao.gov; Moses et al., "Environmental Equity and Pesticide Exposure." Also, a study published in 2001 found that Hispanic farmworkers had higher rates of several cancers than other Hispanics in California. Though the authors did not suggest any explanations, this study provided evidence for those who link pesticide exposure to cancer in the valley. Paul K. Mills and Sandy Kwong, "Cancer Incidence in the United Farmworkers of America (UFW), 1987–1997," *American Journal of Industrial Medicine* 40 (2001): 596–603; Kim Baca, "Hispanic Farmworkers in California Suffer High Cancer Rates," *Houston Chronicle,* 18 March 2002.

62. Quoted in Setterberg and Shavelson, *Toxic Nation,* 71, 136.

63. Testimony of Dr. Leon Hebertson in California Senate, *Childhood Cancer Incidences,* 26–27; "Duke: The Terminator," *California Journal* 19 (February 1988): 85–88; "Dr. Ken Kizer: The State's High-Profile Health Chief," *California Journal* 18 (June 1987): 291–93; Stephen Green, "Glitches, Gremlins and Soap: Staggering Along the Road to Toxic Waste Reform in California," *California Journal* 16 (September 1985): 344–48.

64. Raymond Richard Neutra, "Counterpoint from a Cluster Buster," *Journal of Epidemiology* 132 (July 1990): 1–8; Rothman, "Sobering Start." The most celebrated study was that done of the Woburn leukemia cluster by researchers at the Harvard School of Public Health. The initial study showed a positive correlation between groundwater contamination and cancer incidence; however, later investigators identified several problems with the study and challenged its conclusions. See Steven Lagakos, Barbara J. Wessen, and Marvin Zelen, "An Analysis of Contaminated Well Water and Health Effects in Woburn,

Massachusetts," *Journal of the American Statistical Association* 81 (1986): 583–96. Brown and Mikkelson discuss critiques of that investigation (*No Safe Place*, 24–27).

65. J. P. Vandenbroucke, "Is 'the Causes of Cancer' a Miasma Theory for the End of the Twentieth Century?" *International Journal of Epidemiology* 17 (1988): 708–9; Alvan R. Feinstein, "Scientific Standards in Epidemiologic Studies of the Menace of Everyday Life," *Science* 242 (2 December 1988): 1257–63; Mark Parascandola, "Epidemiology: Second-Rate Science?" *Public Health Reports* 113 (July–August 1998): 312–20. In particular, the tobacco industry had long focused on discrediting epidemiologic methods (David Egilman, Joyce Kim, and Molly Biklen, "Proving Causation: The Use and Abuse of Medical and Scientific Evidence Inside the Courtroom—An Epidemiologist's Critique of the Judicial Interpretation of the Daubert Ruling," *Food and Drug Law Journal* 223 [2003]: 223–50, via LexisNexis).

66. Richard Jackson, "Chemicals and Chromosomes, Children and Cancer, Clusters and Cohorts in a New Century," in *Cancer and the Environment: Gene-Environment Interaction,* ed. Samuel Wilson, Lovell Jones, Christine Coussens, and Kathi Hanna (Washington, DC: National Academy Press, 2002), 92–93; Rothman, "Sobering Start," 6; P. A. Schulte, R. L. Ehrenberg, and M. Singal, "Investigation of Occupational Cancer Clusters: Theory and Practice," *American Journal of Public Health* 77 (1987): 52–56 ; "Why Community Cancer Clusters are often Ignored," *Scientific American* 275 (September 1996): 85. One result is that the field in recent years has emphasized controlled clinical studies (Almeida-Filho, "Epistemological Crisis").

67. Martin, "Cluster: Random or Environmental?"; Aronowitz, *Making Sense of Illness,* 166; Dana Loomis and Steve Wing, "Is Molecular Epidemiology a Germ Theory for the End of the Twentieth Century?" *International Journal of Epidemiology* 19 (1990): 1–3. In epidemiology the influence of the laboratory paradigm is partly evidenced by the turn toward sophisticated mathematics (Susser, "Epidemiology after World War II").

68. Charles W. Schmidt, "Toxicogenomics: An Emerging Discipline," *Environmental Health Perspectives* 110 (December 2002): A750–55; Kenneth Olden, Janet Guthrie, and Sheila Newton, "A Bold New Direction for Environmental Health Research," *American Journal of Public Health* 91 (December 2001): 1964.

69. Proctor, *Cancer Wars,* 171–72; Maren Klawiter, "Chemicals, Cancer, and Prevention: The Synergy of Synthetic Social Movements," in *Synthetic Planet: Chemical Politics and the Hazards of Modern Life,* ed. Monica J. Casper (New York: Routledge, 2003), 155–76; Eileen M. McGurty, "From NIMBY to Civil Rights: The Origins of the Environmental Justice Movement," *Environmental History* 2 (1997): 301–23; Robert Gottlieb, *Forcing the Spring: The Transformation of the American Environmental Movement* (Washington, DC: Island Press, 1993); Philip Shabecoff, *A Fierce Green Fire: The American Environmental Movement,* rev. ed. (Washington, DC: Island Press, 2003), 243–64; Bullard, *Confronting Environmental Racism;* Andrew Szasz, *EcoPopulism: Toxic Waste and the Movement for Environmental Justice* (Minneapolis: University of Minnesota Press, 1994).

70. See, e.g., Penn Loh and Jodi Sugarman-Brozan, "Environmental Justice Organizing for Environmental Health: Case Study on Asthma and Diesel Exhaust in Roxbury, Massachusetts," *Annals of the American Academy of Political and Social Science* 584 (2002): 110–24.

71. On the reappropriation of both the body and differentiated space as a political strategy, see Lefebvre, *Production of Space,* 194–205. Also Neil Smith, "Antimonies of Space and Nature in Henri Lefebvre's *The Production of Space,*" in *Philosophy and Geography II: The Production of Public Space,* ed. Andrew Light and Jonathan M. Smith (Lanham, MD: Rowman & Littlefield, 1998), 49–69.

72. The sociologist Phil Brown has labeled these tactics "popular epidemiology" (Brown and Mikkelson, *No Safe Place*).

73. "You're always walking on eggshells. Anything happens to your child and you run to the doctor because you want to make sure it's just the flu, it's just tonsillitis and not something else." Quote of Irma Alcala in Greg Campbell, "Under the Microscope in McFarland, Again," *BC,* 16 June 1996.

74. "Petitioned Public Health Assessment: McFarland Study Area, McFarland, Kern County, California," at www.atsdr.cdc.gov/HAC/PHA/mcfarland/msa_pl.html (accessed 2 October 2003). In addition, community members specifically requested that the Kern County Health Department and the California Department of Health Services not be involved in the new studies. This and opposition of "Anglo elite" from "Memo from DEODC [Division of Environmental and Occupational Disease Control, CDHS] to Kim Belshe via Genest Munso re: McFarland," 12 April 1996, CDHS-McFarland files; "Notes from Meeting with EPA 8/27/9[5?]," KCDPH files (opposition of KCDPH); CDHS, Environmental Health Investigations Branch, "Update on Cancer Among Children in McFarland," May 1996, KCDPH files.

75. In 1986, a U.S. EPA study on chemicals in human tissue found measurable levels of styrene and ethyl phenol in nearly all adults tested in the United States. Ninety-six percent had clinical levels of chlorobenzene, benzene, and ethyl benzene, 91 percent showed toluene, and 83 percent had detectable levels of polychlorinated biphenols. Jon S. Stanley, *Broad Scan Analysis of Human Adipose Tissue — Executive Summary,* EPA Contract B560/5–86/035 (Springfield, VA: National Technical Information Service, 1986). On the self-refuting nature of modern institutions such as public health, see Beck, *Risk Society.*

76. Dr. Robert Haile of UCLA cited in Setterberg and Shavelson, *Toxic Nation,* 252.

77. Murphy, "The 'Elsewhere within Here'and Environmental Illness." For a similar finding in a somewhat different context, see Martha Balshem, *Cancer in the Community: Class and Medical Authority* (Washington, DC: Smithsonian Institution Press, 1993). For the variety of ways in which bodies are understood even within western biomedicine, see Mol, *Body Multiple.*

78. Raymond Richard Neutra, "Should One Start or Continue a Line of Research? Stakeholders' Interests and Ethical Frameworks Give Different Answers," *Annals of the American Academy of Political and Social Science* 584 (November 2002): 125–34.

79. On persistence of concerns, Martin, "Cluster: Random or Environmental?"

CONCLUSION

1. Stanley, *Broad Scan Analysis of Human Adipose Tissue — Executive Summary.*

2. For more on this point, see Mitman, "In Search of Health."

3. Some scholars have emphasized the role of the state over that of the market in this regard, particularly with respect to the abstraction and modernization of the landscape. However, in the Central Valley, the state has more often worked in tandem with the market; moreover, both interpretations rely on the assumption of a human/nature dichotomy. For state-centered interpretations, see James C. Scott, *Seeing Like a State: Why Certain Schemes to Improve the Human Condition Have Failed* (New Haven, CT: Yale University Press, 1999); Worster, *Rivers of Empire.* For a critique of Scott, see Fernando Coronil, "Smelling Like a Market," *American Historical Review* 106 (February 2001): 119–29. For the human/nature dichotomy in state-sponsored science and its limitations, see Nash, "Changing Experience of Nature."

4. In contrast, the sociologist Ulrich Beck and others have argued that contemporary political conflicts over environmental health are principally arguments about the *institutions* of modernity in industrialized society. What these institutional accounts neglect is the fact that these conflicts are also about particular environments and places. Beck, *Risk Society;* Ulrich Beck, *Ecological Politics in an Age of Risk,* trans. Amos Weisz (Cambridge: Polity Press, 1995); Scott Lash, Bronislaw Szerszynski, and Brian Wynne, eds., *Risk, Environment & Modernity: Towards a New Ecology* (London: Sage, 1996).

5. Lefebvre, *Production of Space,* 200–202.

6. My approach to questions of medicine and disease here is influenced by Mol, *Body Multiple.*

Bibliography

PRIMARY AND TECHNICAL SOURCES
ARCHIVAL AND MANUSCRIPT SOURCES

Bancroft Library, University of California, Berkeley

    John W. H. Baker Letters

    Guy E. Jones Papers

    Elwood Mead Papers

California Department of Health Services, Environmental Health Investigations Branch, Oakland

    McFarland Files

California Historical Society, San Francisco

    William Hammond Hall Papers

California State Archives, Sacramento

    Records of the California Department of Agriculture

    Records of the California Department of Public Health

    Records of the State Engineering Office/William Hammond Hall Papers

California State Library, Sacramento

    Harry Butler Biographical Information File

    Thomas M. Logan Biographical Information File

    Minutes of the Sacramento Society for Medical Improvement

Kern County Department of Public Health, Bakersfield, California

    McFarland Files

Water Resources Control Archives, University of California at Berkeley

California Department of Public Health, Bureau of Sanitary Engineering Records

California State Water Resources Control Board, Bay-Delta Hearings Transcripts

University of California at Berkeley, Public Health Library

California Bureau of Occupational Health, *Community Studies on Pesticides*

Guy P. Jones, Reprints from *California's Health*

SELECTED NEWSPAPERS AND PERIODICALS

American Journal of Public Health

Bakersfield Californian

California Farmer

California's Health

California Journal

California Medical Gazette

California State Journal of Medicine

CSBH/CDPH Monthly Bulletin

CSBH/CDPH Weekly Bulletin

El Malcriado (Delano, CA)

Kern County Weekly Courier (Bakersfield, CA)

Los Angeles Times

Pacific Medical and Surgical Journal

Pacific Rural Press (San Francisco, CA)

Proceedings of the California Mosquito Control Association

San Francisco Chronicle

Sacramento Bee

Southern California Practitioner

Transactions of the Medical Society of the State of California

Transactions of the American Medical Association

BOOKS AND ARTICLES

Abrams, Herbert K. "Public Health Aspects of Agricultural Chemicals." *California's Health* 6 (15 January 1949): 97–102.

Adams, Charles Francis. *The Panama Canal Zone: An Epochal Event in Sanitation.* Boston: Massachusetts Historical Society, 1911.

Agassiz, Louis. "Sketch of the Natural Provinces of the Animal World and Their Relation to the Different Types of Man." In *Types of Mankind*, edited by J. C. Nott and George R. Gliddon, lviii–lxxvi. Philadelphia: Lippincott, Gambo, 1854.

Aitken, Thomas H. C. "The Relationship of the Distribution of Cases of Equine Encephalomyelitis (Human and Equine) and Mosquitoes in California." *Proceedings of the California Mosquito Control Association* 11 (1941): 8–15.

Aldrich, Tim, and Thomas Sinks. "Things to Know and Do about Cancer Clusters." *Cancer Investigations* 20 (2002): 810–16.

Anderson, Henry P. *The Bracero Program in California, with Particular Reference to Health Status, Attitudes, and Practices.* Berkeley: University of California, School of Public Health, 1961.

"Appropriation Needs." *CSBH Monthly Bulletin* 8 (December 1912): 96–101.

Audubon, John W. *Audubon's Western Journal: 1849–1850.* 1906. Reprint, Tucson: University of Arizona Press, 1984.

Ayer, Washington. "Topography and Meteorology." *Transactions of the State of California Medical Society, 1880–1881* [11th]: 41–51.

Babich, H., and D. L. Davis. "Dibromochloropropane (DBCP): A Review." *Science of the Total Environment* 17 (1981): 207–21.

Bailey, Stanley F., D. C. Baerg, and H. A. Cristensen. "Seasonal Distribution and Behavior of California Anopheline Mosquitoes." *Proceedings of the California Mosquito Control Association* 40 (1972): 92–101.

Barnett, Paul G. *Survey of Research on the Impacts of Pesticides on Agricultural Workers and the Rural Environment.* Davis: California Institute for Rural Studies, 1989.

Barry, W. James. *The Central Valley Prairie.* Vol. 1: *California Prairie Ecosystem.* Sacramento: California Department of Parks and Recreation, 1972.

Barthel, W. F., R. T. Murphy, W. G. Mitchell, and Calvin Corley. "The Fate of Heptachlor in the Soil Following Granular Application to the Surface." *Journal of Agriculture and Food Chemistry* 8 (November–December 1960): 445–47.

Belcher, Edward. *Narrative of a Voyage Round the World, Performed in Her Majesty's Ship Sulphur during the Years 1836–1842.* Vol. 2. London: Henry Colburn, 1843.

Benites, José. "California's First Medical Survey: Report of Surgeon-General José Benites." Edited by Sherburne F. Cook. *California and Western Medicine* 45 (October 1931): 352–54.

Berteau, Peter E., and David P. Spath. "The Toxicological and Epidemiological Effects of Pesticide Contamination in California Ground Water." In *Evaluation of Pesticides in Ground Water,* edited by Willa Y. Garner, Richard C. Honeycutt, and Herbert N. Nigg, 423–35. Washington, DC: American Chemical Society, 1986.

Biggs, M. H. "Medical Topography of Santa Barbara." *Transactions of the California Medical Society, 1870–1871* [1st]: 133–37.

Bishop, Joseph Bucklin. *The Panama Gateway.* New York: Scribner, 1913.

Black, J. R. "On the Ultimate Causes of Malarial Disease." *New York Journal of Medicine,* n.s., 14 (March 1854): 197–212.

Blair, Aaron, and Shelia Hoar Zahm. "Cancer among Farmers." *Occupational Medicine: State of the Art Reviews* 6 (1991): 335–54.

Blake, James. "On the Climate and Diseases of California." *American Journal of the Medical Sciences* 24 (1852): 53–64.

———. "On the Climate of California in Its Relations to the Treatment of Pulmonary Consumption." *Pacific Medical and Surgical Journal* 3 (1860): 263–64.

Blodget, Lorin. *Climatology of the United States, and of the Temperate Latitudes of the North American Continent.* Philadelphia: J. B. Lippincott, 1857.

Blot, W. J. "Cancer Mortality in U.S. Counties with Petroleum Industries." *Science* 198 (7 October 1977): 51–53.

Bonte, J. H. C. "The Northerly Winds of California." *Transactions of the California State Agricultural Society, 1881* (1881): 201–13.

Boyland, E. "A Chemist's View of Cancer Prevention." *Proceedings of the Royal Society of Medicine* 60 (1967): 93–99.

Brace, Charles Loring. *The New West: Or, California in 1867–1868.* New York: G. P. Putnam & Son, 1869.

Brown, Philip King. "The Malarial Fevers of the Sacramento and San Joaquin Valleys." *Transactions of the Medical Society of the State of California, 1899,* 29: 268–84.

Brown, Martin. "An Orange Is an Orange." *Environment* 17 (July–August 1975): 6–11.

Bryant, Edwin. *What I Saw in California — Being a Journal of a Tour of the Emigrant Route and South Pass of the Rocky Mountains across the Continent of North American, the Great Desert Basin, and through California, in 1846 and '47.* 1848. Reprint, Santa Ana, CA: Fine Arts Press, 1936.

Buffum, E. Gould. *Six Months in the Gold Mines: From a Journal of Three Years' Residence in Upper and Lower California, 1847-8-9.* Edited by John W. Caughey. [Los Angeles]: Ward Ritchie Press, 1959.

Busvine, James R. *Disease Transmission by Insects.* New York: Springer-Verlag, 1993.

Butler, R. G., G. T. Orlob, and P. H. McGauhey. "Underground Movement of Bacterial and Chemical Pollutants." *Journal of the American Water Works Association* 46 (1954): 97–113.

"The California Sanitation Exhibit." *CSBH Monthly Bulletin* 4 (Mach 1909): 107–11.

Cantor, Kenneth P., Carl M. Shy, and Clair Chilvers. "Water Pollution." In *Cancer Epidemiology and Prevention,* edited by David Schottenfeld Jr. and Joseph F. Fraumeni, 418–37. New York: Oxford University Press, 1996.

Carson, Rachel. *Silent Spring.* Boston: Houghton Mifflin, 1962.

Chambers, Howard W. "Organophosphorus Compounds: An Overview." In *Organophosphates: Chemistry, Fate, Effects,* edited by J. E. Chambers and P. E. Levi, 3–17. San Diego, CA: Academic Press, 1992.

Chapin, Charles V. *The Sources and Modes of Infection.* New York: John Wiley & Sons, 1910.

Chipman, M. M. "Report on the Committee on Medical Topography, Endemics, Etc." *Transactions of the Medical Society of the State of California, 1880–81* [11th]: 128–51.

———. "Importance of Forest Preservation and Timber Cultivation." *Transactions of the Medical Society of the State of California, 1882–1883* [13th]: 240–81.

———. "Government Forest Reservations." *Transactions of the Medical Society of the State of California, 1893* [23d]: 264–80.

———. "Micro-Organisms and Their Relations to Human and Animal Life." *Transactions of the Medical Society of the State of California, 1889* [19th]: 163–83.

———. "Development of Botany in California." *Transactions of the Medical Society of the State of California, 1891* [21st]: 284–90.

"Climate and Disease." *Pacific Medical and Surgical Journal* 14 (February 1872): 420–21.

"The Co-Existence of Health and Sickness." *Pacific Medical and Surgical Journal* 10 (October 1867): 225–26.

Cohen, S. Z., S. M. Creeger, and C. G. Enfield. "Potential Pesticide Contamination of Groundwater from Agricultural Uses." In *Treatment and Disposal of Pesticide Wastes,* edited by Raymond F. Krueger and James N. Seiber, 297–326. Washington, DC: American Chemical Society, 1984.

Cohen, S. Z., C. Eiden, and M. N. Lorber. "Monitoring Ground Water for Pesticides." In *Evaluation of Pesticides in Ground Water,* edited by Willa Y. Garner, Richard C. Honeycutt and Herbert N. Nigg, 170–96. Washington, DC: American Chemical Society, 1986.

Cohen, David B. "Ground Water Contamination by Toxic Substances: A California Assessment." In *Evaluation of Pesticides in Ground Water,* edited by Willa Y. Garner, Richard C. Honeycutt, and Herbert N. Nigg, 499–529. Washington, DC: American Chemical Society, 1986.

Cooper, Elwood. *Forest Culture and Eucalyptus Trees.* San Francisco: Cubery & Co., 1876.

———. *A Treatise on Olive Culture.* San Francisco: Cubery & Co., 1882.

Cort, William W. "Dangers to California from Oriental and Tropical Parasitic Diseases." *CSBH Monthly Bulletin* 14 (July 1918): 6–15.

Cotruvo, Joseph A., and Chieh Wu. "Controlling Organics: Why Now?" *Journal of the American Water Works Association* 70 (November 1978): 590–94.

Cronise, Titus Fey. *The Natural Wealth of California*. San Francisco: H. H. Bancroft & Co., 1868.

Dana, Richard Henry. *Two Years before the Mast; a Personal Narrative of Life at Sea*. 1842. Reprint, Los Angeles: Ward Ritchie Press, 1964.

Davis, Devra Lee, Aaron Blair, and David G. Hoel. "Agricultural Exposures and Cancer Trends in Developed Countries." *Environmental Health Perspectives* 100 (April 1992): 39–44.

Delano, Alonzo. *Alonzo Delano's California Correspondence; Being Letters Hitherto Uncollected from the Ottawa (Illinois) Free Trader and the New Orleans True Delta, 1849–1852*. Edited by Irving McKee. Sacramento, CA: Sacramento Book Collectors Club, 1952.

Derby, George H. "The Topographical Reports of George H. Derby." Parts I and II. Edited by Francis P. Farquhar. *California Historical Society Quarterly* 11 (June 1932): 99–123; (September 1932): 247–65.

Dickie, Walter M. "Public Health Today." *CSBH Weekly Bulletin* 3 (11 October 1924): 137–39.

———. "Migration and the Spread of Disease." *CDPH Weekly Bulletin* 3 (31 January 1925): 201–2.

———. "Health of the Migrant." *CSBH Weekly Bulletin* 17 (18 June 1938): 81–83.

Doane, Rennie W. *Insects and Disease: A Popular Account of the Way in Which Insects May Spread or Cause Some of Our Common Diseases*. New York: Henry Holt and Company, 1910.

Drake, Daniel. *A Systematic Treatise Historical, Etiological, and Practical on the Principal Diseases of the Interior Valley of North America as They Appear in the Caucasian, African, Indian, and Esquimaux Varieties of Its Population*. 2 vols. New York: Burt Franklin, 1850–54.

Dubos, René. *Mirage of Health: Utopias, Progress, and Biological Change*. New York: Harper & Brothers, 1959.

———. *So Human an Animal: How We Are Shaped by Surroundings and Events*. New York: Charles Scribner, 1968. Reprint, with an introduction by Jill Cooper and David Mechanic, New Brunswick, NJ: Transaction, 1998.

Dubos, René, and Jean Dubos. *The White Plague: Tuberculosis, Man and Society*. Boston: Little, Brown, 1952.

Dupuytren, J. B. Pigne. "Notes on Miasmatic Diseases." *Pacific Medical and Surgical Journal* 1 (1858): 63–69.

Ebright, George E. "Plans for Malaria Control under the New Mosquito Abatement Act." *CSBH Monthly Bulletin* 11 (December 1916): 251–54.

Emmons, C. W. "A Reservoir of Coccidioidomycosis in Wild Rodents." *Journal of Bacteriology* 45 (1943): 306.

Epstein, Samuel. *Politics of Cancer.* New York: Anchor, 1979.

"Equable Climates." *Pacific Medical and Surgical Journal* 15 (February 1873): 447.

The Family Health Annual. Oakland, CA: Pacific Press, 1878.

Farinholt, Mary K. *The New Masked Man in Agriculture: Pesticides and the Health of Agricultural Users.* Cleveland, OH: National Consumers Committee for Research and Education, 196[2].

Farnsworth, R. W. C., ed. *A Southern California Paradise.* Pasadena, CA: R. W. C. Farnsworth, 1883.

Farnsworth, Stanford F. "Malaria and the Migratory Problem in California." *Proceedings and Papers of the Annual Conference of Mosquito Abatement Officials in California* 9 (1938): 39–46.

Faust, Ernest Carroll. "Malaria Incidence in North America." In *Malariology: A Comprehensive Survey of All Aspects of This Group of Diseases from a Global Standpoint,* edited by Mark F. Boyd, 749–63. Philadelphia: W. B. Saunders, 1949.

Feinstein, Alvan R. "Scientific Standards in Epidemiologic Studies of the Menace of Everyday Life." *Science* 242 (2 December 1988): 1257–63.

Fenske, R. A., et. al. "Fluorescent Tracer Evaluation of Chemical Protective Clothing during Pesticide Applications in Central Florida Citrus Groves." *Journal of Agricultural Safety and Health* 8 (2002): 319–31.

Forry, Samuel. *The Climate of the United States and Its Endemic Influences.* New York: J. & H. G. Langley, 1842.

Fourgeaud, Victor J. "Two Fourgeaud Letters." Edited by John Francis McDermott. *California Historical Society Quarterly* 20 (June 1941): 117–22.

Freeborn, Stanley B. "Rice, Mosquitoes and Malaria." *CSBH Monthly Bulletin* 12 (November 1916): 247–52.

———. "The Rice Fields as a Factor in the Control of Malaria." *Journal of Economic Entomology* 10 (June 1917): 354–59.

———. "The Malaria Problem in the Rice Fields." *California State Journal of Medicine* 15 (October 1917): 412–14.

———. "Malaria Control: A Report of Demonstration Studies at Anderson, California." *CSBH Monthly Bulletin* 15 (March 1920): 279–88.

Freeman, Allen W. "The Farm: The Next Point of Attack in Sanitary Progress." *JAMA* 55 (27 August 1910): 736–38.

Frémont, John Charles. *The Exploring Expedition to the Rocky Mountains, Oregon and California.* Auburn, NY: Derby and Miller, 1854.

Garnier, Pierre. *A Medical Journey in California by Dr. Pierre Garnier.* Translated by L. Jay Oliva. Edited by Doyce B. Nunis Jr. Los Angeles: Zeitlin & Ver Brugge, 1967.

Geiger, Maynard, and Clement W. Meighan. *As the Padres Saw Them: California Indian Life and Customs as Reported by the Franciscan Missionaries, 1813–1815.* Santa Barbara, CA: Santa Barbara Mission Archive Library, 1976.

Gibbons, Henry. "Epidemic Cholera." *Pacific Medical and Surgical Journal* 9 (June 1866): 80–91.

———. "Medical Novelties and Improvements." *Pacific Medical and Surgical Journal* 10 (June 1867): 17–23.

———. "Insanity and Disease in California." *Pacific Medical and Surgical Journal* 14 (April 1872): 496–503.

———. "Report on Practical Medicine." *Transactions of the Medical Society of the State of California, 1875–76* [6th]: 29–39.

Gibbons, William P. "On Forest Culture as a Prophylactic to Miasmatic Diseases." *Pacific Medical and Surgical Journal* 18 (August 1875): 115–24.

———. "Report on Indigenous Botany." *Transactions of the Medical Society of the State of California, 1878–1879* [9th]: 129–41.

———. "Notes on Topography, and on the Distribution of Plants in California." In *Tenth Biennial Report of the California State Board of Health*, 184–87. Sacramento, 1888.

Gibbs, Lois. "Social Policy and Social Movements." *Annals of the American Academy of Political and Social Science* 97 (November 2002): 98–109.

Gill, B. M. "The Medical Topography of Northern California." *Transactions of the Medical Society of the State of California, 1895* [25th]: 174–76.

Gillet, Felix. *Fragariculture; or the Culture of the Strawberry.* San Francisco: Spaulding & Barto, Steam Book and Job Printers, 1876.

Glotfelty, Dwight E., Michael S. Majewski, and James N. Seiber. "Distribution of Several Organophosphorus Insecticides and Their Oxygen Analogues in a Foggy Atmosphere." *Environmental Science and Technology* 24 (1990): 353–57.

Glotfelty, D. E., J. N. Seiber, and L. A. Liljedahl. "Pesticides in Fog." *Nature* 325 (12 February 1987): 602–5.

Goldman, Lynn R., Donald Mengle, David M. Epstein, Delois Fredson, Kem Kelly, and Richard J. Jackson. "Acute Symptoms in Persons Residing near a Field Treated with the Soil Fumigants Methyl Bromide and Chloropicrin." *Western Journal of Medicine* 147 (1987): 95–98.

Goldsmith, John R., and Lester Breslow. "Epidemiological Aspects of Air Pollution." *Journal of the Air Pollution Control Association* 9 (November 1959): 129–32.

Gorgas, W. C. "The Conquest of the Tropics for the White Race." *JAMA* 52 (19 June 1909): 1967–69.

Gray, Harold Farnsworth. "Malaria Control in California." *American Journal of Public Health* 2 (1912): 452–55.

———. "The Cost of Malaria: A Study of Economic Loss Sustained by the Anderson-Cottonwood Irrigation District, Shasta County, Calif." *JAMA* 72 (24 May 1919): 1533–35.

———. "Which Way Now?" *Proceedings of the California Mosquito Control Association* 18 (1950): 3–4.

———. "Historical Highlights of 'Permanent' Mosquito Control in California." *Proceedings of the California Mosquito Control Association* 20 (1952): 61–62.

———. "The Confusing Epidemiology of Malaria in California." *American Journal of Tropical Medicine and Hygiene* 5 (May 1956): 411–18.

Gray, Harold Farnsworth, J. J. Rosenthal, E. P. Felt, Ralph H. Hunt, Robert A. Rutherford, Kenneth Allen, W. E. Britton, Thomas H. Means, and Spencer Miller. "Discussion on 'Prevention of Mosquito Breeding' by Spencer Miller." *Transactions of the American Society of Civil Engineers* 76 (1913): 767–84.

Gray, Sean, Zev Ross, and Bill Walker. *Every Breath You Take: Airborne Pesticides in the San Joaquin Valley.* Washington, DC: Environmental Working Group, 2001.

Green, Stephen. "Glitches, Gremlins, and Soap: Staggering along the Road to Toxic Waste Reform in California." *California Journal* 16 (September 1985): 344–48.

Green, Will S. *The History of Colusa County, California and the General History of the State, with Supplement.* Sacramento: Sacramento Lithograph Co., 1950.

Greenberg, Michael, and Daniel Wartenberg. "Communicating to an Alarmed Community about Cancer Clusters: A Fifty State Survey." *Journal of Community Health* 16 (April 1991): 71–82.

———. "Newspaper Coverage of Cancer Clusters." *Health Education Quarterly* 18 (Fall 1991): 363–74.

Griffin, John S. *A Doctor Comes to California: The Diary of John S. Griffin, Assistant Surgeon with Kearny's Dragoons, 1846–1847.* Edited by George Walcott Ames Jr. San Francisco: California Historical Society, 1943.

Grob, D., W. L. Garlick, and A. M. Harvey. "The Toxic Effects in Man of the Anti-Cholinesterase Insecticide Parathion (P-Nitrophenyl Diethylthionophosphate)." *Johns Hopkins Hospital Bulletin* 81 (1950): 106–29.

Groh, George W. *Gold Fever: Being a True Account, Both Horrifying and Hilar-*

ious, of the Art of Healing (So-Called) during the California Gold Rush. New York: William Morrow, 1966.

Gunn, Herbert. "Uncinariasis in California, Based on Observation of Sixty-two Cases." *California State Journal of Medicine* 3 (July 1905): 212–14.

Gunther, F. A. "Insecticide Residues in California Citrus Fruits and Products." *Residue Reviews* 28 (1968): 1–120.

Hall, Harvey Monroe. "Walnut Pollen as a Cause of Hay Fever." *Science* 47 (24 May 1918): 516–17.

Hall, S. Hastings. *Health and Disease: How to Obtain One and Avoid the Other.* San Francisco: Towne & Bacon, 1867.

Harkness, H. W. "Northers, or North-West Winds of California." *Pacific Medical and Surgical Journal* 11 (May 1869): 545–50.

Hastings, Lansford W. *The Emigrant's Guide to Oregon and California.* Cincinnati, OH: George Conclin, 1845.

Hatch, F. W. "Insolation or Sunstroke." *Pacific Medical and Surgical Journal* 20 (July 1877): 49–60.

———. "On the Climate of the Valley of the Sacramento, California." *New York Journal of Medicine,* n.s., 15 (July 1855): 9–50.

———. "Report on the Epidemics of California in 1868." *Transactions of the American Medical Association* 20 (1869): 513–42.

———. "Report on Climatology and Diseases of California." *Transactions of the American Medical Association* 23 (1872): 335–67.

———. *Sixth Anniversary Address before the Sacramento Society for Medical Improvement.* San Francisco: Joseph Winterburn & Company, 1874.

———. "Some of the Health Resorts of the Coast Range of Mountains." *Transactions of the Medical Society of the State of California, 1881–1882* [12th]: 225–43.

———. "The Seaside Health Resorts of California." In *Seventh Biennial Report of the California State Board of Health,* 85–93. Sacramento, 1882.

Hays, W. W. "The Connection of Ozone with Malarial Diseases." *Pacific Medical and Surgical Journal* 9 (August 1866): 124–34.

Heath, Clark W., Jr. "Environmental Pollutants and the Epidemiology of Cancer." *Environmental Health Perspectives* 27 (1978): 7–10.

Herms, William B. "Medical Entomology, Its Scope and Method." *Journal of Economic Entomology* 2 (August 1909): 265–68.

———. "How to Control the Common House Fly." *CSBH Monthly Bulletin* 5 (May 1910): 269–77.

———. *The House Fly in Its Relation to Public Health.* California Agricultural Experiment Station Bulletin No. 215. Sacramento, 1911.

———. *Malaria: Cause and Control.* New York: Macmillan, 1913.

——. "Health on the Farm." *University of California Journal of Agriculture* 2 (December 1914): 136–37.

——. "Successful Methods of Attack on Malaria in California." *California State Journal of Medicine* 13 (May 1915): 185–89.

——. "Flies—Their Habits and Control." *CSBH Monthly Bulletin* 11 (March 1916): 448–64.

——. "Malaria Control." *CSBH Monthly Bulletin* 16 (November 1920): 75–78.

——. "Rural Hygiene and Sanitation." *CSBH Monthly Bulletin* 15 (February 1920): 247–54.

——. *Medical and Veterinary Entomology.* New York: Macmillan, 1923.

——. *Medical Entomology, with Special Reference to the Health and Well-Being of Man and Animals.* 3d ed. New York: Macmillan, 1939.

Herms, William B., and Harold Farnsworth Gray. *Mosquito Control: Practical Methods for Abatement of Disease Vectors and Pests.* New York: Commonwealth Fund, 1940.

Higginson, John. "Population Studies in Cancer." *Acta Unio Internalis Contra Cancerum* 16 (July 1958): 1667–70.

Higginson, John, and Calum S. Muir. "The Role of Epidemiology in Elucidating the Importance of Environmental Factors in Human Cancer." *Cancer Detection and Prevention* 1 (1976): 79–105.

Hill, Hibbert Winslow. *The New Public Health.* New York: Macmillan, 1916.

——. "What Is the New Public Health?" *CSBH Weekly Bulletin* 2 (3 November 1923): 149–50.

Hittel, J. S. *Resources of California.* 3d ed. San Francisco: A. Roman and Co., 1867.

Hostetter's California Almanac for Merchants, Mechanics, Miners, Farmers, Planters, and General Family Use. San Francisco: Hostetter, Smith & Dean, 1866.

"The Hot Air Bath in Therapeutics: Sun-Stroke." *Pacific Medical and Surgical Journal* 11 (September 1868): 174–76.

Howard, L. O. *The House Fly—Disease Carrier: An Account of Its Dangerous Activities and of the Means of Destroying It.* 2d ed. New York: Frederick A. Stokes Co., 1911.

——. "A Fifty Year Sketch History of Medical Entomology and Its Relation to Public Health." In *A Half Century of Public Health: Jubilee Historical Volume of the American Public Health Association,* edited by Mazÿck P. Ravenel, 412–38. New York: American Public Health Association, 1921.

——. *A History of Applied Entomology (Somewhat Anecdotal).* Washington, DC: Smithsonian Institution, 1930.

————. *Mosquitoes: How They Live; How They Carry Disease; How They Are Classified; How They May Be Destroyed.* New York: McClure, Phillips & Co., 1901.

[Howland, J. L.]. *The Olive in California.* Pomona, CA: Pomona Nursery, 1892.

Hueper, W. C. "Cancer Hazards from Natural and Artificial Water Pollutants." In *Conference on Physiological Aspects of Water Quality,* edited by Harry A. Faber and Lena J. Bryson, 181–94. Washington, DC: U.S. Public Health Service, 1960.

Illustrated Family Medical Almanac. San Francisco: Park & White, 1855.

"An Intensive Anti-mosquito Campaign." *CSBH Monthly Bulletin* 7 (January 1912): 177.

"The Insane in California." *Pacific Medical and Surgical Journal* 11 (August 1868): 138.

"Irrigation—Effect on Health." *Pacific Medical and Surgical Journal* 14 (July 1871): 139.

Jackson, Richard. "Chemicals and Chromosomes, Children and Cancer, Clusters and Cohorts in a New Century." In *Cancer and the Environment: Gene-Environment Interaction,* edited by Samuel Wilson, Lovell Jones, Christine Coussens, and Kathi Hanna, 92–93. Washington, DC: National Academy Press, 2002.

Jefferson, Thomas. *Notes on the State of Virginia.* Edited by William Peden. Chapel Hill: University of North Carolina Press, 1954.

Jenkins, H. O. "A Traveling Sanitation Exhibit Directed by the State Board of Health of California in 1909." Massachusetts Institute of Technology, Boston, 1910. Mimeograph.

Johnson, Arthur T. *California: An Englishman's Impressions of the Golden State.* London: Stanley Paul & Co., [1913].

Johnson, James, and James Ranald Martin. *The Influence of Tropical Climates on European Constitutions.* 6th ed. New York: Samuel S. and William Wood, 1846.

Jones, Guy P. "Typhus Fever in California." *CSBH Monthly Bulletin* 12 (October 1916): 180–85.

Jones, W. H. S. *Malaria, a Neglected Factor in the History of Greece and Rome.* Cambridge: Macmillan and Bowes, 1907.

Kahn, Ephraim. "Pesticide Related Illness in California Farm Workers." *Journal of Occupational Medicine* 18 (October 1976): 693–96.

Karalliedde, Lakshman, Stanley Feldman, John Henry, and Timothy Marrs, eds. *Organophosphates and Health.* London: Imperial College Press, 2001.

Kent, George F. "Life in California in 1849, as Described in the 'Journal' of George F. Kent." Edited by John Walton Caughey. *California Historical Society Quarterly* 20 (March 1941): 26–46.

Kerr, Thomas. "An Irishman in the Gold Rush: The Journal of Thomas Kerr." Edited by Charles L. Camp. *California Historical Society Quarterly* 8 (June 1929): 167–82.

King, T. Butler. *Report of Hon. T. Butler King, on California.* Washington, DC: Gideon and Co., 1850.

Kinney, Abbot. *Eucalyptus.* Los Angeles: B. R. Baumgardt & Co., 1895.

Kloos, Helmut. "1,2-Dibromo-3-Chloropropane (DBCP) and Ethylene Dibromide (EDB) in Well Water in the Fresno/Clovis Metropolitan Area, California." *Archives of Environmental Health* 51 (July–August 1996): 291–99.

Knox, Robert. *The Races of Men: A Fragment.* 1850. Reprint, Miami, FL: Mnemosyne, 1969.

Lagakos, Steven, Barbara J. Wessen, and Marvin Zelen. "An Analysis of Contaminated Well Water and Health Effects in Woburn, Massachusetts." *Journal of the American Statistical Association* 81 (1986): 583–96.

Lane, L. C. "Notes of Travel in the Interior." *San Francisco Medical Press* 3 (1862): 223–24.

Langsdorff, Georg H. von. *Langsdorff's Narrative of the Rezanov Voyage to Nueva California in 1806.* Fairfield, WA: Ye Galleon Press, 1988.

Lawrence, George W. "Report of the Committee on Climatology, Etc., of Arkansas." *Transactions of the American Medical Association* 23 (1872): 387–400.

Leonard, J. P. "Health in California." *Boston Medical and Surgical Journal* 41 (1850): 323–24.

———. "Letter from California." *Boston Medical and Surgical Journal* 41 (1850): 394–99.

LePrince, Joseph A., and A. J. Orenstein. *Mosquito Control in Panama: The Eradication of Malaria and Yellow Fever in Cuba and Panama.* New York: G. P. Putnam's Sons, 1916.

Lind, James. *An Essay on Diseases Incidental to Europeans in Hot Climates, with the Method of Preventing Their Fatal Consequences.* 1st Amer. ed. Philadelphia: W. Duane, 1811.

Lippmann, Morton. *Environmental Toxicants: Human Exposures and Their Health Effects.* New York: Wiley-Interscience, 2000.

Logan, Thomas M. "On the Climate and Health of Charleston." *Southern Literary Journal* 2 (July 1836): 348–56.

———. "Letters from California." Parts 1 and 2. *New York Journal of Medicine*, n.s., 13 (March 1851): 278–83; (May 1851): 421–26.

———. "Contributions to the Medical History of California." *California Medical Gazette* 1 (October 1856): 166–97.

———. *Circular.* Sacramento, 1856.

———. "Report of Dr. T. M. Logan, Corresponding Secretary of the Medical Society of the State of California." *California Medical Gazette* 1 (April 1857): 436–39.

———. "Report on the Medical Topography and Epidemics of California." *Transactions of the American Medical Association* 12 (1859): 80–127.

———. "Contributions to the Physics, Hygiene and Thermology of the Sacramento River." *Pacific Medical and Surgical Journal* 7 (1864): 145–51.

———. *Medical History of the Year 1868, in California.* Sacramento: Sacramento Society for Medical Improvement, 1868.

———. *Address of Thomas M. Logan, M.D., President of the American Medical Association, Delivered in St. Louis, May 6th, 1873.* St. Louis, 1873.

———. *Valedictory Address in Behalf of the Faculty of the Medical Department of the University of California.* San Francisco, 1874.

———. "Malarial Fevers and Consumption in California." In *Third Biennial Report of the State Board of Health of California,* 117–21. Sacramento, 1875.

[Logan, Thomas M., et. al.] "Report on the Medical Topography and Epidemics of California." *Transactions of the American Medical Association* 16 (1865): 497–567.

Longo, Lawrence D. "Environmental Pollution and Pregnancy: Risks and Uncertainties for the Fetus and Infant." *American Journal of Obstetrics and Gynecology* 137 (15 May 1980): 162–73.

M'Collum, William. *California as I Saw It.* 1850. Reprint, Los Gatos, CA: Talisman Press, 1960.

MacCulloch, John. *An Essay on the Remittent and Intermittent Diseases, Including, Generically Marsh Fever and Neuralgia.* Philadelphia: Carey and Lea, 1830.

"The Malaria Problem." *Transactions of the Commonwealth Club of California* 9 (March 1916): 1–40.

Markos, Basil G. "Distribution and Control of Mosquitoes in Rice Fields." *Journal of the National Malaria Society* 10 (1951): 233–47.

Marsh, John. "Unpublished Letters of Dr. Marsh." *Overland Monthly* 15 (February 1890): 213–20.

Martin, James Ranald. *The Influence of Tropical Climates on European Constitutions.* 7th ed. London: John Churchill, 1856.

Martin, Joan C. "Drugs of Abuse during Pregnancy: Effects upon Offspring Structure and Function." *Signs: Journal of Women and Culture in Society* 2 (Winter 1976): 357–68.

"Medical Topography in California." *California Medical Gazette* 2 (October 1869): 33.

"Medicine among the Mormons and the Indians of North America." *Pacific Medical and Surgical Journal* 4 (May 1861): 343–47.

Merrill, Malcolm H. "Recent Trends in Public Health." *California Medicine* 72 (January 1950): 22–25.

———. "Health Conditions and Services in California for Domestic Seasonal Agricultural Workers and Their Families," Parts 1 and 2. *California's Health* 18 (1 February 1961): 113–17; (15 February 1961): 121–24.

———. "The Sanitarian in Our Changing Environment." *California's Health* 19 (15 May 1962): 161–64.

Meyer, K. F. "Source of Malaria in California." *Transactions of the Commonwealth Club of California* 11 (March 1916): 22–26.

———. "Why Epidemics." *CDPH Weekly Bulletin* 16 (5 June 1937): 73–75; (12 June 1937): 77–79; (19 June 1937): 81–83; (3 July 1937): 89–91; (10 July 1937): 95.

———. "The Prevention of Plague in Light of Newer Knowledge." *Annals of the New York Academy of Sciences* 48 (1947): 429–66.

Milby, Thomas H., Fred Ottoboni, and Howard W. Mitchell. "Parathion Residue Poisoning among Orchard Workers." *JAMA* 189 (3 August 1964): 351–56.

Miller, P. B. M. "Epidemic Relapsing Fever among the Chinese at Oroville." *Pacific Medical and Surgical Journal* 16 (January 1875): 370–75.

Miller, Robert W. "Areawide Chemical Contamination: Lessons from Case Histories." *JAMA* 245 (17 April 1981): 1548–51.

Miller, Spencer. "Prevention of Mosquito Breeding." *Transactions of the American Society of Civil Engineers* 76 (1913): 759–66.

Mills, Paul K., and Sandy Kwong. "Cancer Incidence in the United Farmworkers of America (UFW), 1987–1997." *American Journal of Industrial Medicine* 40 (2001): 596–603.

Moore, C. W. "Water: Its Impurities, Gathered from the Air and Earth." *Pacific Record of Medicine and Surgery* (15 March 1888). Offprint.

Morse, E. Malcolm. "Conservative Medicine." *California Medical Gazette* 1 (October 1868): 80–83.

———. "Something about the Small-Pox Epidemic." *California Medical Gazette* 1 (January 1869): 130–31.

Morton, Samuel George. *Illustrations of Pulmonary Consumption, Its Anatomical Characters, Symptoms, and Treatment.* Philadelphia: Key and Biddle, 1834.

Moses, Marion. "Pesticide-related Health Problems and Farmworkers." *AAOHN Journal* 37 (March 1989): 115–30.

Moses, Marion, Eric S. Johnson, W. Kent Anger, Virlyn W. Burse, Sanford W. Horstman, Richard J. Jackson, Robert G. Lewis, Keith T. Maddy, Rob McConnell, William J. Meggs, and Sheila Hoar Zahm. "Environmental

Equity and Pesticide Exposure." *Toxicology and Industrial Health* 9 (1993): 913–59.

"The Mosquito Pest Can Be Stamped Out." *CSBH Monthly Bulletin* 14 (February 1919): 273–74.

"Nature of Malaria." *Pacific Medical and Surgical Journal* 8 (April 1866): 50.

Neff, Johnson A. "Impressions of Mosquito Control vs. Wildlife." *Proceedings and Papers of the Annual Conference of Mosquito Abatement Officials in California* 10 (1939): 13–18.

Neutra, Raymond Richard. "Counterpoint from a Cluster Buster." *Journal of Epidemiology* 132 (July 1990): 1–8.

———. "Should One Start or Continue a Line of Research? Stakeholders' Interests and Ethical Frameworks Give Different Answers." *Annals of the American Academy of Political and Social Science* 584 (November 2002): 125–34.

Neutra, Raymond, Shanna Swan, and Thomas Mack. "Clusters Galore: Insights about Environmental Clusters from Probability Theory." *Science of the Total Environment* 127 (1992): 187–200.

Newman, Penny. "Cancer Clusters among Children: The Implications of McFarland." *Journal of Pesticide Reform* 9 (Fall 1989): 10–13.

Nixon, A. B. "Random Thoughts on 'Conservative Medicine.'" *California Medical Gazette* 2 (June 1870): 207–8.

———. "Nature in Disease, or Conservative Medicine." *Pacific Medical and Surgical Journal* 15 (April 1873): 528–35.

Nordhoff, Charles. *Nordhoff's West Coast: California, Oregon, and Hawaii.* 1874. Reprint, London: KPI, 1987.

Nuttall, George H. F. "On the Role of Insects, Arachnids and Myriapods as Carriers in the Spread of Bacterial and Parasitic Diseases of Man and Animals: A Critical and Historical Study." *Johns Hopkins Hospital Reports* 8 (1900): 1–154.

Oakley, Godfrey P., Jr., and Clark W. Heath Jr. "Cancer, Environmental Health, and Birth Defects—Examples of New Directions in Public Health Practice." *American Journal of Epidemiology* 144, Suppl. (15 October 1996): S58–S64.

Ogden, Henry. *Rural Hygiene.* New York: Macmillan, 1913.

Olden, Kenneth, Janet Guthrie, and Sheila Newton. "A Bold New Direction for Environmental Health Research." *American Journal of Public Health* 91 (December 2001): 1964.

Orme, H. S. "Topography, Climate, and Diseases of Los Angeles County, Valley, and City." *Transactions of the Medical Society of the State of California, 1874–1875* [5th]: 84–89.

———. "Irrigation—Its Influence on Health, Etc." In *Eighth Biennial Report of the California State Board of Health,* 51–59. Sacramento, 1884.

———. "Irrigation and Forestry Considered in Connection with Malarial Diseases." In *Tenth Biennial Report of the California State Board of Health,* 224–27. Sacramento, 1888.

"The Outbreak of Typhoid Fever at Hanford." *CSBH Monthly Bulletin* 9 (May 1914): 269–72.

"Pacific Coast Diseases." *California Medical Gazette* 1 (January 1857): 323–44.

Parascandola, Mark. "Epidemiology: Second-Rate Science?" *Public Health Reports* 113 (July–August 1998): 312–20.

Paynter, Orville E. "Worker Reentry Safety: Viewpoint and Program of the Environmental Protection Agency." *Residue Reviews* 62 (1976): 13–20.

Pease, William S., James Liebman, Dan Landy, and David Albright. *Pesticide Use in California: Strategies for Reducing Environmental Health Impacts.* Berkeley: University of California, California Policy Seminar, 1996.

Peoples, S. A., K. T. Maddy, W. Cusick, T. Jackson, C. Cooper, and A. S. Frederickson. "A Study of Samples of Well Water Collected from Selected Areas in California to Determine the Presence of DBCP and Certain Other Pesticide Residues." *Bulletin of Environmental Contamination and Toxicology* 24 (1989): 611–18.

Phelan, Henry du R. "Trypansome and Its Relation to Certain Diseases." *California State Journal of Medicine* 3 (November 1905): 351–52.

Pickering, Charles. *The Races of Man: And Their Geographical Distribution.* London: John Chapman, 1849.

Pierce, William Dwight. *Sanitary Entomology: The Entomology of Disease, Hygiene and Sanitation.* Boston: Richard G. Badger, 1921.

"The Place of Nature in Therapeutics." *Pacific Medical and Surgical Journal* 15 (May 1873): 573–92.

"Plague in Squirrels." *CSBH Monthly Bulletin* 4 (July 1908): 15–16.

Popendorf, William J. "Exploring Citrus Harvesters' Exposure to Pesticide Contaminated Foliar Dust." *American Industrial Hygiene Association Journal* 41 (September 1980): 652–59.

Popendorf, William J., and John T. Leffingwell. "Regulating OP Pesticide Residues for Farmworker Protection." *Residue Reviews* 82 (1982): 125–202.

Popendorf, William J., and Robert C. Spear. "Preliminary Survey of Factors Affecting the Exposure of Harvesters to Pesticide Residues." *American Industrial Hygiene Association Journal* 35 (June 1974): 374–80.

Pottenger, Francis Marion. *The Fight against Tuberculosis: An Autobiography.* New York: H. Schuman, 1952.

Pottenger, Francis Marion, and Bernard Krohn. "Poisoning from DDT and Other Chlorinated Hydrocarbon Pesticides: Pathogenesis, Diagnosis, and Treatment." *Journal of Applied Nutrition* 14 (1961): 126–39.

Powers, Stephen. *Tribes of California.* 1877. Reprint, Berkeley: University of California Press, 1976.

Praslow, Dr. J. *The State of California: A Medico-Geographical Account.* Translated by Frederick C. Cordes. 1857. Reprint, San Francisco: John J. Newbegin, 1939.

Quinby, Griffith E., and Allen B. Lemmon. "Parathion Residues as a Cause of Poisoning in Crop Workers." *JAMA* 166 (15 February 1958): 740–46.

Randolph, Theron G. *Environmental Medicine: Beginnings and Bibliographies of Clinical Ecology.* Fort Collins, CO: Clinical Ecology Publications, 1987.

Remondino, Peter C. *Longevity and Climate.* San Francisco: Woodward & Co., [1890].

Reynolds, Peggy, Daniel F. Smith, Enid Satariano, David O. Nelson, Lynn R. Goldman, and Raymond R. Neutra. "The Four County Study of Childhood Cancer: Clusters in Context." *Statistics in Medicine* 15 (1996): 683–97.

Robinson, David. "Cancer Clusters: Findings vs. Feelings." *Medscape General Medicine* 4 (6 November 2002) [electronic resource].

Rogers, Justus H. *Colusa County, Its History Traced from a State of Nature through the Early Period of Settlement and Development, to the Present Day.* Orland, CA, 1891.

Rollins, Robert Z. "Federal and State Regulation of Pesticides." *American Journal of Public Health* 53 (1963): 1427–31.

Ross, Edward Halford. *The Reduction of Domestic Mosquitoes: Instructions for the Use of Municipalities, Town Councils, Health Officers, Sanitary Inspectors, and Residents in Warm Climates.* Philadelphia: P. Blakiston's Son & Co., 1911.

Ross, Ronald. *Mosquito Brigades and How to Organize Them.* London: George Philip & Son, 1902.

———. "Malaria in Greece." *Annual Report of the Smithsonian Institution for 1908,* 697–710. Washington, DC: U.S. GPO, 1909.

Ross, Zev, and Jonathon Kaplan. *Poisoning the Air: Airborne Pesticides in California.* San Francisco: California Public Interest Research Group and Californians for Pesticide Reform, 1998.

Rothman, Kenneth J. "Causes." *American Journal of Epidemiology* 104 (1976): 587–92.

———. "A Sobering Start for the Cluster Busters' Conference." *American Journal of Epidemiology* 132, Suppl. (July 1990): S6-S13.

Rowell, Chester H. "Chinese and Japanese Immigrants—A Comparison." *Annals of the American Academy of Political and Social Science* 34 (September 1909): 223–30.

Rudkin, Charles N., ed. *The First French Expedition to California: Laperouse in 1768.* Los Angeles: G. Dawson, 1959.

Rupp, Henry R. "Adverse Assessments of *Gambusia Affinis:* An Alternate View for Mosquito Control Practitioners." *Journal of the American Mosquito Control Association* 12 (June 1996): 155–66.

Rush, Benjamin. "An Inquiry into the Causes of the Increase of Bilious and Intermitting Fevers in Pennsylvania." In *Medical Inquiries and Observations,* new ed., 2:265–76. Philadelphia: T. Dobson, 1797.

———. *Medical Inquiries and Observations.* New ed. Vol. 2. Philadelphia: T. Dobson, 1797.

Sacramento Valley Reclamation Company. *Tule Lands of the Sacramento Valley in Yolo and Colusa Counties, California.* Louisville: John P. Morton and Co., 1872.

"Sanitary Inspections." *CSBH Monthly Bulletin* 10 (June 1915): 318–41.

"Sanitation in Country Districts." *CSBH Monthly Bulletin* 6 (November 1910): 346–47.

Sargent, A. A. "Irrigation and Drainage." *Overland Monthly,* 2d ser., 8 (1886): 19–32.

Schmidt, Charles W. "Childhood Cancer: A Growing Problem." *Environmental Health Perspectives* 106 (January 1998): A18–A23.

———. "Toxicogenomics: An Emerging Discipline." *Environmental Health Perspectives* 110 (December 2002): A750–55.

Schulte, P. A., R. L. Ehrenberg, and M. Singal. "Investigation of Occupational Cancer Clusters: Theory and Practice." *American Journal of Public Health* 77 (1987): 52–56.

Schwartz, Stephen P., Paul E. White, and Robert G. Hughes. "Environmental Threats, Communities, and Hysteria." *Journal of Public Health Policy* 6 (March 1985): 58–77.

Settlers' Experience in Kern County, California as Related by Themselves with Advice to Newcomers. Bakersfield, CA, 1894.

Shorb, J. Campbell. "The Miasmatic Diseases of California." Parts 1–6. *California Medical Gazette* 1 (July 1868): 5–6; (August 1868): 31–32; (September 1868): 53–55; (October 1868): 77–78; (November 1868): 101–3; (January 1869): 125–27.

Shy, Carl M., and Robert J. Struba. "Air and Water Pollution." In *Cancer Epidemiology and Prevention,* edited by David Schottenfeld Jr. and Joseph F. Fraumeni, 336–63. Philadelphia: W. B. Saunders, 1982.

Snow, William F. "Malaria, the Minotaur of California." *CSBH Monthly Bulletin* 5 (August 1909): 109–12.

———. "Some of California's New Health Laws." *CSBH Monthly Bulletin* 9 (October 1913): 53–59.

———. "Some Other Tropical Diseases Occasionally Seen in California." *CSBH Monthly Bulletin* 5 (December 1909): 117–19.

———. "When Commerce and Health Unite." *CSBH Monthly Bulletin* 6 (February 1911): 520–21.

Spear, Robert C. "Report of the Status of Research into the Pesticide Residue Intoxication Problem in the Central Valley of California." In *Pesticide Residue Hazards to Farm Workers: Proceedings of a Workshop Held February 9–10, 1976*, 43–62. Salt Lake City: U.S. Department of Health, Education and Welfare, National Institute for Occupational Health and Safety, 1976.

Spear, Robert C., David L. Jenkins, and Thomas H. Milby. "Pesticide Residues and Field Workers." *Environmental Science and Technology* 9 (April 1975): 308–13.

Spiller, D. *Mosquito Problems in California's Central Valley.* Berkeley: University of California, Division of Agricultural Science, 1968.

Stallard, J. H. *Female Health and Hygiene on the Pacific Coast.* San Francisco: Bonnard & Daly, 1876.

Stanley, William E., and Rolf Eliassen. *Status of Knowledge of Ground Water Contaminants.* Cambridge, MA: Department of Civil and Sanitary Engineering, MIT, 1960.

Stanley, Jon S. *Broad Scan Analysis of Human Adipose Tissue — Executive Summary.* EPA Contract No. B560/5–86/035. Springfield, VA, 1986.

Stead, Frank M. "Historical Concept of Environmental Sanitation." In *Managing Man's Environment in the San Francisco Bay Area*, 3–17. Berkeley: University of California, Institute of Governmental Studies, 1963.

Stillman, J. D. B. "Observations on the Medical Topography and Diseases (Especially Diarrhoea) of the Sacramento Valley, California, during the Years 1849–50." *New York Journal of Medicine*, n.s., 13 (July 1851). Offprint.

———. *The Gold Rush Letters of J. D. B. Stillman.* Palo Alto, CA: Lewis Osborne, 1967.

Susser, Mervyn. "Epidemiology Today: 'A Thought-Tormented World.'" *International Journal of Epidemiology* 18 (September 1989): 481–88.

Taubes, Gary. "Epidemiology Faces Its Limits." *Science* 269 (14 July 1995): 164–69.

Taylor, A. L. "Progress in Chemical Control of Nematodes." In *Plant Pathology: Problems and Progress, 1908–1958*, compiled by Charles S. Holton, 427–34. Madison: University of Wisconsin Press, 1959.

Taylor, Bayard. *Eldorado, or Adventures in the Path of Empire.* 1850. Reprint, New York: Alfred A. Knopf, 1949.

Todd, Frank Morton. *Eradicating Plague from San Francisco: Report of the Citizens' Health Committee and an Account of Its Work.* San Francisco: C. A. Murdock & Co., 1909.

Toner, J. M. "Life and Professional Labors of Thomas Muldrup Logan, M.D., of

California." *Transactions of the Medical Society of the State of California, 1875–76* [6th]: 136–43.

Truman, Benjamin C. *Semi-Tropical California: Its Climate, Healthfulness, Productiveness, and Scenery.* San Francisco: A. L. Bancroft, 1874.

Trumbo, Craig W. "Public Requests for Cancer Cluster Investigations: A Survey of State Health Departments." *American Journal of Public Health* 90 (August 2000): 1300–1302.

"A Typhoid Outbreak and Clean-up Campaign." *CSBH Monthly Bulletin* 12 (August 1916): 82–83.

Tyson, James L. *Diary of a Physician in California; Being the Results of Actual Experience, Including Notes of the Journey by Land and Water, and Observations on the Climate, Soil, Resources of the Country, Etc.* 1850. Reprint, Oakland, CA: Biobooks, 1955.

Vandenbroucke, J. P. "Is 'The Causes of Cancer' a Miasma Theory for the End of the Twentieth Century?" *International Journal of Epidemiology* 17 (1988): 708–9.

Weir, Walter W. "The Drainage Situation in the Rice Growing Areas of the Sacramento Valley." 1921. Typescript. WRCA.

Wellman, Creighton. "Comments on Tropical Medicine." *California State Journal of Medicine* 8 (January 1910): 23–24; (February 1910): 46–47; (March 1910): 66–67; (April 1910): 102–3; (September 1910): 312–13; (October 1910): 346–47.

West, Irma. "Occupational Disease of Farm Workers." *Archives of Environmental Health* 9 (July 1964): 92–98.

———. "Biological Effects of Pesticides in the Environment." In *Organic Pesticides in the Environment,* edited by Aaron A. Rosen and H. F. Kraybill, 38–53. Washington, DC: American Chemical Society, 1966.

"What Is Malaria?" *Pacific Medical and Surgical Journal* 14 (November 1871): 283.

Whorton, M. D., R. M. Krauss, S. Marshall, and T. H. Milby. "Infertility in Male Pesticide Workers." *Lancet* 2 (1977): 1259–61.

Whorton, M. Donald, Robert W. Morgan, Otto Wong, Suzanne Larson, and Nancy Gordon. "Problems Associated with Collecting Drinking Water Quality Data for Community Studies: A Case Example, Fresno County, California." *American Journal of Public Health* 78 (January 1988): 47–51.

Widess, Ellen. *Neurotoxic Pesticides and the Farmworker: A Report.* Washington, DC: Office of Technology Assessment, 1988.

Widney, J. P. "Medical Topography, Meteorology, Endemics and Epidemics." *Transactions of the Medical Society of the State of California, 1880–1881* [11th]: 213–17.

———. "Irrigation and Drainage." In *Seventh Biennial Report of the State Board of Health of California,* 104–6. Sacramento, 1882.

———. "The Anglo-Teuton in a New Home." *Southern California Practitioner* 1 (January 1886): 1–3.

———. "Climatic Changes Which Man Is Working in Southern California." *Southern California Practitioner* 1 (October 1886): 389–93.

———. "Report of the Committee on Medical Topography, Meteorology, Endemics, and Epidemics." *Transactions of the Medical Society of the State of California, 1889* [19th]: 213–17.

Wolfe, Homer R., and John F. Armstrong. "Exposure of Workers to Pesticides." *Archives of Environmental Health* 14 (April 1967): 622–33.

Work, John. "Letter of 24 February 1834." *Washington Historical Quarterly* 2 (1908): 163–64.

———. *Fur Brigade to the Bonaventura: John Work's California Expedition, 1832–1833, for the Hudson's Bay Company.* Edited by Alice Bay Maloney. San Francisco: California Historical Society, 1945.

Worthington, H. "Fevers in Southern California." In *Fifth Biennial Report of the State Board of Health of California,* 16–17. Sacramento, 1879.

Wrensch, Margaret, Shanna Swan, Peter J. Murphy, Jane Lipscomb, Kathleen Claxton, David Epstein, and Raymond Neutra. "Hydrogeologic Assessment of Exposure to Solvent-contaminated Drinking Water: Pregnancy Outcomes in Relation to Exposure." *Archives of Environmental Health* 45 (1990): 210–16.

Wynns, Harlin L. "The Danger to Civilian Populations on the Pacific Coast from Mosquito-transmitted Infections in Returning Military Personnel." *Proceedings of the California Mosquito and Vector Control Association* 13 (1944): 22–29.

Zabik, J. M., and J. N. Seiber. "Atmospheric Transport of Organophosphate Pesticides from California's Central Valley to the Sierra Nevada Mountains." *Journal of Environmental Quality* 22 (March 1993): 80–90.

Zaki, Mahfouz H., Dennis Moran, and David Harris. "Pesticides in Groundwater: The Aldicarb Story in Suffolk County, NY." *American Journal of Public Health* 72 (December 1982): 1391–94.

GOVERNMENT DOCUMENTS

Biennial Reports of the California State Board of Health. Sacramento, 1871–1926.

California Assembly. Public Health Committee. *A Research Report on Encephalitis in California.* Sacramento, 1953.

California Assembly, Interim Research Committee, Subcommittee on Pesticides. *Hearings.* Davis, 6 February 1964.

California Assembly. Interim Committee on General Research, Subcommittee on Pesticides. *Government and Pesticides in California.* Sacramento, 1965.

California Assembly. Office of Research. *California's Pesticide Regulatory Program and Farmworker Safety Issues Related to the Use of Organophosphate Pesticides: A Background Report.* By Steve Lewis. Sacramento, 1977.

California Department of Fish and Game. *Pesticides: Their Use and Toxicity in Relation to Wildlife.* By Robert L. Rudd and Richard E. Genelly. Game Bulletin No. 7. Sacramento, 1956.

California Department of Food and Agriculture. *Pesticide Movement to Ground Water.* Vol. 1. *Survey of Ground Water Basins for DBCP, EDB, Simazine, and Carbofuran.* By D. J. Weaver et al. Sacramento, 1983.

California Department of Health Services. *Literature Review on the Toxicological Aspects of DBCP and an Epidemiological Comparison of Patterns of DBCP Drinking Water Contamination with Mortality Rates from Selected Cancers in Fresno County, California 1970–1979.* By Richard J. Jackson, Carolyn J. Greene, Jeffrey T. Thomas, Edward L. Murphy, and John Kaldor. Berkeley, 1982.

California Department of Health Services. *Epidemiologic Study of Adverse Health Effects in Children in McFarland, California — Draft Phase II Report.* Berkeley, 1988.

California Department of Public Health. *Occupational Disease in California Attributed to Pesticides and Other Agricultural Chemicals.* Berkeley, 1950–73.

California Department of Water Resources. *Water Well Standards: San Joaquin County.* Bulletin No. 74–5. Sacramento, 1965.

———. *The Fate of Pesticides Applied to Irrigated Agricultural Land.* Bulletin No. 174–1. Sacramento, 1968.

California Engineer's Office. *Report of the State Engineer to the Legislature, Session of 1880–1881.* 2 vols. Sacramento, 1881.

California Governor's Special Committee on Public Policy Regarding Agricultural Chemicals. *Report on Agricultural Chemicals and Recommendations for Public Policy.* Sacramento, 1960.

California Legislature. Joint Legislative Committee on Agriculture and Livestock Problems. *Special Report on Agricultural Use of Aircraft, Regulation of Pest Control Operators, and Use and Application of Hazardous Materials.* Sacramento, 1949.

———. *Special Report on Enforcement of State Laws Relating to Agricultural Pest Control Operators and Their Use of Injurious Materials.* Sacramento, 1953.

California Senate. Fact Finding Committee on Agriculture. *Hearings* [Sacramento, 23 October 1963; El Centro, 16 June 1964; Riverside, 17 June 1964; San Francisco, 19 June 1964]. Sacramento, 196[4?].

California Senate. Toxics and Public Safety Management Committee. *Childhood Cancer Incidences — McFarland* [Hearings: 23 July 1985, McFarland, CA]. Sacramento, 1986.

California State Board of Health. *Malaria and Mosquito Control.* By William B. Herms. Sacramento, 1923.

―――. *Malaria in California: A Detailed Survey of Nine Widely Separated California Communities with Recommendations and Estimates.* By Louva G. Lennert and Edward T. Ross. Sacramento, 1923.

California Surveyor-General. *Annual Report.* Sacramento, 1856, 1861–62.

California Water Resources Control Board. *Groundwater Contamination by Pesticides: A California Assessment.* By Yoram J. Litwin, Norman N. Hantzache, and Nancy A. George. Sacramento, 1983.

Council on Environmental Quality. *Environmental Quality: The Eleventh Annual Report, 1980.* Washington, DC: GPO, 1980.

Howard, L. O. *The House Fly and How to Suppress It.* Washington, DC: U.S. Department of Agriculture, 1924.

Kern County Health Department. *Epidemiologic Study of Cancer in Children in McFarland, California, 1985–1986; Phase 1; Statistical Considerations, Current Environment.* By Leon M. Hebertson, Richard W. Whitfield, Daphne Washington, Richard Casagrande, Jan Libby, Guy Shaw, David Alva, Chris Burger, Mary Ann Ante, Thomas Lazar, and Henry Mayrsohn. Bakersfield, CA, 1986.

National Cancer Institute. *Atlas of Cancer Mortality for U.S. Counties: 1950–1969.* By Thomas J. Mason, Frank W. McKay, Robert Hoover, William J. Blot Jr., and Joseph F. Fraumeni. Bethesda, MD: U.S. Department of Health, Education, and Welfare, 1975.

New York State Department of Health. *Bibliography of Organic Pesticide Publications Having Relevance to Public Health and Water Pollution Problems.* By Patrick R. Dugan, Robert M. Pfister, and Margaret L. Sprague. [Albany], 1963.

Pesticide Residue Hazards to Farm Workers: Proceedings of a Workshop Held February 9–10, 1976. Salt Lake City: U.S. Department of Health, Education and Welfare, National Institute for Occupational Safety and Health, 1976.

U.S. Commissioner of Agriculture. *Report of the Commissioner of Agriculture for the Year 1874.* Washington, DC: U.S. GPO, 1875.

U.S. Congress. *Report of the Joint Special Committee to Investigate Chinese Immigration.* 44th Cong., 2d sess., 1877. S. Rept. 689.

U.S. Congress. House. *Mr. Kelley's Memoir.* 25th Cong., 3d sess., 1839. H. Rept. 101.

―――. Committee on Appropriations, Subcommittee on Departments of Labor and Health, Education and Welfare, and Related Agencies. *Report on Environmental Health Problems.* 86th Cong., 2d sess., 1960.

———. Committee on Education and Labor. *Occupational Safety and Health Act of 1969: Hearings on H.R. 843, H.R. 3809, H.R. 4294, H.R. 13373.* 91st Cong., 1st sess., 1969.

———. Committee on Energy and Commerce, Subcommittee on Health and the Environment and Committee on Interstate and Foreign Commerce, Subcommittee on Transportation and Commerce. *Hazardous Waste and Drinking Water: Joint Hearings.* 96th Cong., 2d sess., 1980.

U.S. Congress. Senate. *Statistical Report on the Sickness and Mortality in the Army of the United States.* 34th Cong., 1st sess., 1856. S. Exec. Doc. No. 96.

———. *Statistical Report on the Sickness and Mortality in the Army of the United States.* 36th Cong., 1st sess., 1860. S. Exec. Doc. No. 52.

———. *Report of the Special Committee of the U.S. Senate on the Irrigation and Reclamation of Arid Lands.* Part 2, *The Great Basin Region and California.* 51st Cong., 1st sess., 1890. S. Rept. 928.

———. *Report of the Country Life Commission.* 60th Cong., 2d sess., 1909. S. Doc. 705.

U.S. Congress. Senate. Committee on Agriculture, Nutrition and Forestry. *Circle of Poison: Impact of U.S. Pesticides on Third World Workers.* 102d Cong., 1st sess., 1991.

U.S. Congress. Senate. Committee on Labor and Public Welfare. *Hearings on Migratory Labor.* Part 2. 86th Cong., 2d sess., 1960.

———. *Amending Migratory Labor Laws.* 89th Cong., 1st and 2d sess., 1965–66.

U.S. Congress. Senate. Subcommittee on Migratory Labor. *Hearings on Migrant and Seasonal Farmworker Powerlessness.* Part 6A–B, *Pesticides and the Farmworker.* 91st Cong., 1st and 2d sess., 1969–71.

U.S. Environmental Protection Agency, Office of Water Programs. *The Effects of Agricultural Pesticides in the Agricultural Environment, Irrigated Croplands, San Joaquin Valley.* Washington, DC: U.S. GPO, 1972.

U.S. Federal Water Pollution Control Administration. *Effects of the San Joaquin Master Drain on Water Quality of the San Francisco Bay and Delta.* San Francisco, 1967.

U.S. Geological Survey. *Irrigation near Bakersfield, California.* By Carl Ewald Grunsky. Water-Supply and Irrigation Papers No. 17. Washington, DC: U.S. GPO, 1898.

———. *Irrigation near Fresno, California.* By Carl Ewald Grunsky. Water-Supply and Irrigation Papers No. 18. Washington, DC: U.S. GPO, 1898.

———. *Ground Water in the Central Valley, California — A Summary Report.* By G. L. Bertholdi, R. H. Johnston and K. D. Evenson. USGS Professional Paper 1401A. Sacramento, CA, 1991.

———. *Regional Assessment of Nonpoint-Source Pesticide Residues in Ground*

Water, San Joaquin Valley, California. By Joseph L. Domagalski and Neil M. Dubrovsky. Water-Resources Investigations Report No. 91–4027. Sacramento, CA, 1991.

———. *Ground Water Atlas of the United States.* Vol. 1, *California, Nevada.* By Michael Planert and John W. Williams. Reston, VA: USGS, 1995.

———. *Pesticides in the Atmosphere: Distribution, Trends, and Governing Factors.* By Michael S. Majewski, and Paul D. Capel. Open-File Report No. 94–506. Sacramento, CA, 1995.

———. *Environmental Setting of the San Joaquin-Tulare Basins, California.* By Jo Ann M. Gronberg, Neil M. Dubrovsky, Charles R. Kratzer, Joseph L. Domagalski, Larry R. Brown, and Karen R. Burow. Water-Resources Investigations Report No. 97–4205. Sacramento, CA, 1998.

———. *Nitrate and Pesticides in Ground Water in the Eastern San Joaquin Valley, California: Occurrence and Trends.* By Karen R. Burow, Sylvia V. Stork, and Neil M. Dubrovsky. Water-Resources Investigations Report No. 98–4040. Sacramento, CA, 1998.

———. *Occurrence of Nitrate and Pesticides in Ground Water beneath Three Agricultural Land-Use Settings in the Eastern San Joaquin Valley, California, 1993–1995.* By Karen R. Burow, Jennifer L. Shelton, and Neil M. Dubrovsky. Water Resources Investigations Report No. 97–4284. Sacramento, CA, 1998.

U.S. General Accounting Office. *Pesticides on Farms: Limited Capability Exists to Monitor Occupational Illnesses and Injuries: Report to the Chairman, Committee on Agriculture, Nutrition, and Forestry, U.S. Senate.* Washington, D.C.: U.S. GPO, 1993.

U. S. Government Accounting Office. *Sizing of Hazardous Waste Landfills and Their Correlation with Racial and Economic Status of Surrounding Communities.* Washington, D.C.: U.S. GPO, 1983.

U.S. Public Health Service. *A Study of Medical Problems Associated with Transients.* By C. F. Blankenship and F. Safer. Public Health Bulletin No. 258. Washington, DC: U.S. GPO, 1940.

———. *Environment and Health: Problems of Environmental Health in the United States and the Public Health Service Programs Which Aid States and Communities in Their Efforts to Solve Such Problems.* Washington, DC: U.S. GPO, 1951.

———. *Report of the Committee on Environmental Health Problems to the Surgeon General.* Washington, DC: U.S. GPO, 1962.

U.S. War Department. *Reports of Explorations and Surveys, to Ascertain the Most Practicable and Economical Route for a Railroad from the Mississippi River to the Pacific Ocean. Made under the Direction of the Secretary of War, in 1853–[6].* Washington, DC: A. O. P. Nicolson, 1855–61.

West, Irma. *Pesticides and Other Agricultural Chemicals as a Public Health Prob-*

lem with Special Reference to Occupational Disease in California. Berkeley: California Department of Public Health, 1963.

SECONDARY SOURCES

Ackerknecht, Erwin H. *Malaria in the Upper Mississippi Valley, 1760–1900.* Supplement No. 4 to the *Bulletin of the History of Medicine.* 1945.

———. "Anticontagionism between 1821 and 1867." *Bulletin of the History of Medicine* 22 (1954): 562–93.

Adams, Frank. *The Historical Background of California Agriculture.* Berkeley: California Agricultural Experiment Station, 1946.

Allen, Barbara L. *Uneasy Alchemy: Citizens and Experts in Louisiana's Chemical Corridor Disputes.* Cambridge: MIT Press, 2003.

Almaguer, Tomás. *Racial Fault Lines: The Historical Origins of White Supremacy in California.* Berkeley: University of California Press, 1994.

Almeida-Filho, Naomar. "The Epistemological Crisis of Contemporary Epidemiology: Paradigms in Perspective." *Sante Culture Health* 8 (1991): 145–66.

Anderson, Warwick. "Excremental Colonialism: Public Health and the Poetics of Pollution." *Critical Inquiry* 21 (Spring 1995): 640–69.

———. "The Trespass Speaks: White Masculinity and Colonial Breakdown." *American Historical Review* 102 (December 1997): 1343–70.

———. "The Third-World Body." In *Medicine in the Twentieth Century,* edited by Roger Cooter and John Pickstone, 235–45. Amsterdam: Harwood Academic Publishers, 2000.

———. "Going through the Motions: American Public Health and Colonial 'Mimicry.'" *American Literary History* (2002): 686–719.

———. *The Cultivation of Whiteness: Science, Health and Racial Destiny in Australia.* New York: Basic Books, 2003.

———. "Natural Histories of Infectious Disease: Ecological Vision in Twentieth-Century Biomedical Science." *Osiris* 19 (2004): 39–61.

Arnold, David. *The Problem of Nature: Environment, Culture and European Expansion.* Oxford: Blackwell, 1996.

———, ed. *Warm Climates and Western Medicine: The Emergence of Tropical Medicine, 1500–1900.* Amsterdam: Rodopi, 1996.

———. "An Ancient Race Outworn: Malaria and Race in Colonial India, 1860–1930." In *Race, Science and Medicine, 1700–1960,* edited by Waltrud Ernst and Bernard Harris, 123–43. London: Routledge, 1999.

Aronowitz, Robert A. *Making Sense of Illness: Science, Society, and Disease.* New York: Cambridge University Press, 1998.

Balshem, Martha. *Cancer in the Community: Class and Medical Authority.* Washington, DC: Smithsonian Institution Press, 1993.

Barnes, David S. *The Making of a Social Disease: Tuberculosis in Nineteenth-Century France.* Berkeley: University of California Press, 1995.

Barnett, Harold C. *Toxic Debts and the Superfund Dilemma.* Chapel Hill: University of North Carolina Press, 1994.

Barrett, Frank A. "Daniel Drake's Medical Geography." *Social Science of Medicine* 42 (1996): 791–800.

Baur, John E. *The Health Seekers of Southern California, 1870–1900.* San Marino, CA: The Huntington Library, 1959.

Beals, Ralph L., Jr., and Joseph A. Hester, eds. *California Indians.* Vol. 1. New York: Garland Publishing, 1974.

Bean, Walton. *California: An Interpretive History.* New York: McGraw-Hill, 1978.

Beck, Ulrich. *Risk Society: Toward a New Modernity.* Translated by Mark Ritter. London: Sage, 1992.

———. *Ecological Politics in an Age of Risk.* Translated by Amos Weisz. Cambridge: Polity Press, 1995.

Bewell, Alan. *Romanticism and Colonial Disease.* Baltimore, MD: Johns Hopkins University Press, 1999.

Bosso, Christopher J. *Pesticides and Politics: The Life Cycle of a Public Issue.* Pittsburgh: University of Pittsburgh Press, 1987.

Bowers, William L. *The Country Life Movement in America, 1900–1920.* Port Washington, NY: Kennikat Press, 1974.

Boyd, Robert. *The Coming of the Spirit of Pestilence: Introduced Infectious Diseases and Population Decline among Northwest Coast Indians, 1774–1874.* Seattle: University of Washington Press, 1999.

Briggs, Charles L., and Clara Mantini-Briggs. *Stories in the Time of Cholera: Racial Profiling during a Medical Nightmare.* Berkeley: University of California Press, 2003.

Brown, Jerald. "The United Farmworkers Grape Strike and Boycott, 1965–1970: An Evaluation of the Culture of Poverty Theory." Ph.D. diss., Cornell University, 1972.

Brown, Phil, and Edwin J. Mikkelsen. *No Safe Place: Toxic Waste, Leukemia, and Community Action.* Berkeley: University of California Press, 1990.

Bruce-Chwatt, Leonard J. "Ronald Ross, William Gorgas, and Malaria Eradication." *American Journal of Tropical Medicine and Hygiene* 26 (1977): 1071–79.

Buell, Lawrence. "Toxic Discourse." *Critical Inquiry* 24 (1998): 639–65.

Bullard, Robert D. *Dumping in Dixie: Race, Class, and Environmental Quality.* 3d ed. Boulder: Westview Press, 2000.

Bynum, W. F. "'Reasons for Contentment': Malaria in India, 1900–1920." *Parassitologia* 40 (1998): 19–27.

Camacho, David E., ed. *Environmental Injustices, Political Struggles: Race, Class, and the Environment.* Durham, NC: Duke University Press, 1998.

Carter, Paul. *The Road to Botany Bay: An Essay in Spatial History.* London: Faber and Faber, 1987.

Casey, Edward S. *The Fate of Place: A Philosophical Inquiry.* Berkeley: University of California Press, 1997.

Cassedy, James H. *Charles V. Chapin and the Public Health Movement.* Cambridge, MA: Harvard University Press, 1962.

———. *Medicine and American Growth, 1800–1860.* Madison: University of Wisconsin Press, 1986.

Castillo, Edward D. "The Impact of Euro-American Exploration and Settlement." In *Handbook of North American Indians.* Vol. 2, *California,* edited by Robert F. Heizer, 99–127. Washington, DC: Smithsonian Institution Press, 1978.

Cavanaugh, Dan C. "K. F. Meyer's Work on Plague, Biographical Notes." *Journal of Infectious Diseases* 129, Suppl. (May 1974): 411–13.

Cayleff, Susan E. *Wash and Be Healed: The Water-Cure Movement and Women's Health.* Philadelphia: Temple University Press, 1987.

Cetina, Karin Knorr. *Epistemic Cultures: How the Sciences Make Knowledge.* Cambridge, MA: Harvard University Press, 1999.

Chakrabarty, Dipesh. "Open Space/Public Place: Garbage, Modernity and India." *South Asia* 14 (1991): 15–31.

Chan, Sucheng. *This Bittersweet Soil: The Chinese in California Agriculture, 1869–1910.* Berkeley: University of California Press, 1986.

Chaplin, Joyce. *Subject Matter: Technology, the Body, and Science on the Anglo-American Frontier, 1500–1676.* Cambridge, MA: Harvard University Press, 2001.

Chinard, Gilbert. "Eighteenth-Century Theories on America as Human Habitat." *Proceedings of the American Philosophical Society* 91 (February 1947): 27–57.

Coclanis, Peter A. *The Shadow of a Dream: Economic Life and Death in the South Carolina Low Country, 1670–1920.* New York: Oxford University Press, 1989.

Colten, Craig. "A Historical Perspective on Industrial Wastes and Groundwater Contamination." *Geographical Review* 81 (April 1991): 215–28.

Comaroff, John, and Jean Comaroff. "Medicine, Colonialism, and the Black

Body." In *Ethnography and the Historical Imagination,* 215–34. Boulder: Westview Press, 1993.

Cook, Sherburne F. "The Monterey Surgeons during the Spanish Period in California." *Bulletin of the History of Medicine* 5 (1937): 43–72.

———. "Smallpox in Spanish and Mexican California, 1770–1845." *Bulletin of the History of Medicine* 7 (1939): 153–91.

———. *The Conflict between the California Indian and White Civilization.* Berkeley: University of California Press, 1943.

———. "The Epidemic of 1830–33 in California and Oregon." *University of California Publications in American Archaeology and Ethnology* 43 (1955): 303–26.

———. "Historical Demography." In *Handbook of North American Indians.* Vol. 2, *California,* edited by Robert F. Heizer, 91–98. Washington, DC: Smithsonian Institution Press, 1978.

Cooper, Frederick. *Colonialism in Question: Theory, Knowledge, History.* Berkeley: University of California Press, 2005.

Cooper, Frederick, and Ann Stoler, eds. *Tensions of Empire: Colonial Cultures in a Bourgeois World.* Berkeley: University of California Press, 1997.

Cooper, R., and R. David. "The Biological Concept of Race and Its Application to Public Health and Epidemiology." *Journal of Health Politics, Policy, and Law* 11 (1986): 97–116.

Corn, Jacqueline Karnell. *Response to Occupational Health Hazards: A Historical Perspective.* New York: Van Nostrand Reinhold, 1992.

Coronil, Fernando. "Smelling Like a Market." *American Historical Review* 106 (February 2001): 119–29.

Courtwright, David T. "Disease, Death, and Disorder on the American Frontier." *Journal of the History of Medicine* 46 (October 1991): 457–92.

Cox, William E. "Evolution of the Safe Drinking Water Act: A Search for Effective Quality Assurance Strategies and Workable Concepts of Federalism." *William and Mary Law and Policy Review* 21 (Winter 1997): 69–165.

Craddock, Susan. *City of Plagues: Disease, Poverty, and Deviance in San Francisco.* Minneapolis: University of Minnesota Press, 2000.

Crary, Jonathan, and Sanford Kwinter, eds. *Incorporations.* New York: Zone, 1992.

Cronon, William. *Changes in the Land: Indians, Colonists, and the Ecology of New England.* New York: Hill and Wang, 1983.

———. *Nature's Metropolis: Chicago and the Great West.* New York: Norton, 1991.

———. "The Trouble with Wilderness; or, Getting Back to the Wrong Nature." In *Uncommon Ground: Toward Reinventing Nature,* edited by William Cronon, 69–90. New York: Norton, 1995.

———, ed. *Uncommon Ground: Toward Reinventing Nature.* New York: Norton, 1995.

Crosby, Alfred W. "Virgin Soil Epidemics as a Factor in the Aboriginal Depopulation in America." *William and Mary Quarterly,* 3d ser., 33 (April 1976): 289–99.

———. *Ecological Imperialism: The Biological Expansion of Europe, 900–1900.* Cambridge: Cambridge University Press, 1986.

Cunningham, Andrew, and Perry Williams, eds. *The Laboratory Revolution in Medicine.* Cambridge: Cambridge University Press, 1992.

Curtin, Philip D. *Death by Migration: Europe's Encounter with the Tropical World in the Nineteenth Century.* Cambridge: Cambridge University Press, 1989.

Dally, Ann. "Thalidomide: Was the Tragedy Preventable?" *Lancet* 351 (1998): 1197–99.

Danbom, David B. *The Resisted Revolution: Urban American and the Industrialization of Agriculture, 1900–1930.* Ames: University of Iowa Press, 1979.

Daniel, Cletus. *Bitter Harvest: A History of California Farmworkers, 1870–1941.* Berkeley: University of California Press, 1981.

Daniels, Roger. *The Politics of Prejudice: The Anti-Japanese Movement in California and the Struggle for Japanese Exclusion.* Berkeley: University of California Press, 1977.

Dasmann, Raymond. *The Destruction of California.* New York: Collier Books, 1966.

Davis, Devra. *When Smoke Ran Like Water: Tales of Environmental Deception and the Battle against Pollution.* New York: Basic Books, 2002.

Deacon, Harriet. "Racial Segregation and Medical Discourse in Nineteenth-Century Cape Town." *Journal of Southern African Studies* 22 (June 1996): 287–308.

Deitering, Cynthia. "The Postnatural Novel." In *The Ecocriticism Reader: Landmarks in Literary Ecology,* edited by Cheryll Glotfelty and Harold Fromm, 196–203. Athens: University of Georgia Press, 1996.

Dewey, Scott Hamilton. *Don't Breathe the Air: Air Pollution and U.S. Environmental Politics, 1945–1970.* College Station: Texas A&M University Press, 2000.

Diamond, Jared. *Guns, Germs, and Steel: The Fates of Human Societies.* New York: Norton, 1997.

Douglas, Mary, and Aaron Wildavsky. *Risk as Culture: An Essay on the Selection of Technical and Environmental Dangers.* Berkeley: University of California Press, 1982.

Duden, Barbara. *The Woman beneath the Skin: A Doctor's Patients in Eigh-*

teenth-Century Germany. Translated by Thomas Dunlap. Cambridge, MA: Harvard University Press, 1991.

Duffy, John, ed. *The Rudolph Matas History of Medicine in Louisiana.* Vol. 2. Baton Rouge: Louisiana State University Press, 1962.

———. *The Sanitarians: A History of American Public Health.* Urbana: University of Illinois Press, 1990.

Dunlap, Thomas R. *DDT: Scientists, Citizens, and Public Policy.* Princeton, NJ: Princeton University Press, 1981.

Edelstein, Michael R. *Contaminated Communities: Coping with Residential Toxic Exposure.* 2d ed. Boulder: Westview Press, 2004.

Eden, Trudy. "Food, Assimilation, and the Malleability of the Human Body." In *A Centre of Wonders: The Body in Early America,* edited by Janet Moore Lindman and Michele Lise Tarter, 29–42. Ithaca, NY: Cornell University Press, 2001.

Egilman, David, Joyce Kim, and Molly Biklen. "Proving Causation: The Use and Abuse of Medical and Scientific Evidence inside the Courtroom—An Epidemiologist's Critique of the Judicial Interpretation of the Daubert Ruling." *Food and Drug Law Journal* 58 (2003): 223–50.

Elenes, Victoria. "Farmworker Pesticide Exposures: Interplay of Science and Politics in the History of Regulation (1947–1988)." Master's thesis, University of Wisconsin–Madison, 1991.

Elias, Norbert. *The Civilizing Process.* Vol. 1, *The History of Manners.* Translated by E. Jephcott. New York: Pantheon Press, 1978.

Erlandson, Jon M., and Kevin Bartoy. "Protohistoric California: Paradise or Pandemic?" *Proceedings of the Society for California Archaeology* 9 (1996): 304–9.

Etheridge, Elizabeth W. *Sentinel for Health: A History of the Centers for Disease Control.* Berkeley: University of California Press, 1992.

Ettling, John. *The Germ of Laziness: Rockefeller Philanthropy and Public Health in the New South.* Cambridge, MA: Harvard University Press, 1981.

Fee, Elizabeth, and Dorothy Porter. "Public Health, Preventive Medicine and Professionalization: England and America in the Nineteenth Century." In *Medicine in Society: Historical Essays,* edited by Andrew Wear, 249–75. Cambridge: Cambridge University Press, 1992.

Feher, Michel, Ramona Nadaff, and Nadia Tazi, eds. *Fragments for a History of the Human Body.* New York: Zone, 1989.

Fiege, Mark. *Irrigated Eden: The Making of an Agricultural Landscape in the American West.* Seattle: University of Washington Press, 1999.

Fisher, Lloyd H. *The Harvest Labor Market in California.* Cambridge, MA: Harvard University Press, 1953.

Fitzgerald, Deborah. "Accounting for Change: Farmers and the Modernizing

State." In *The Countryside and the Age of the Modern State: Political Histories of Rural America*, edited by Catherine Stock and Robert D. Johnson, 189–212. Ithaca, NY: Cornell University Press, 2001.

Fleming, James Rodger. *Meteorology in America, 1800–1870*. Baltimore, MD: Johns Hopkins University Press, 1990.

Flynn, James, Paul Slovic, and C. K. Mertz. "Gender, Race, and Perception of Environmental Health Risks." *Risk Analysis* 14 (December 1994): 1101–8.

Fredrickson, George M. *The Black Image in the White Mind: The Debate on Afro-American Character and Destiny, 1817–1914*. New York: Harper and Row, 1971.

Frey, Sharon. "DBCP: A Lesson in Groundwater Management." *UCLA Journal of Law and Policy* 5 (1985): 81–99.

Galarza, Ernesto. *Merchants of Labor: The Mexican Bracero Story*. Santa Barbara, CA: McNally and Loftin, 1964.

———. *Farm Workers and Agri-Business in California, 1947–1960*. Notre Dame, IN: University of Notre Dame Press, 1977.

Galishoff, Stuart. *Newark: The Nation's Unhealthiest City, 1832–1895*. New Brunswick, NJ: Rutgers University Press, 1988.

Garr, Daniel J. "A Rare and Desolate Land: Population and Race in Hispanic California." *Western Historical Quarterly* 6 (April 1975): 133–48.

Gayton, A. H. "Culture-Environment Integration: External References in Yokuts Life." *Southwestern Journal of Anthropology* 2 (Autumn 1946): 252–68.

Geong, Hae-Gyung. "Exerting Control: Biology and Bureaucracy in the Development of American Entomology, 1870–1930." Ph.D. diss., University of Wisconsin–Madison, 1999.

Glacken, Clarence J. *Traces on the Rhodian Shore: Nature and Culture in Western Thought from Ancient Times to the End of the Eighteenth Century*. Berkeley: University of California Press, 1967.

Goldschmidt, Walter. *As You Sow*. New York: Harcourt, Brace, 1947.

Gordon, Robert. "Poisons in the Fields: The United Farm Workers, Pesticides and Environmental Politics." *Pacific Historical Review* 68 (February 1999): 51–77.

Gottlieb, Robert. *Forcing the Spring: The Transformation of the American Environmental Movement*. Washington, DC: Island Press, 1993.

Goubert, Jean-Pierre. *The Conquest of Water: The Advent of Health in the Industrial Age*. Princeton, NJ: Princeton University Press, 1989.

Graf, Michael W. "Regulating Pesticide Pollution in California under the 1986 Safe Drinking Water and Toxic Exposure Act (Proposition 65)." *Ecology Law Quarterly* 28 (September–October 2001): 663–754.

Gray, Harold Farnsworth, and Russell E. Fontaine. "A History of Malaria in

California." *Proceedings and Papers of the California Mosquito Control Association* 25 (1957): 18–37.

Gregory, James N. *American Exodus: The Dust Bowl Migration and Okie Culture in California.* New York: Oxford University Press, 1989.

Grosz, Elizabeth. "Space, Time, and Bodies." In *Places through the Body*, edited by Heidi J. Nast and Steve Pile. New York: Routledge, 1998.

Gurian, Patrick, and Joel A. Tarr. "The First Federal Drinking Water Standards and Their Evolution: A History from 1914 to 1974." In *Improving Regulation: Cases in Environment, Health, and Safety*, edited by Paul S. Fischbeck and R. Scott Farrow, 43–69. Washington, DC: Resources for the Future, 2001.

Guthman, Julie. *Agrarian Dreams: The Paradox of Organic Farming in California.* Berkeley: University of California Press, 2004.

Hagen, Joel B. *An Entangled Bank: The Origins of Ecosystem Ecology.* New Brunswick, NJ: Rutgers University, 1992.

Hamlin, Christopher. *A Science of Impurity: Water Analysis in Nineteenth-Century Britain.* Berkeley: University of California Press, 1990.

Hannaway, Caroline. "Environment and Miasmata." In *Companion Encyclopedia of the History of Medicine*, edited by W. F. Bynum and Roy Porter, 292–308. New York: Routledge, 1993.

Haraway, Donna J. *Simians, Cyborgs, and Women: The Reinvention of Nature.* New York: Routledge, 1991.

———. "The Politics of Postmodern Bodies: Constitutions of Self in Immune System Discourse." In *Simians, Cyborgs, and Women*, 203–30. New York: Routledge, 1991.

Harding, Sandra. "Women's Standpoints on Nature: What Makes Them Possible?" *Osiris* 12 (1997): 186–200.

Harr, Jonathan. *A Civil Action.* New York: Random House, 1995.

Harris, Henry. *California's Medical Story.* San Francisco: J. W. Stacey, 1932.

Harrison, Mark. "'The Tender Frame of Man': Disease, Climate and Racial Difference in India and the West Indies, 1760–1869." *Bulletin of the History of Medicine* 70 (1996): 68–93.

———. *Climates and Constitutions: Health, Race, Environment, and British Imperialism in India, 1600–1850.* New York: Oxford University Press, 1999.

Hays, Samuel P. *Beauty, Health, and Permanence: Environmental Politics in the United States, 1955–1985.* New York: Cambridge University Press, 1987.

Hazlett, Maril. "The Story of *Silent Spring* and the Ecological Turn." Ph.D. diss., University of Kansas, 2003.

———. "Voices from the *Spring*: *Silent Spring* and the Ecological Turn in American Health." In *Seeing Nature through Gender*, edited by Virginia J. Scharff, 103–28. Lawrence: University Press of Kansas, 2003.

Hildreth, Martha L., and Bruce T. Moran, eds. *Disease and Medical Care in the Mountain West: Essays on Region, History, and Practice*. Reno: University of Nevada Press, 1998.

Holt, Marilin Irvin. *Linoleum, Better Babies and the Modern Farm Woman, 1890–1930*. Albuquerque: University of New Mexico Press, 1995.

Horkheimer, Max, and Theodor W. Adorno. *Dialectic of Enlightenment*. Translated by John Cumming. 1944. Reprint, New York: Continuum Books, 1972.

Horsman, Reginald. *Race and Manifest Destiny: The Origins of American Racial Anglo-Saxonism*. Cambridge, MA: Harvard University Press, 1981.

Hoy, Suellen. *Chasing Dirt: The American Pursuit of Cleanliness*. New York: Oxford University Press, 1995.

Humphreys, Margaret. "Kicking a Dying Dog: DDT and the Demise of Malaria in the American South, 1942–1950." *Isis* 87 (March 1996): 1–17.

———. *Malaria: Poverty, Race, and Public Health in the United States*. Baltimore, MD: Johns Hopkins University Press, 2001.

Igler, David. *Industrial Cowboys: Miller & Lux and the Transformation of the Far West, 1850–1920*. Berkeley: University of California Press, 2001.

Illeto, Reynaldo. "Cholera and Origins of American Sanitary Order in the Philippines." In *Discrepant Histories: Translocal Essays on Filipino Cultures*, edited by Vincente Rafael, 51–82. Philadelphia: Temple University Press, 1995.

Irwin, Alan, and Brian Wynne, eds. *Misunderstanding Science? The Public Reconstruction of Science and Technology*. Cambridge: Cambridge University Press, 1996.

Jacobsen, Matthew Frye. *Whiteness of a Different Color: European Immigrants and the Alchemy of Race*. Cambridge, MA: Harvard University Press, 1998.

Jelinek, Lawrence J. *Harvest Empire: A History of California Agriculture*. San Francisco: Boyd & Fraser, 1979.

Johnson, Stephen, Robert Dawson, and Gerald Haslam. *The Great Central Valley: California's Heartland*. Berkeley: University of California Press, 1993.

Johnson, Susan Lee. *Roaring Camp: The Social World of the California Gold Rush*. New York: Norton, 2000.

Johnston, Warren E., and Alex F. McCalla. *Whither California Agriculture: Up, Down, or Out?* Giannini Foundation Special Report No. 04–1. Davis: University of California, Giannini Foundation, 2004.

Jones, Guy P. "Thomas M. Logan, M.D., Organizer of California State Board of Health." *California's Health* 2 (15 March 1945): 129–33.

———. "Early Public Health in California." *California's Health* 2 (15 April 1945): 145–52; (30 April 1945): 153–60; (15 June 1945): 177–79.

Jones, J. Roy. *Memories, Men and Medicine: A History of Medicine in Sacramento, California*. Sacramento: Sacramento Society for Medical Improvement, 1950.

Jordanova, L. J. "Earth Science and Environmental Medicine: The Synthesis of the Late Enlightenment." In *Images of the Earth: Essays in the History of the Environmental Sciences,* edited by L. J. Jordanova and Roy S. Porter, 119–46. Bucks: British Society for the History of Science, 1979.

Juettner, Otto. *Daniel Drake and His Followers: Historical and Biographical Sketches.* Cincinnati, OH: Harvey Publishing Company, 1909.

Kalof, Linda, Thomas Dietz, Gregory Guagnano, and Paul C. Stern. "Race, Gender and Environmentalism: The Atypical Beliefs of White Men." *Race, Gender and Class* 9 (2002): 112–30.

Katz, Cindi. "Major/Minor: Theory, Nature, and Politics." *Annals of the Association of American Geographers* 85 (1995): 164–68.

Katz, Cindi, and Andrew Kirby. "In the Nature of Things: The Environment and Everyday Life." *Transactions of the Institute of British Geographers* 16 (1991): 259–71.

Kelley, Robert. *Gold vs. Grain: California's Hydraulic Mining Controversy.* Glendale, CA: Arthur H. Clark, 1959.

———. *Battling the Inland Sea: American Political Culture, Public Policy, and the Sacramento Valley, 1850–1986.* Berkeley: University of California Press, 1989.

Kennedy, Dane. "The Perils of the Midday Sun: Climatic Anxieties in the Colonial Tropics." In *Imperialism and the Natural World,* edited by John M. MacKenzie, 118–40. Manchester: University of Manchester Press, 1990.

Kiple, Kenneth F., and Stephen V. Beck, eds. *Biological Consequences of European Expansion, 1450–1800.* Aldershot: Ashgate, 1997.

Klaus, Alisa. "Depopulation and Race Suicide: Maternalism and Pronatalist Ideologies in France and the United States." In *Mothers of a New World: Maternalist Politics and the Origins of Welfare States,* edited by Seth Koven and Sonya Michel, 188–212. New York: Routledge, 1993.

Klawiter, Maren. "Chemicals, Cancer, and Prevention: The Synergy of Synthetic Social Movements." In *Synthetic Planet: Chemical Politics and the Hazards of Modern Life,* edited by Monica J. Casper, 155–76. New York: Routledge, 2003.

Kloos, Helmut. "Valley Fever (*Coccidioidomycosis*): Changing Concepts of a 'California Disease.'" *Southern California Quarterly* 55 (Spring 1973): 59–88.

Kluger, James R. *Turning on Water with a Shovel: The Career of Elwood Mead.* Albuquerque: University of New Mexico Press, 1992.

Kohler, Robert E. *Landscapes and Labscapes: Exploring the Lab-Field Border in Biology.* Chicago: University of Chicago Press, 2002.

Kolodny, Annette. *The Land before Her: Fantasy and Experience of the American Frontiers, 1630–1860.* Chapel Hill: University of North Carolina Press, 1984.

Krauss, Celene. "Challenging Power: Toxic Waste Protests and the Politicization of White, Working-Class Women." In *Community Activism and Feminist Politics: Organizing across Race, Class, and Gender,* edited by Nancy A. Naples, 129–50. New York: Routledge, 1998.

Kraut, Alan M. *Silent Travelers: Germs, Genes, and the "Immigrant Menace."* New York: Basic Books, 1994.

Krieger, Nancy. "Epidemiology and the Web of Causation: Has Anyone Seen the Spider?" *Social Science and Medicine* 39 (1994): 887–903.

Krissman, Fred. "California Agribusiness and Mexican Farm Workers (1942–1992): A Bi-National Agricultural System of Production/Reproduction." Ph.D. diss., University of California, Santa Barbara, 1996.

———. "Cycles of Deepening Poverty in Rural California: The San Joaquin Valley Towns of McFarland and Farmersville." In *The Dynamics of Hired Farm Labour: Constraints and Community Responses,* edited by Jill L. Findeis, Ann M. Vandeman, Janelle M. Larson, and Jack L. Runyan, 183–96. Oxon: CABI Publishing, 2002.

Kroeber, A. L. *Handbook of the Indians of California.* Bureau of American Ethnology Bulletin 78. Smithsonian Institution, Washington, DC, 1925.

Kroll-Smith, Steve, and H. Hugh Floyd. *Bodies in Protest: Environmental Illness and the Struggle over Medical Knowledge.* New York: New York University Press, 1997.

Kunitz, Stephen J. "Explanations and Ideologies of Mortality Patterns." *Population and Development Review* 13 (1987): 379–408.

Kupperman, Karen Ordahl. "Fear of Hot Climates in the Anglo-American Colonial Experience." *William and Mary Quarterly* 41 (1984): 213–40.

Kushner, Sam. *Long Road to Delano.* New York: International Publishers, 1975.

Langston, Nancy. "Gender Transformed: Endocrine Disruptors in the Environment." In *Seeing Nature through Gender,* edited by Virginia J. Scharff, 129–66. Lawrence: University Press of Kansas, 2003.

Laqueur, Thomas. *Making Sex: Body and Gender from the Greeks to Freud.* Cambridge, MA: Harvard University Press, 1990.

Lash, Scott, Bronislaw Szerszynski, and Brian Wynne, eds. *Risk, Environment & Modernity: Towards a New Ecology.* London: Sage, 1996.

Latour, Bruno. *The Pasteurization of France.* Translated by Alan Sheridan. Cambridge, MA: Harvard University Press, 1988.

———. *We Have Never Been Modern.* Translated by Catherine Porter. Cambridge, MA: Harvard University Press, 1993.

Latour, Bruno, and Steven Woolgar. *Laboratory Life: The Social Construction of Scientific Facts.* Beverly Hills, CA: Sage, 1979.

Lear, Linda. *Rachel Carson: Witness for Nature.* New York: Henry Holt, 1997.

Leavitt, Judith Walzer. *The Healthiest City: Milwaukee and the Politics of Health Reform*. Princeton, NJ: Princeton University Press, 1982.

———. *Typhoid Mary: Captive to the Public's Health*. Boston: Beacon Press, 1996.

Lefebvre, Henri. *The Production of Space*. Translated by Donald Nicholson. Oxford: Blackwell, 1991.

Levine, Adeline. *Love Canal: Science, Politics, and People*. Lexington, MA: Lexington Books, 1982.

Limerick, Patricia Nelson. *The Legacy of Conquest: The Unbroken Past of the American West*. New York: Norton, 1987.

Livingstone, David N. "Human Acclimatization: Perspectives on a Contested Field of Inquiry in Science, Medicine and Geography." *History of Science* 25 (1987): 359–94.

Lock, Margaret. *Encounters with Aging: Mythologies of Menopause in Japan and North America*. Berkeley: University of California Press, 1993.

Loh, Penn, and Jodi Sugerman-Brozan. "Environmental Justice Organizing for Environmental Health: Case Study on Asthma and Diesel Exhaust in Roxbury, Massachusetts." *Annals of the American Academy of Political and Social Science* 584 (2002): 110–24.

Loomis, Dana, and Steve Wing. "Is Molecular Epidemiology a Germ Theory for the End of the Twentieth Century?" *International Journal of Epidemiology* 19 (1990): 1–3.

Lupton, Deborah. *Risk*. London: Routledge, 1999.

Lutts, Ralph H. "Chemical Fallout: Rachel Carson's *Silent Spring*, Radioactive Fallout, and the Environmental Movement." *Environmental Review* 9 (1985): 210–25.

Lyons, Maryinez. *The Colonial Disease: A Social History of Sleeping Sickness in Northern Zaire, 1900–1940*. Cambridge: Cambridge University Press, 1992.

Maass, Arthur, and Raymond L. Anderson. . . . *And the Desert Shall Rejoice: Conflict, Growth, and Justice in Arid Environments*. Cambridge, MA: MIT Press, 1978.

Majka, Linda C., and Theo J. Majka. *Farm Workers, Agribusiness, and the State*. Philadelphia: Temple University Press, 1982.

Marcus, Alan I. "Physicians Open a Can of Worms: American Nationality and Hookworm in the United States, 1893–1909." *American Studies* 30 (Fall 1989): 103–22.

Martin, Emily. *Flexible Bodies: The Role of Immunity in American Culture from the Days of Polio to the Age of AIDS*. Boston: Beacon Press, 1994.

———. "The Body at Work: Boundaries and Collectivities in the Late Twentieth Century." In *The Social and Political Body*, edited by Theodore R. Schatzki and Wolfgang Natter, 145–59. London: Guilford Press, 1996.

Mazur, Allan. *A Hazardous Inquiry: The Rashoman Effect at Love Canal*. Cambridge, MA: Harvard University Press, 1998.

McClung, L. S., and K. F. Meyer. "The Beginnings of Bacteriology in California." *Bacteriological Reviews* 38 (September 1974): 251–71.

McGowan, Joseph A. *History of the Sacramento Valley*. 3 vols. New York: Lewis Historical Publishing Company, 1961.

McGurty, Eileen M. "From NIMBY to Civil Rights: The Origins of the Environmental Justice Movement." *Environmental History* 2 (1997): 301–23.

McKeown, Thomas. *The Modern Rise of Population*. New York: Academic Press, 1976.

McNeill, William H. *Plagues and Peoples*. Garden City, NY: Anchor, 1976.

McWilliams, Carey. *Factories in the Field: The Story of Migratory Farm Labor in California*. Boston: Little, Brown, 1939.

Melosi, Martin V., ed. *Pollution and Reform in American Cities, 1870–1930*. Austin: University of Texas Press, 1980.

———. *The Sanitary City: Urban Infrastructure in America from Colonial Times to the Present*. Baltimore, MD: Johns Hopkins University Press, 2000.

Merchant, Carolyn. *The Death of Nature: Women, Ecology, and the Scientific Revolution*. San Francisco: Harper and Row, 1980.

Merleau-Ponty, Maurice. *Phenomenology of Perception*. Translated by Colin Smith. London: Routledge & Kegan Paul, 1962.

Merrens, H. Roy, and George D. Terry. "Dying in Paradise." *Journal of Southern History* 50 (November 1984): 533–50.

Milkis, Sidney M., and Jerome M. Mileur, eds. *Progressivism and the New Democracy*. Amherst: University of Massachusetts Press, 1999.

Mitchell, Don. *The Lie of the Land: Migrant Workers and the California Landscape*. Minneapolis: University of Minnesota Press, 1996.

Mitchell, Timothy. "The Stage of Modernity." In *Questions of Modernity*, edited by Timothy Mitchell, 1–34. Minneapolis: University of Minnesota Press, 2000.

———. *Rule of Experts: Egypt, Techno-Politics, Modernity*. Berkeley: University of California Press, 2002.

Mitman, Gregg. "Hay Fever Holiday: Health, Leisure and Place in Gilded Age America." *Bulletin of the History of Medicine* 77 (2003): 600–635.

———. "Natural History and the Clinic: The Regional Ecology of Allergy in America." *Studies in the History and Philosophy of Biology and Biomedical Science* 34 (2003): 491–510.

———. "In Search of Health: Landscape and Disease in American Environmental History." *Environmental History* 10 (April 2005): 184–210.

Mitman, Gregg, Michelle Murphy, and Christopher Sellers. "Introduction: A Cloud over History." *Osiris* 19 (2004): 1–20.

———, eds. *Landscapes of Exposure: Knowledge and Illness in Modern Environments. Osiris* 19 (2004).

Moes, Robert J. "Manuel Quijano and Waning Spanish California." *California History* 67 (June 1988): 78–93.

Mol, Annemarie. "Missing Links, Making Links: The Performance of Some Atheroscleroses." In *Differences in Medicine: Unraveling Practices, Techniques, and Bodies,* edited by Marc Berg and Annemarie Mol, 144–65. Durham, NC: Duke University Press, 1998.

———. *The Body Multiple: Ontology in Medical Practice.* Durham, NC: Duke University Press, 2002.

Montejano, David. *Anglos and Mexicans in the Making of Texas, 1836–1986.* Austin: University of Texas Press, 1987.

Morris, David B. "Environment: The White Noise of Health." *Literature and Medicine* 15 (Spring 1996): 1–15.

———. *Illness and Culture in the Postmodern Age.* Berkeley: University of California Press, 1998.

Mortenson, Earl W. "A Historical Review of Mosquito Prevention in California, Part I (1904–1946)." *Proceedings of the California Mosquito and Vector Control Association* 53 (1985): 41–46.

Moses, Marion. "Farmworkers and Pesticides." In *Confronting Environmental Racism: Voices from the Grassroots,* edited by Robert D. Bullard, 161–78. Boston: South End Press, 1993.

Murphy, Michelle. "The 'Elsewhere within Here' and Environmental Illness; or, How to Build Yourself a Body in a Safe Space." *Configurations* 8 (Winter 2000): 87–120.

———. "Uncertain Exposures and the Privilege of Imperception: Activist Scientists and Race at the U.S. Environmental Protection Agency." *Osiris* 19 (2004): 266–82.

Nash, Linda. "The Changing Experience of Nature: Historical Encounters with a Northwest River." *Journal of American History* 86 (March 2000): 1600–1629.

———. "Finishing Nature: Harmonizing Bodies and Landscapes in Late Nineteenth-Century California." *Environmental History* 8 (January 2003): 25–52.

———. "The Fruits of Ill-Health: Pesticides and Workers' Bodies in Post–World War II California." *Osiris* 19 (2004): 203–19.

Nast, Heidi J., and Steve Pile, eds. *Places through the Body.* New York: Routledge, 1998.

Nelkin, Dorothy. *Selling Science: How the Press Covers Science and Technology.* Rev. ed. New York: W. H. Freeman, 1995.

Nicolson, M. "Alexander von Humboldt, Humboldtian Science and the Origins of the Study of Vegetation." *History of Science* 25 (1987): 167–94.

Nunis, Doyce B., Jr. "Medicine in Spanish California." *Southern California Quarterly* 76 (Spring 1994): 31–58.

Nutton, Vivian. "Humoralism." In *Companion Encyclopedia to the History of Medicine,* edited by W. F. Bynum and Roy Porter, 281–91. New York: Routledge, 1993.

Nye, Ronald Loren. "Visions of Salt: Salinity and Drainage in the San Joaquin Valley, California, 1870–1970." Ph.D. diss., University of California, Santa Barbara, 1986.

Okun, Daniel A. "Drinking Water and Public Health Protection." In *Drinking Water Regulation and Health,* edited by Frederick W. Pontius, 3–24. New York: John Wiley, 2003.

Orsi, Robert J. "Selling the Golden State: A Study of Boosterism in Nineteenth-Century California." Ph.D. diss., University of Wisconsin–Madison, 1973.

Parsons, James J. "A Geographer Looks at the San Joaquin Valley." *Geographical Review* 76 (1986): 371–89.

Paterson, Alan M. *Land, Water, and Power: A History of the Turlock Irrigation District, 1887–1987.* Glendale, CA: Arthur H. Clark, 1987.

Patterson, James T. *The Dread Disease: Cancer and Modern American Culture.* Cambridge, MA: Harvard University Press, 1987.

Paull, Jeffrey M. "The Origin and Basis of Threshold Limit Values." *American Journal of Industrial Medicine* 5 (1984): 227–38.

Pearce, Neil. "Traditional Epidemiology, Modern Epidemiology, and Public Health." *American Journal of Public Health* 86 (May 1996): 678–83.

Pelling, Margaret. "Contagion/Germ Theory/Specificity." In *Companion Encyclopedia of the History of Medicine,* edited by W. F. Bynum and Roy Porter, 309–34. New York: Routledge, 1993.

Pellow, David Naguib, and Lisa Sun-Hee Park. *The Silicon Valley of Dreams: Environmental Injustice, Immigrant Workers, and the High-Tech Global Economy.* New York: New York University Press, 2002.

Perkins, John H. *Insects, Experts, and the Insecticide Crisis: The Quest for New Pest Management Strategies.* New York: Plenum Press, 1982.

Peters, Richard F. "Development of Mosquito Control in California through the Window of the Department of Health Services." *Proceedings of the California Vector Control Association* 52 (1984): 35–38.

Petersen, Alan, and Deborah Lupton. *The New Public Health: Health and Self in the Age of Risk.* London: Sage, 1996.

Phillips, George Harwood. *Indians and Intruders in Central California, 1769–1849.* Norman: University of Oklahoma Press, 1993.

Pick, Daniel. *Faces of Degeneration: A European Disorder, c. 1848–1918.* Cambridge: Cambridge University Press, 1989.

Pickering, Andy, ed. *Science as Practice and Culture.* Chicago: University of Chicago Press, 1992.

Pisani, Donald J. *From the Family Farm to Agribusiness: The Irrigation Crusade in California and the West, 1850–1931.* Berkeley: University of California Press, 1984.

Porter, Dorothy. *The History of Public Health and the Modern State.* Amsterdam: Rodopi, 1994.

Prakash, Gyan. *Another Reason: Science and the Imagination of India.* Princeton, NJ: Princeton University Press, 1999.

Preston, William L. *Vanishing Landscapes: Land and Life in the Tulare Lake Basin.* Berkeley: University of California Press, 1981.

———. "Serpent in Eden: Dispersal of Foreign Diseases into Pre-Mission California." *Journal of California and Great Basin Anthropology* 18 (1996): 2–37.

Preston, William L., Jon M. Erlandson, and Kevin Bartoy. "Protohistoric California: Paradise or Pandemic?" *Proceedings of the Society for California Archaeology* 9 (1996): 304–9.

Proctor, Robert. *Cancer Wars: How Politics Shapes What We Know and Don't Know About Cancer.* New York: Basic Books, 1995.

Pulido, Laura. "Perceptions of Agricultural Chemicals in Kern County, California." Master's thesis, University of Wisconsin–Madison, 1987.

———. *Environmentalism and Economic Justice: Two Chicano Struggles in the Southwest.* Tucson: University of Arizona Press, 1996.

Pulido, Laura, and Devon Pena. "Environmentalism and Positionality: The Early Pesticide Campaign of the United Farm Workers' Organizing Committee, 1965–1971." *Race, Gender & Class* 6 (1998): 33–50.

Rabinbach, Anson. *The Human Motor: Energy, Fatigue, and the Origins of Modernity.* Berkeley: University of California Press, 1992.

Reisler, Mark. *By the Sweat of Their Brow: Mexican Immigrant Labor in the United States, 1900–1940.* Westport, CT: Greenwood Press, 1976.

Rhode, Paul W. "Learning, Capital Accumulation, and the Transformation of California Agriculture." *Journal of Economic History* 55 (December 1995): 773–800.

Rice, Richard B., William A. Bullough, and Richard J. Orsi. *The Elusive Eden: A New History of California.* Boston: McGraw-Hill, 2002.

Risse, Guenter B. "'A Long Pull, a Strong Pull, and All Together': San Francisco and Bubonic Plague, 1907–1908." *Bulletin of the History of Medicine* 66 (Summer 1992): 260–82.

Roberts, Brian. *American Alchemy: The California Gold Rush and Middle-Class Culture.* Chapel Hill: University of North Carolina Press, 2000.

Rogaski, Ruth. *Hygienic Modernity: Meanings of Health and Disease in Treaty-Port China.* Berkeley: University of California Press, 2004.

Rogers, Frank B. "The Rise and Decline of the Altitude Therapy of Tuberculosis." *Bulletin of the History of Medicine* 43 (1969): 1–16.

Rogers, Naomi. "Germs with Legs: Flies, Disease, and the New Public Health." *Bulletin of the History of Medicine* 63 (1989): 599–617.

Rome, Adam. *The Bulldozer in the Countryside: Suburban Sprawl and the Rise of American Environmentalism.* New York: Cambridge University Press, 2001.

Rosen, George. *A History of Public Health.* New York: MD Publications, 1958.

Rosenberg, Charles E. "The Cause of Cholera: Aspects of Etiological Thought in Nineteenth-Century America." *Bulletin of the History of Medicine* 34 (1960): 331–54.

———. "Body and Mind in Nineteenth-Century Medicine: Some Clinical Origins of the Neurosis Concept." In *Explaining Epidemics and Other Studies in the History of Medicine.* New York: Cambridge University Press, 1992.

———. "Pathologies of Progress: The Idea of Civilization as Risk." *Bulletin of the History of Medicine* 72 (1998): 714–30.

Rosenberg, Charles E., and Janet Golden, eds. *Framing Disease: Studies in Cultural History.* New Brunswick, NJ: Rutgers University Press, 1992.

Rosenkrantz, Barbara Gutmann. *Public Health and the State: Changing Views in Massachusetts, 1842–1936.* Cambridge, MA: Harvard University Press, 1972.

Rosner, David, and Gerald Markowitz. *Dying for Work: Workers' Safety and Health in Twentieth-Century America.* Bloomington: Indiana University Press, 1987.

Roth, Mitchel. "Cholera, Community, and Public Health in Gold Rush Sacramento and San Francisco." *Pacific Historical Review* 66 (1997): 527–51.

Rothman, Sheila M. *Living in the Shadow of Death: Tuberculosis and the Social Experience of Illness in American History.* New York: Basic Books, 1994.

Rupke, Nicolaas A. "Humboldtian Medicine." *Medical History* 40 (July 1996): 293–310.

———, ed. *Medical Geography in Historical Perspective.* Supplement No. 20 to *Medical History.* London: Wellcome Trust Center for the History of Medicine, 2000.

Russell, Edmund P. "The Strange Career of DDT: Experts, Federal Capacity, and Environmentalism after World War II." *Technology and Culture* 40 (1999): 770–96.

———. *War and Nature: Fighting Humans and Insects with Chemicals from World War I to Silent Spring.* New York: Cambridge University Press, 2001.

Rutman, Darrett B., and Anita H. Rutman. "Of Agues and Fevers: Malaria in the Early Chesapeake." In *Biological Consequences of European Expansion, 1450–1800,* edited by Kenneth F. Kiple and Stephen V. Beck, 203–32. Aldershot: Ashgate, 1997.

Sackman, Douglas Cazaux. *Orange Empire: California and the Fruits of Eden.* Berkeley: University of California Press, 2005.

Sagan, Dorion. "Metametazoa: Biology and Multiplicity." In *Incorporations,* edited by Jonathan Crary and Sanford Kwinter, 362–85. New York: Zone, 1992.

Saunders, J. B. de C. M. *Humboldtian Physicians in California.* Davis: University of California, Davis, 1971.

Schiebinger, Londa. *Nature's Body: Gender and the Making of Modern Science.* Boston: Beacon Press, 1993.

Scott, James L. *Seeing Like a State: Why Certain Schemes to Improve the Human Condition Have Failed.* New Haven, CT: Yale University Press, 1999.

Scott, Joan W. "The Evidence of Experience." *Critical Inquiry* 18 (Spring 1991): 773–97.

Sellers, Christopher. "The Public Health Service's Office of Industrial Hygiene and the Transformation of Industrial Medicine." *Bulletin of the History of Medicine* 65 (Spring 1991): 42–73.

———. *Hazards of the Job: From Industrial Disease to Environmental Health Science.* Chapel Hill: University of North Carolina Press, 1997.

———. "Discovering Environmental Cancer: Wilhelm Hueper, Post–World War II Epidemiology, and the Vanishing Clinician's Eye." *American Journal of Public Health* 87 (November 1997): 1824–35.

———. "Body, Place, and the State: The Making of an 'Environmentalist' Imaginary in the Post–World War II U.S." *Radical History Review* 74 (1999): 31–64.

———. "Thoreau's Body: Towards an Embodied Environmental History." *Environmental History* 4 (October 1999): 486–515.

Setterberg, Fred, and Lonny Shavelson. *Toxic Nation: The Fight to Save Our Communities from Chemical Contamination.* New York: John Wiley & Sons, 1993.

Shabecoff, Philip. *A Fierce Green Fire: The American Environmental Movement.* Rev. ed. Washington, DC: Island Press, 2003.

Shah, Nayan. *Contagious Divides: Epidemics and Race in San Francisco's Chinatown.* Berkeley: University of California Press, 2001.

Shapin, Steven, and Simon Schaffer. *Leviathan and the Air-Pump: Hobbes, Boyle, and the Experimental Life.* Princeton, NJ: Princeton University Press, 1985.

Shapin, Steven. "The House of Experiment in Seventeenth-Century England." *Isis* 79 (1988): 373–404.

Showalter, Elaine. *Hystories: Hysterical Epidemics and Modern Media.* New York: Columbia University Press, 1997.

Shryock, Richard Harrison. *The Development of Modern Medicine: An Interpretation of the Social and Scientific Factors Involved.* London: Victor Gollancz, 1948.

———. *Medical Licensing in America, 1650–1965.* Baltimore, MD: Johns Hopkins University Press, 1967.

Smith, Dale C. "The Rise and Fall of Typhomalarial Fever." *Journal of the History of Medicine and Allied Sciences* 37 (1982): 182–220, 287–321.

Smith, Neil. "Antimonies of Space and Nature in Henri Lefebvre's *The Production of Space.*" In *Philosophy and Geography II: The Production of Public Space,* edited by Andrew Light and Jonathan M. Smith, 49–69. Lanham, MD: Rowman & Littlefield, 1998.

Smith, Wallace. *Garden of the Sun: A History of the San Joaquin Valley, 1772–1939.* Edited by William B. Secrest Jr. 2d ed. Fresno, CA: Linden Publishing, 2004.

Snyder, Lynne Page. "'The Death-dealing Smog over Donora, Pennsylvania': Industrial Air Pollution, Public Health Policy, and the Politics of Expertise, 1948–1949." *Environmental History Review* 18 (Spring 1994): 117–39.

Sorensen, W. Conner. *Brethren of the Net: American Entomology, 1840–1880.* Tuscaloosa: University of Alabama Press, 1995.

Spence, Clark C. "The Golden Age of Dredging: The Development of an Industry and Its Environmental Impact." *Western Historical Quarterly* 11 (October 1980): 401–14.

Starr, Paul. *The Social Transformation of American Medicine: The Rise of the Sovereign Profession and the Making of a Vast Industry.* New York: Basic Books, 1982.

Steinberg, Ted. *Down to Earth: Nature's Role in American History.* New York: Oxford University Press, 2002.

Steingraber, Sandra. *Living Downstream: A Scientist's Personal Investigation of Cancer and the Environment.* New York: Vintage, 1998.

Stemen, Mark L. "Genetic Dreams: An Environmental History of the California Cotton Industry, 1902–1953." Ph.D. diss., University of Iowa, 1999.

Stepan, Nancy. "Biological Degeneration: Races and Proper Places." In *Degeneration: The Dark Side of Progress,* edited by J. Edward Chamberlain and Sander L. Gilman, 97–120. New York: Columbia University Press, 1985.

———. *Picturing Tropical Nature.* Ithaca, NY: Cornell University Press, 2001.

Stern, Alexandra Minna. "Buildings, Boundaries, and Blood: Medicalization and

Nation-building on the U.S.-Mexico Border, 1910–1930." *Hispanic American Historical Review* 79 (February 1999): 41–82.

———. *Eugenic Nation: Faults and Frontiers of Better Breeding in Modern America.* Berkeley: University of California Press, 2005.

Stoler, Ann. "Sexual Affronts and Racial Frontiers: European Identities and the Cultural Politics of Exclusion in Colonial Southeast Asia." *Comparative Studies in Society and History* 34 (July 1992): 514–51.

Stoll, Steven. *The Fruits of Natural Advantage: Making the Industrial Countryside in California.* Berkeley: University of California Press, 1998.

Susser, Mervyn. "Epidemiology in the United States after World War II: The Evolution of Technique." *Epidemiologic Reviews* 7 (1985): 147–77.

———. "Choosing a Future for Epidemiology: I. Eras and Paradigms." *American Journal of Public Health* 86 (May 1996): 668–73.

Szasz, Andrew. *Ecopopulism: Toxic Waste and the Movement for Environmental Justice.* Minneapolis: University of Minnesota Press, 1994.

Szerszynski, Bronislaw, Scott Lash, and Brian Wynne, eds. *Risk, Environment and Modernity: Towards a New Ecology.* London: Sage, 1996.

Tarr, Joel A. *The Search for the Ultimate Sink: Urban Pollution in Historical Perspective.* Akron: University of Ohio Press, 1996.

Taylor, Herbert C., Jr., and Lester L. Hoaglin Jr. "The 'Intermittent Fever' Epidemic of the 1830's on the Lower Columbia River." *Ethnohistory* 9 (1962): 160–78.

Temkin, Owsei. "Health and Disease." In *The Double Face of Janus and Other Essays in the History of Medicine,* 419–40. Baltimore, MD: Johns Hopkins University Press, 1977.

Tesh, Sylvia Noble. *Hidden Arguments: Political Ideology and Disease Prevention Policy.* New Brunswick, NJ: Rutgers University Press, 1988.

Thompson, Kenneth. "Insalubrious California: Perception and Reality." *Annals of the Association of American Geographers* 59 (March 1969): 50–64.

———. "Irrigation as a Menace to Health in California." *Geographical Review* 59 (April 1969): 195–214.

———. "The Australian Fever Tree in California: Eucalypts and Malaria Prophylaxis." *Annals of the Association of American Geographers* 60 (June 1970): 230–44.

———. "Climatotherapy in California." *California Historical Quarterly* 50 (1971): 111–30.

———. "Wilderness and Health in the Nineteenth Century." *Journal of Historical Geography* 2 (1976): 145–61.

Tilley, Helen. "Ecologies of Complexity: Tropical Environments, African Trypanosomiasis, and the Science of Disease Control Strategies in British Colonial Africa, 1900–1940." *Osiris* 19 (2004): 21–38.

Tomes, Nancy. *The Gospel of Germs: Men, Women, and the Microbe in American Life*. Cambridge, MA: Harvard University Press, 1998.

Tomes, Nancy, and John Harley Warner, eds. "Introduction to Special Issue on *Rethinking the Germ Theory of Disease: Comparative Perspectives*." *Journal of the History of Medicine and the Allied Sciences* 52 (January 1997): 7–16.

Tuan, Yi-Fu. *Space and Place: The Perspective of Experience*. Minneapolis: University of Minnesota Press, 1977.

Tyrell, Ian. *True Gardens of the Gods: Californian-Australian Environmental Reform, 1860–1930*. Berkeley: University of California Press, 1999.

Valenčius, Conevery Bolton. *The Health of the Country: How American Settlers Understood Themselves and Their Land*. New York: Basic Books, 2002.

Valle, Rosemary Keupper. "Prevention of Smallpox in Alta California during the Franciscan Mission Period (1769–1833)." *California Medicine* 119 (July 1973): 73–77.

Vaught, David. *Cultivating California: Growers, Specialty Crops, and Labor, 1875–1920*. Baltimore, MD: Johns Hopkins University Press, 1999.

Vogel, Virgil J. *American Indian Medicine*. Norman: University of Oklahoma Press, 1970.

Waddell, Craig, ed. *And No Birds Sing: Rhetorical Analyses of Rachel Carson's "Silent Spring."* Carbondale: Southern Illinois University Press, 2000.

Wargo, John. *Our Children's Toxic Legacy: How Science and Law Fail to Protect Us from Pesticides*. 2d ed. New Haven, CT: Yale University Press, 1998.

Warner, John Harley. "'The Nature-Trusting Heresy': American Physicians and the Concept of the Healing Power of Nature in the 1850s and 1860s." *Perspectives in American History* 11 (1977–78): 295–96.

———. *The Therapeutic Perspective: Medical Practice, Knowledge, and Identity in America, 1820–1885*. Cambridge, MA: Harvard University Press, 1986.

———. "Medical Sectarianism, Therapeutic Conflict, and the Shaping of Orthodox Professional Identity in Antebellum American Medicine." In *Medical Fringe and Medical Orthodoxy, 1750–1850*, edited by W. F. Bynum and Roy Porter, 234–60. London: Croom Helm, 1987.

———. "The Idea of Southern Medical Distinctiveness: Medical Knowledge and Practice in the Old South." In *Science and Medicine in the Old South*, edited by Ronald L. Numbers and Todd L. Savitt, 206–25. Baton Rouge: Louisiana State University, 1989.

Warren, Louis S. "Buffalo Bill Meets Dracula: William F. Cody, Bram Stoker, and the Frontiers of Racial Decay." *American Historical Review* 107 (October 2002): 1124–57.

Wasserstrom, Robert F., and Richard Wiles. *Field Duty: U.S. Farmworkers and Pesticide Safety*. Washington, DC: World Resources Institute, 1985.

Weber, Devra. *Dark Sweat, White Gold: California Farm Workers, Cotton, and the New Deal.* Berkeley: University of California Press, 1996.

Wells, Miriam J. *Strawberry Fields: Politics, Class, and Work in California Agriculture.* Ithaca, NY: Cornell University Press, 1996.

White, Richard. *Land Use, Environment, and Social Change: The Shaping of Island County, Washington.* Seattle: University of Washington Press, 1980.

———. "Are You an Environmentalist or Do You Work for a Living?" In *Uncommon Ground: Toward Reinventing Nature,* edited by William Cronon, 171–85. New York: Norton, 1995.

———. *The Organic Machine: The Making and Unmaking of the Columbia River.* New York: Hill and Wang, 1995.

White, Suzanne Rebecca. "Chemistry and Controversy: Regulating the Use of Chemicals in Foods, 1883–1959." Ph.D. diss., Emory University, 1994.

White, Luise. "Tsetse Visions: Narratives of Blood and Bugs in Colonial Northern Rhodesia, 1931–9." *Journal of African History* 36 (1995): 219–45.

Whorton, James. *Before Silent Spring: Pesticides and Public Health in Pre-DDT America.* Princeton, NJ: Princeton University Press, 1974.

Wiebe, Robert. *The Search for Order.* New York: Hill and Wang, 1967.

Willems, Dick. "Inhaling Drugs and Making Worlds: The Proliferation of Lungs and Asthma." In *Differences in Medicine: Unraveling Practices, Techniques, and Bodies,* edited by Marc Berg and Annemarie Mol, 105–18. Durham, NC: Duke University Press, 1998.

Williams, Ralph Chester. *The United States Public Health Service, 1798–1950.* Washington, DC: U.S. Public Health Service, 1951.

Wing, Steve. "Limits of Epidemiology." *Medicine and Global Survival* 1 (1994): 74–86.

Woo-Sam, Anne Marie. "Domesticating the Immigrant: California's Commission of Immigration and Housing and the Domestic Immigration Policy Movement, 1910–1945." Ph.D. diss., University of California, Berkeley, 1999.

Worboys, Michael. "Germs, Malaria, and the Invention of Mansonian Tropical Medicine: From 'Disease in the Tropics' to 'Tropical Diseases.'" In *Warm Climates and Western Medicine: The Emergence of Tropical Medicine, 1500–1900,* edited by David Arnold, 181–207. Amsterdam: Rodopi, 1996.

———. *Spreading Germs: Disease Theories and Medical Practice in Britain, 1865–1900.* Cambridge: Cambridge University Press, 2000.

Worster, Donald. *Dust Bowl: The Southern Plains in the 1930s.* New York: Oxford University Press, 1979.

———. *Rivers of Empire: Water, Aridity, and the Growth of the American West.* New York: Pantheon, 1985.

Wrobel, David M. *Promised Lands: Promotion, Memory, and the Creation of the American West.* Lawrence: University Press of Kansas, 2002.

Index

Page numbers in *italics* refer to illustrations.

Text: 10/13 Sabon
Display: Akzidenz
Compositor: BookMatters, Berkeley
Printer and binder: Sheridan Books, Inc.